Carbon Capture and Environmental Sustainability

Carbon Capture and Environmental Sustainability

Karan Sotoodeh
Baker Hughes, Oslo, Norway

WILEY

This edition first published 2026

Registered Offices
John Wiley & Sons, Inc., 111 River Street, Hoboken, NJ 07030, USA
John Wiley & Sons Ltd. New Era House, 8 Oldlands Way, Bognor Regis, West Sussex, PO22 9NQ, UK

For details of our global editorial offices, customer services, and more information about Wiley products visit us at www.wiley.com.

The manufacturer's authorized representative according to the EU General Product Safety Regulation is Wiley-VCH GmbH, Boschstr. 12, 69469 Weinheim, Germany, e-mail: Product_Safety@wiley.com.

Wiley also publishes its books in a variety of electronic formats and by print-on-demand. Some content that appears in standard print versions of this book may not be available in other formats.

Library of Congress Cataloging-in-Publication Data
Names: Sotoodeh, Karan author | John Wiley & Sons publisher
Title: Carbon capture and environmental sustainability / Karan Sotoodeh, Oslo, Norway.
Description: Hoboken, NJ : Wiley, 2026.
Identifiers: LCCN 2026020403 (print) | LCCN 2026020404 (ebook) | ISBN 9781394436378 hardback | ISBN 9781394436392 adobe pdf | ISBN 9781394436385 epub
Subjects: LCSH: Carbon sequestration | Carbon sequestration–Social aspects | Sustainable development | Sustainable engineering
Classification: LCC TD885.5.C3 S675 2026 (print) | LCC TD885.5.C3 (ebook)
LC record available at https://lccn.loc.gov/2026020403
LC ebook record available at https://lccn.loc.gov/2026020404

Cover Design: Wiley
Cover Image: tanit boonruen/Alamy Stock Photo

Set in 9.5/12.5pt STIXTwoText by Straive, Pondicherry, India

Contents

Preface

This book represents a deliberate departure from my previous work. While my academic and professional background is rooted in engineering and technical disciplines, *Carbon Capture and Environmental Sustainability: Science, Engineering, Economics, Political, and Governance Pathways to a Low-Carbon Future* is my first nontechnical and interdisciplinary book. The motivation for writing this book arises from the realization that carbon capture can no longer be meaningfully discussed within purely technical boundaries.

Carbon capture has become a central topic in recent years due to the growing urgency of environmental protection, climate change mitigation, and the need to remove or manage carbon emissions from industrial systems and the atmosphere. However, carbon capture does not exist in isolation. It cannot be implemented, scaled, or sustained through engineering solutions alone. Without supportive policies, economic incentives, social acceptance, institutional frameworks, and effective governance, even the most advanced carbon capture technologies will remain limited in impact.

Many engineers, scientists, and technical professionals tend to view carbon capture narrowly through the lenses of capture efficiency, separation technologies, transportation, and storage. While these aspects are essential, they represent only one dimension of a much broader challenge. Carbon capture is equally shaped by political decision-making, regulatory structures, public perception, economic feasibility, and global governance mechanisms. Ignoring these dimensions risks oversimplifying a complex problem and undermining real-world implementation.

The purpose of this book is therefore to broaden the perspective of readers, engineers, and non-engineers alike, by examining carbon capture as a multidisciplinary and societal challenge. By stepping beyond my traditional technical domain, I aim to explore how science and engineering intersect with economics, policy, governance, and social systems in shaping low-carbon pathways. The intention is not to replace technical analysis but to complement it by situating carbon capture within the wider context in which it must operate.

This book is written for readers who seek a more holistic understanding of carbon capture and environmental sustainability, including engineers wishing to expand their viewpoints, policymakers seeking technical awareness without excessive detail, and students and researchers interested in the broader implications of low-carbon transitions. Ultimately, addressing climate change and achieving a low-carbon future requires not only better technologies but also broader thinking, interdisciplinary collaboration, and an openness to perspectives beyond traditional disciplinary boundaries.

1

Introduction to Carbon Capture and Environmental Sustainability

Climate change is no longer a distant concern. It is a present-day challenge that shapes how we live, govern, and develop. Rising levels of carbon dioxide (CO_2) in the atmosphere are a major driver of global warming, with consequences that reach across ecosystems, economies, and communities [1]. Consequently, carbon capture has emerged as a key tool in our environmental strategy. This technology is not intended to replace renewable energy, but rather to complement it, particularly in hard-to-decarbonize sectors, such as cement and chemical industries, transportation, and heavy energy usage [2].

Figure 1.1 illustrates the central role of carbon dioxide removal in protecting natural ecosystems. The integration of nature (tree and moss) with carbon symbols and technology icons reflects the multidisciplinary nature of carbon capture strategies aimed at sustainability.

This chapter introduces the core concepts of carbon capture and explains how they fit within the broader idea of environmental sustainability. Sustainability is not just about protecting nature; it also involves social equity, economic resilience, and ethical responsibility [3]. The idea of a low-carbon future is not merely technical; it is political, moral, and deeply human.

We begin by exploring what carbon capture actually means, why CO_2 is such a central issue in climate science, and how this relates to the global frameworks for sustainable development. This chapter also provides a brief historical overview of how carbon capture technologies have evolved. It also discusses why they are increasingly seen as essential even by industrial solutions critics, people who often doubt or criticize standard engineering or industrial approaches. They might think some solutions are too costly, inefficient, outdated, or not sustainable enough.

Finally, this chapter presents the structure of the book, outlining how each chapter addresses a unique dimension of carbon capture: from policy and ethics to public perception to integration with renewable energy and global equity.

1.1 What is Carbon Capture?

Carbon capture refers to a set of technologies and processes that remove carbon dioxide (CO_2) from industrial emissions or directly from the atmosphere [4]. The goal is to prevent CO_2 from reaching the atmosphere and contributing to global warming. Once captured, the

Figure 1.1 Visual symbol for CO_2 reduction and environmental harmony. (Courtesy of Shutterstock.com).

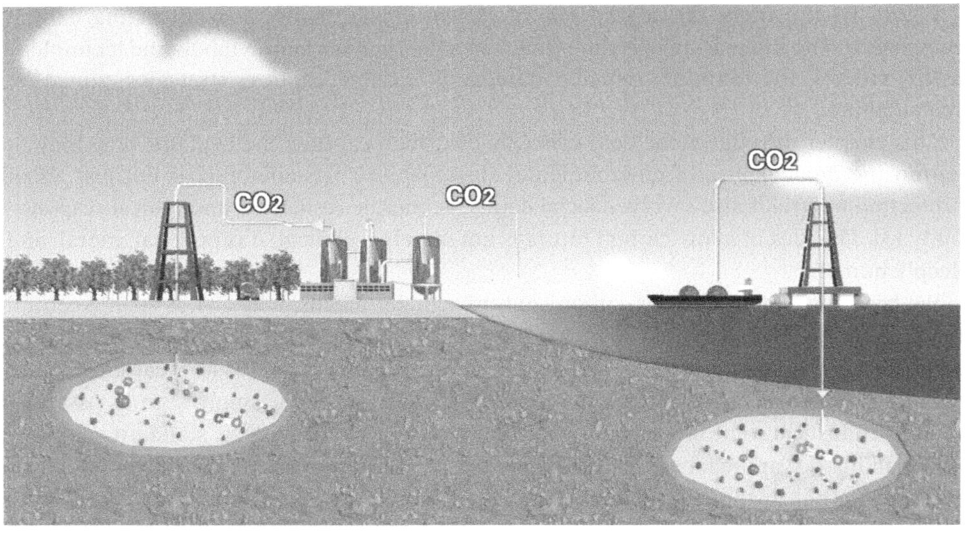

Figure 1.2 Schematic representation of carbon capture and geological storage (CCS). (Courtesy of Shutterstock.com).

CO_2 can either be stored underground in geological formations (carbon capture and storage, or CCS) or used in industrial applications (carbon capture and utilization, or CCU) [4, 5]. Figure 1.2 shows the basic process of capturing carbon dioxide from industrial sources, transport with ship, and inject into the underground storage sites. The image illustrates both onshore and offshore injection into deep geological formations, a key strategy for long-term CO_2 separation and climate mitigation.

Carbon capture is not a single technology, but a category that includes several methods. These range from post-combustion capture, where CO_2 is removed after fuel is burned, to direct air capture, which filters CO_2 directly from ambient air. Each method has its own advantages, challenges, and environmental implications [4].

The idea behind carbon capture is not new. For decades, CO_2 has been captured in the oil and gas industry for use in enhanced oil recovery (EOR) [6]. What's new is the urgency of deploying it at a wide range, across multiple sectors, as part of a comprehensive response to the climate crisis.

Today, carbon capture is being discussed not just in technical or engineering terms, but as a political, economic, and environmental issue. It represents both a climate solution and a subject of debate: Who should pay for it? Who benefits? And can it truly be sustainable?

1.2 The Role of CO_2 in Global Climate Change

Carbon dioxide (CO_2) is one of the most well-known greenhouse gases and for good reason. It plays a central role in the warming of our planet by trapping heat in the Earth's atmosphere. While it occurs naturally and is essential for processes like plant photosynthesis, human activities have drastically increased the concentration of CO_2 in the atmosphere over the past two centuries [7].

The burning of fossil fuels such as coal, oil, and natural gas is the primary source of anthropogenic (human-caused) CO_2 emissions. Industrial processes, transportation, and electricity generation contribute heavily. Deforestation also plays a major role by reducing the Earth's capacity to absorb CO_2 naturally [7].

According to the Intergovernmental Panel on Climate Change (IPCC), atmospheric CO_2 levels have now exceeded 420 parts per million (ppm), compared to preindustrial levels of about 280 ppm [8]. This increase is closely linked to rising global temperatures, more extreme weather events, melting polar ice, and rising sea levels.

CO_2 has a long atmospheric lifetime, often remaining for hundreds of years. This means that even if emissions stopped today, their climate effects would continue for generations. That's why reducing emissions alone is not enough; we must also actively remove CO_2 from the atmosphere, which is where carbon capture technologies become crucial [8].

In short, carbon dioxide is not just one of many greenhouse gases; it is the driver of long-term climate instability. Any serious effort to combat climate change must directly address CO_2, both at the source and in the atmosphere.

1.3 Definitions and Pillars of Environmental Sustainability

There is a term called environmental sustainability that is frequently used but not always understood. At its core, sustainability refers to meeting the needs of today without compromising the ability of future generations to meet their own needs. This involves protecting the natural systems that support life—air, water, soil, and climate—as well as establishing fair and sustainable economic and social structures [9].

Most experts agree that sustainability rests on three main pillars:

1) **Environmental Protection**: This means taking care of natural resources and ecosystems. It includes reducing pollution, protecting biodiversity, and managing resources like water, energy, and land in ways that don't cause long-term harm.
2) **Social Equity**: A sustainable world must also be fair. This pillar focuses on justice and human well-being. It means ensuring that all people, especially vulnerable communities, have access to clean air, safe water, healthy food, education, and opportunities to thrive [10].
3) **Economic Resilience**: Sustainability doesn't mean stopping economic growth, but it does mean changing how we grow. A resilient and sustainable economy creates jobs, supports innovation, and reduces poverty without destroying the environment or increasing inequality [11].

These three pillars are deeply connected. We cannot solve environmental problems without addressing social and economic issues at the same time. For example, carbon capture technology may reduce emissions, but to be truly sustainable, it must also be affordable, fairly distributed, and developed in ways that do not harm communities or ecosystems.

In this book, we will return to these three pillars often. They help us evaluate whether carbon capture solutions are not only technically effective but also ethically and socially responsible.

1.4 Historical Context and Evolution of Carbon Capture Technologies

The idea of capturing carbon dioxide (CO_2) is not new. In fact, industries have been separating and managing CO_2 for decades, long before climate change became a major public concern. For example, in the oil and gas industry, CO_2 has been captured from oil and gas and used since the 1970s for a process called EOR, where it is injected into underground wells to extract more oil. However, this early use was focused on improving efficiency and profits and not reducing emissions [6].

The shift toward using carbon capture as a climate solution began in the late 1990s and early 2000s, when scientists and policymakers started to recognize the urgent need to limit greenhouse gas emissions. Research and pilot projects were launched to explore how captured CO_2 could be stored safely underground, rather than being released into the atmosphere. This marked the beginning of carbon capture and storage (CCS) as a climate strategy [6].

Over time, the field has grown and changed. Newer technologies have emerged [12], such as:

- **Post-combustion capture**, which removes CO_2 after fossil fuels are burned.
- **Precombustion capture**, where CO_2 is removed before burning, during fuel processing.
- **Oxy-fuel combustion**, in which oxygen is used instead of air to make CO_2 easier to capture
- **Direct air capture (DAC)**, a more recent innovation that pulls CO_2 straight from the atmosphere.

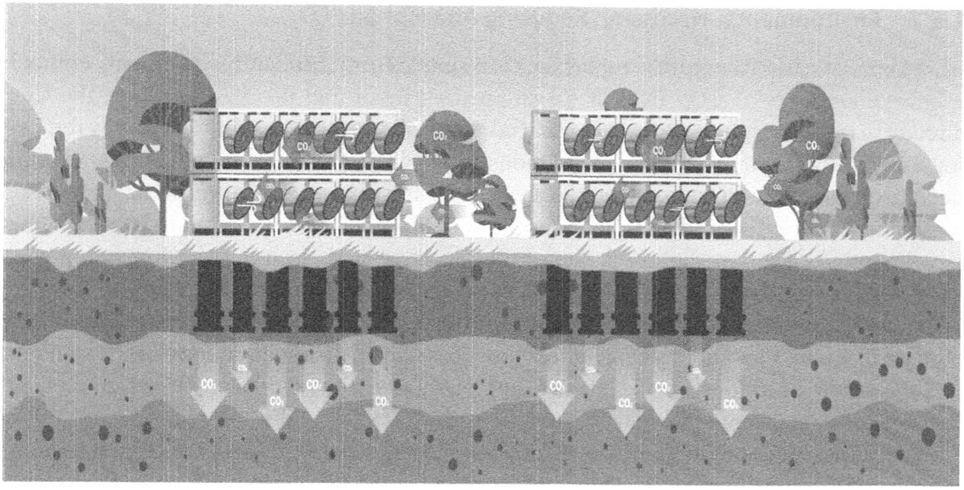

Figure 1.3 Direct air capture (DAC) with subsurface carbon storage. (Courtesy of Shutterstock.com).

These technologies are at different stages of development and deployment [12]. Some are already used in industrial settings; others are still in the research or demonstration phase. As climate goals become more ambitious, there is growing interest in scaling up carbon capture technologies and combining them with renewable energy, hydrogen production, and sustainable industry.

Figure 1.3 shows a conceptual DAC facility that captures CO_2 directly from the ambient air using large fan systems and filters. The captured CO_2 is then injected underground into geological formations for long-term storage [13].

Despite these advances, carbon capture remains a contentious issue in some communities. Critics worry that it could be used as an excuse to delay the transition away from fossil fuels. Some supporters, however, assert that carbon capture is a necessary tool, especially in sectors where there are no zero-emission alternatives [14]. We can gain a better understanding of carbon capture's role in the current climate landscape by examining its historical context. It is not a magic bullet, but it is a rapidly evolving technology that can play a significant role when used responsibly and alongside other solutions.

1.5 Why Carbon Capture is Needed: A Multidimensional Perspective

To understand why carbon capture is increasingly viewed as essential, we need to look beyond just science. The case for carbon capture is not only environmental; it is also economic, political, social, and even ethical. In this section, we explore the many reasons why carbon capture technologies are becoming a necessary part of the global response to climate change.

1.5.1 Environmental Necessity: Reducing Atmospheric CO_2

At the core of carbon capture's importance is a simple truth: humanity has already emitted excessive amounts of CO_2. Even if we stopped all fossil fuel use by tomorrow, the CO_2 already in the atmosphere would keep warming the planet for centuries. That's why removing CO_2 is just as vital as reducing emissions [1]. Even if we cut future emissions, we still need to clean up the CO_2 already in the air because it will continue to warm the planet. So, removing CO_2 (pulling it out) is just as important maybe even more so than just reducing new emissions. Carbon capture is one of the few large-scale technologies that can physically remove CO_2 from industrial processes or even directly from the air [1].

Figure 1.4 contrasts two possible futures: one driven by unchecked carbon emissions and climate inaction, the other by sustainable practices and environmental recovery. It symbolizes the urgent need for carbon capture technologies as part of a broader strategy to prevent irreversible damage and promote a livable future.

Renewable energy reduces future emissions but does not eliminate existing emissions already in the atmosphere [15]. This is especially relevant for sectors such as cement, steel, and chemical manufacturing. These sectors emit CO_2 not only from fuel but also from chemical reactions in the production process. These sectors are hard to decarbonize with renewables alone [16].

1.5.2 Economic Logic: Supporting Industry Without Shutting It Down

For many countries, heavy industries are vital for jobs, income, and exports. Abruptly removing these sectors could lead to economic disruption and social unrest. Carbon

Figure 1.4 Climate crisis vs. sustainability: a visual contrast. (kwest/Shutterstock.com).

capture provides a solution, allowing these industries to continue operating during the shift to greener alternatives while reducing their climate impact.

In this way, carbon capture contributes to what is often referred to as a "just transition," a strategy that is intended to move economies toward sustainability without adversely affecting workers and local communities. Carbon capture can contribute to the preservation of employment in regions that are dependent on fossil fuels while new clean energy solutions are being developed [17].

1.5.3 Political Realities: Bridging Gaps Between Nations

Globally, climate politics is shaped by a complex balance of three key factors: responsibility, capability, and historical emissions. Responsibility refers to the question of who caused the most climate damage, especially through decades of burning fossil fuels during early industrialization [18]. Capability means who has the financial resources, technology, and institutional strength to take meaningful climate action. Finally, historical emissions refer to the total amount of greenhouse gases released over time, not just what's emitted today.

High-income countries, mainly those that industrialized early, such as the United States and many European nations score, high on all three counts. They are largely responsible for past emissions, they have the means to act, and they are expected to lead the global response to climate change. However, many low- and middle-income countries are still focused on urgent priorities like poverty reduction, industrial growth, and ensuring access to reliable energy. For these nations, rapidly cutting emissions can conflict with their development goals [19].

In this context, carbon capture technologies offer a realistic middle ground. By allowing continued use of fossil fuels while capturing CO_2 from power plants and industrial processes, these technologies make it possible for developing countries to grow their economies and improve energy access without abandoning climate goals [17]. Carbon capture, therefore, provides a strategic pathway to balance climate responsibility with development needs, especially in parts of the world where renewable energy infrastructure is not yet fully established or economically viable.

Carbon capture is politically attractive as a technology-based solution. It does not require lifestyle changes like using less energy or flying less. This makes it easier for governments to support, especially in politically sensitive countries. It focuses on technology, not personal sacrifice, gaining support across the political spectrum. Even cautious governments include it in their strategies. They see it as a way to cut emissions without slowing economic activity.

1.5.4 Time Pressure: Acting Quickly on Multiple Fronts

Because climate change is so urgent, we cannot depend on just one solution. We need to use all available tools to reduce emissions. Major organizations like the International Energy Agency (IEA) and the IPCC both agree that carbon capture, utilization, and storage (CCUS) is one of these important tools. According to their research, if we do not use carbon capture, it will be much more expensive or even impossible to reach our net-zero targets by the middle of this century [20].

This is why carbon capture is often called a "bridge technology." It may not be the final answer, but it is something we can use right now to reduce emissions while we continue to develop cleaner and more permanent solutions like renewable energy and new infrastructure. In short, carbon capture helps us stay within the safe limits for global temperature rise, buying us time to build a more sustainable future.

1.5.5 Social and Ethical Dimensions: Responsibility and Justice

Climate change creates big ethical questions. For example, Who caused the problem? Who is being hurt the most? And who should take responsibility to fix it? Carbon capture can help answer some of these questions but only if it is used in a fair and equal way.

Many low-income countries did not cause much of the pollution, but they are often the ones suffering the most from floods, droughts, and heatwaves. On the other hand, many wealthy countries produced more emissions in the past, and they have more money and technology today [19]. This means they may have a moral duty to help pay for solutions like carbon capture, to protect everyone, not just themselves.

Also, carbon capture can help people who live near factories or power plants, where the air is often dirty. If carbon capture is done with care and fairness, and if local communities are involved in decisions, it can reduce both carbon emissions and local pollution. In this way, carbon capture can support environmental justice, not just climate action [21].

Figure 1.5 shows the social and ethical dimensions of carbon capture and climate justice [21]. In the middle, there's a big scale—like a balance. On the one side is a dark cloud

Figure 1.5 Balancing the scales: carbon capture as a pathway to climate justice, bridging the gap between responsibility, equity, and environmental protection.

labeled "carbon," and on the other side is the Earth. This means we need to balance pollution with protecting the planet. The scale is held up by the hands of different people, showing that people from all over the world should work together. On the left, there's a woman standing near houses in a flood. This shows that people in poor countries often suffer the most from climate change, even though they didn't cause much of the pollution. On the right, there's a man in a suit with money, standing near big factories with smoke. This shows that rich countries and companies have caused more pollution in the past, and now they have the money to help fix it. At the bottom, there's a building labeled "carbon capture." This is a mechanism that helps take carbon out of the air. If used fairly, it can help clean the air and protect both the planet and the people who live near polluting factories. The visual composition emphasizes the need for fairness, inclusion, and shared responsibility in addressing climate change.

1.5.6 Criticism and Caution

Of course, carbon capture is not without its critics. Some argue that it may delay the transition away from fossil fuels by giving polluting industries a "license to operate." Others point out that some carbon capture projects have failed to deliver promised reductions or have been used more for commercial oil production than climate goals [21, 22].

These concerns are valid and deserve attention, but they shouldn't lead to abandoning carbon capture. Instead, they emphasize the need to use this technology carefully and responsibly. To achieve this, we need clear regulations and policies that ensure carbon capture genuinely reduces emissions while protecting people and the environment [21, 23].

Independent oversight is equally important. Trusted experts, researchers, and organizations monitor and evaluate how carbon capture is being implemented. This helps prevent misuse, ensures safety, and builds public trust [21, 22].

Carbon capture projects should align directly with climate goals. Their primary purpose should be to help countries and industries meet real targets for reducing greenhouse gas emissions. By focusing on areas where they can have the most positive impact such as heavy industries like cement, steel, and chemicals, or places where they can also improve local air quality and public health, carbon capture can be a valuable part of the climate solution [24].

However, it's crucial to remember that carbon capture is not a substitute for larger changes in our energy and economic systems. It cannot replace the shift to renewable energy, energy efficiency improvements, or changes in unsustainable consumption habits. Instead, it should be seen as one piece of a broader, long-term strategy to build a cleaner, more equitable future.

1.5.7 Conclusion

In summary, carbon capture is needed not just for environmental reasons also but because of the complex political, economic, and social challenges of transitioning to a low-carbon future. It is one of several tools we need and, like all tools, it must be used with care, purpose, and transparency.

In the chapters that follow, we will continue to examine how carbon capture fits into this broader picture. We will ask questions about who controls it, who benefits, and how it can truly support a sustainable future for all.

1.6 Bridging Policy, Environment, and Ethics

Addressing climate change is not only a scientific or technological challenge but also deeply political and ethical [25]. Carbon capture technologies sit at the intersection of three major domains: public policy, environmental protection, and moral responsibility. If carbon capture is to play a meaningful role in building a sustainable future, it must be developed and governed through fair and effective policies that reflect broader social values.

1.6.1 The Role of Policy

Public policy plays a central role in determining how carbon capture technologies are developed, funded, and implemented [26]. Without strong policy support, carbon capture will likely remain limited in scale and slow to progress. Governments can shape this field by offering financial support through mechanisms such as subsidies or tax incentives. They can also set firm emission targets that require industry to invest in capture systems [27]. In addition, they can build a regulatory environment that ensures safety and long-term reliability. Furthermore, policies that promote international cooperation can enable the transport and storage of CO_2 across borders, which may be necessary in some regions [26, 27]. However, if policies are poorly designed or overly influenced by special interests, they can lead to public distrust, inefficient investments, or even environmental catastrophes. That is why effective carbon capture policies must be guided by sound science, transparency, and the public good—not just market priorities.

1.6.2 Environmental Goals and Trade-Offs

Environmentally, the primary goal of carbon capture is clear: to reduce the concentration of CO_2 in the atmosphere and limit global warming [2]. But achieving this goal is not without trade-offs. Carbon capture processes require significant energy such as water or electricity, and depending on the source of that energy, the net environmental benefit can vary [28]. If fossil fuels are used to power the capture system, emissions reductions might be lower than expected. Additionally, infrastructure needs such as pipelines and storage sites can impact ecosystems, land use, and water resources. To understand the full picture, carbon capture systems must be evaluated using life cycle assessments (LCAs), which examine both direct and indirect environmental impacts across the system's entire lifespan. This kind of analysis helps avoid unintended consequences, such as solving one problem by creating another. Environmental planning must be embedded at every stage of carbon capture development to ensure that the solutions are genuinely sustainable, not just technically impressive [29].

Figure 1.6 A comparison of carbon capture systems powered by renewable energy versus fossil fuels. Life cycle assessment (LCA) helps reveal the true environmental impact by considering both inputs and outputs across the system's lifespan.

Figure 1.6 shows two different environmental outcomes of carbon capture based on the type of energy used. On the left side, the system is powered by renewable energy sources like wind turbines and solar panels, resulting in clean air, green landscapes, and a healthy environment, representing a sustainable approach. On the right side, the carbon capture facility is powered by fossil fuels, with smoke-filled skies, dry and damaged land, and environmental harm, showing that using dirty energy can reduce the overall benefit of using carbon capture. In the center, a sign labeled "Life Cycle Assessment" with arrows pointing between "Inputs" and "Outputs" reminds us that we must examine the full environmental impact of carbon capture systems from the resources they consume to the emissions they produce. This helps avoid solving one problem while creating another.

1.6.3 Ethics and Climate Responsibilities

Ethical concerns are a key part of the carbon capture debate [30]. Climate change itself is a question of fairness. Many of the people and communities who are most affected by climate change such as low-income populations and climate-vulnerable regions have done the least to cause the problem. This makes it essential to think carefully about who benefits from carbon capture, who pays the price, and who gets to make the decisions.

For example, should carbon capture facilities be built in areas that are already facing pollution, health risks, and environmental damage, especially in poor or marginalized communities? Should rich countries be allowed to keep producing emissions, as long as they invest in carbon capture projects in other parts of the world? These are not just technical or economic questions. They are deeply moral questions about justice, fairness, and responsibility.

A fair and ethical approach to carbon capture should prioritize genuine community involvement. It's essential to listen to and respect the voices of local people, especially those most impacted. Projects must be transparent, with all plans and decisions shared openly. They should also be accessible, ensuring that everyone, including Indigenous peoples and historically marginalized groups, is treated with dignity and given a meaningful role in the process [31].

1.6.4 Connecting the Dots

When policy, environmental science, and ethics are brought together, carbon capture has the potential to serve not just as a technical solution [26, 30]. It can also serve as a pathway toward a more welcoming and sustainable future [4]. The right governance ensures that carbon capture is developed responsibly [26]; environmental assessment confirms that it does not harm ecosystems or waste resources; and ethical frameworks verify that it contributes to climate justice [29]. For example, a government might support carbon capture only in industries without current low-emission alternatives. Regulators might require full environmental impact studies before permitting new carbon storage sites. And ethical guidelines might be used to ensure communities are fully informed and fairly treated [30]. Ultimately, carbon capture is most effective when decisions are guided not only by scientific research and economic calculations but also by core values such as justice, equity, public health, and environmental protection; values that prioritize both people's well-being and the long-term health of the planet [32].

Figure 1.7 shows a triangular framework that highlights the three fundamental dimensions influencing the success of carbon capture systems. Environmental integrity ensures

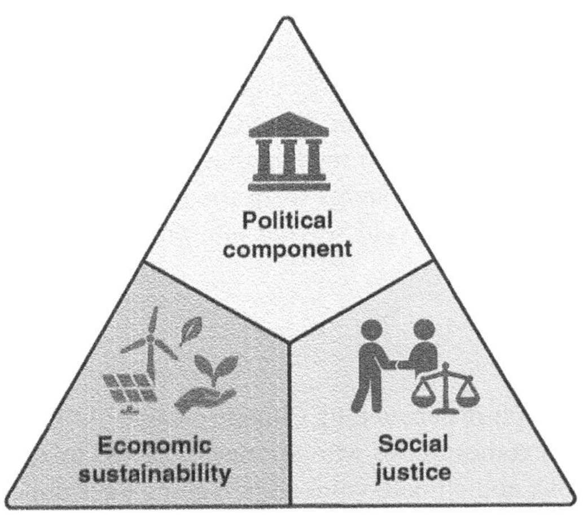

Figure 1.7 Core dimensions of carbon capture systems: environmental, political, and economic.

sustainability, political will enables regulation and governance, and economic viability supports real-world implementation. Effective carbon capture must balance all three to achieve meaningful, long-term climate impact.

1.7 Structure and Objectives of This Book

This book offers a broad yet focused exploration of carbon capture and its relationship with environmental sustainability. It integrates multiple perspectives from a scientific, social, economic, political, and ethical perspective rather than looking at it solely from a technical or engineering perspective. The goal is to present carbon capture not just as a set of technologies, but as a complex system that interacts with real-world challenges and choices. Readers from diverse backgrounds, including students, policymakers, environmentalists, and industry professionals, will find accessible analysis, practical insights, and critical questions throughout.

The book's structure reflects this interdisciplinary approach. Following this introductory chapter, each subsequent chapter explores a distinct but interconnected theme. Some chapters focus on the technical and environmental aspects of carbon capture—how it works, how it interacts with energy systems, and what its ecological impacts may be. Others examine broader aspects, including public perception, international policy, economic implications, and global equity concerns. Special attention is given to justice, ethics, and community engagement, topics often overlooked in traditional engineering discussions.

One of the central objectives of this book is to show that carbon capture is not a one-size-fits-all solution. It is a flexible tool whose outcomes depend on how, where, and why it is used. The book emphasizes the importance of governance, transparency, and sustainability metrics in determining whether carbon capture efforts truly benefit society and the environment. It also highlights the risks of misuse or relying too heavily on these technologies instead of more systemic changes.

One of the goals of this book is to make communication easier between technical experts and the general public. Carbon capture is often explained in very technical language, which makes it hard for many people to fully understand or join the conversation. This book tries to make the topic clear and accessible without making it too simple or leaving out important details.

By explaining complex ideas in an understandable way, the book helps readers think more deeply about climate change and the role carbon capture can play. It invites everyone not just scientists or engineers to take part in shaping the future of climate solutions.

By the end of the book, readers should have a strong understanding of the various dimensions of carbon capture: its potential, its limitations, and its place within a sustainable and socially responsible climate strategy. Whether you are approaching this subject from the standpoint of science, policy, advocacy, or industry, this book aims to provide a balanced and compelling foundation for further exploration and action.

1.8 Key Highlights (Chapter Summary)

◇ Chapter 1—Key Highlights:

1) Climate change is a current and urgent global issue, largely driven by rising CO_2 levels.
2) Carbon capture is a complementary tool to renewable energy, especially for hard-to-decarbonize sectors.
3) Environmental sustainability involves three pillars: environmental protection, social equity, and economic resilience.
4) Carbon capture technologies have evolved from industrial use to a broader climate mitigation strategy.
5) CO_2 has a long atmospheric lifetime, making both emissions reduction and removal essential.
6) Carbon capture is supported by major climate models (e.g. IPCC, IEA) as part of achieving net-zero targets.
7) Policy frameworks are crucial to ensure carbon capture is deployed effectively and fairly.
8) Environmental impact of carbon capture projects must be assessed using life cycle analysis to avoid unintended harm.
9) Ethical considerations such as justice, public participation, and transparency must shape deployment.
10) The book takes an interdisciplinary approach to explore carbon capture through policy, ethics, science, and society.

Keywords: Carbon Capture and Storage; Direct Air Capture; Post-Combustion Capture; Greenhouse Gas Mitigation; Fossil Fuels; Climate Justice; Just Transition; Life Cycle Assessment; Policy Frameworks; Public Participation; Energy Systems; Sustainable Infrastructure; Global Equity; Carbon Markets; Industrial Emissions; Climate Policy; Technological Innovation; Environmental Trade-Offs; Ethical Responsibility; Long-Term Storage.

References

1 K. Abbass, M. Z. Qasim, H. Song, *et al.*, "A review of the global climate change impacts, adaptation, and sustainable mitigation measures," *Environmental Science and Pollution Research*, vol. 29, pp. 42539–42559, 2022. DOI: 10.1007/s11356-022-19718-6.

2 A. Dubey and A. Arora, "Advancements in carbon capture technologies: A review," *Journal of Cleaner Production*, vol. 373, 133932, Nov., 2022. DOI: 10.1016/j.jclepro.2022.133932.

3 A. R. Pearson, C. G. Tsai, and S. Clayton, "Ethics, morality, and the psychology of climate justice," *Current Opinion in Psychology*, vol. 42, pp. 36–42, Dec., 2021. DOI: 10.1016/j.copsyc.2021.03.001.

4 A. Dey, S. K. Dash, and B. Mandal, "Introduction to carbon capture," in *Emerging Carbon Capture Technologies: Towards a Sustainable Future*, M. Khalid, S. A. Dharaskar, M.

Sillanpää, and H. Siddiqui, Eds, 1st ed. Elsevier, 2022, pp. 1–31, DOI: 10.1016/B978-0-323-89782-2.00005-3.

5 M. Al-Shafi, O. Massarweh, A. S. Abushaikha, and Y. Bicer, "A review on underground gas storage systems: Natural gas, hydrogen and carbon sequestration," *Energy Reports*, vol. 9, pp. 6251–6266, Dec., 2023. DOI: 10.1016/j.egyr.2023.05.236.

6 J. Ma *et al.*, "Carbon capture and storage: History and the road ahead," *Engineering*, vol. 14, pp. 33–43, Jul., 2022. DOI: 10.1016/j.eng.2021.11.024.

7 M. Kabir *et al.*, "Climate change due to increasing concentration of carbon dioxide and its impacts on environment in 21st century: a mini review," *Journal of King Saud University, Science*, vol. 35, no. 5, 102693, Jul., 2023. DOI: 10.1016/j.jksus.2023.102693.

8 L. Yu, S. D. M. Tkachenko, I. S. Sherbakova, and A. S. Rovnyagina, "Why we need to purify the air from carbon dioxide," *IOP Conference Series: Earth and Environmental Science*, vol. 815, no. 1, p. 012006, 2021. DOI: 10.1088/1755-1315/815/1/012006.

9 F. Ahmad, Q. Saeed, S. M. U. Shah, M. A. Gondal, and S. Mumtaz, "Environmental sustainability: Challenges and approaches," in *Natural Resources Conservation and Advances for Sustainability*, M. K. Jhariya, R. S. Meena, A. Banerjee, and S. N. Meena, Eds, 1st ed. Elsevier, 2022, ch. 11, pp. 243–270, DOI: 10.1016/B978-0-12-822976-7.00019-3.

10 D. Fiack, J. Cumberbatch, M. Sutherland, and N. Zerphey, "Sustainable adaptation: Social equity and local climate adaptation planning in U.S. cities," *Cities*, vol. 115, 103235, Aug., 2021. DOI: 10.1016/j.cities.2021.103235.

11 H. Qudrat-Ullah and C. M. Nevo, "The impact of renewable energy consumption and environmental sustainability on economic growth in Africa," *Energy Reports*, vol. 7, pp. 3877–3886, Nov., 2021. DOI: 10.1016/j.egyr.2021.05.083.

12 T. Wilberforce, A. G. Olabi, E. T. Sayed, K. Elsaid, and M. A. Abdelkareem, "Progress in carbon capture technologies," *Science of the Total Environment*, vol. 761, 143203, Mar., 2021. DOI: 10.1016/j.scitotenv.2020.143203.

13 M. Ozkan, S. P. Nayak, A. D. Ruiz, and W. Jiang, "Current status and pillars of direct air capture technologies," *iScience*, vol. 25, no. 4, 103990, Apr., 2022. DOI: 10.1016/j.isci.2022.103990.

14 K. T. Raimi, K. S. Wolske, P. S. Hart, and S. Choi, "Exploring public perceptions of carbon capture and utilization in the U.S.," *Sustainable Production and Consumption*, vol. 50, pp. 314–326, Oct., 2024. DOI: 10.1016/j.spc.2024.08.003.

15 Q. Zhao, "A review of pathways to carbon neutrality from renewable energy and carbon capture," *E3S Web of Conferences*, vol. 245, 01018,, 2021. DOI: 10.1051/e3sconf/202124501018.

16 S. Paltsev, J. Morris, H. Kheshgi, and H. Herzog, "Hard-to-Abate sectors: The role of industrial carbon capture and storage (CCS) in emission mitigation," *Applied Energy*, vol. 300, 117322, Oct., 2021. DOI: 10.1016/j.apenergy.2021.117322.

17 A. G. Olabi *et al.*, "Large scale application of carbon capture to process industries—a review," *Journal of Cleaner Production*, vol. 362, 132300, Aug., 2022. DOI: 10.1016/j.jclepro.2022.132300.

18 M. Goodhart, "Climate change and the politics of responsibility," *Perspectives on Politics*, vol. 21, no. 2, pp. 550–568, 2023. DOI: 10.1017/S153759272200319X.

19 C. B. Milindi and R. Inglesi-Lotz, "The role of green technology on carbon emissions: Does it differ across countries' income levels?," *Applied Economics*, vol. 54, no. 29, pp. 3309–3339, 2022. DOI: 10.1080/00036846.2021.1998331.

20 K. Radunsky and T. Cadman, "Addressing climate change risks: Importance and urgency," in *Handbook of Climate Change Management*, J. M. Luetz and D. Ayal, Eds. Cham: Springer, 2021, pp. 1405–1431, DOI: 10.1007/978-3-030-57281-5_288.

21 J. A. E. Nielsen, K. Stavrianakis, and Z. Morrison, "Community acceptance and social impacts of carbon capture, utilization and storage projects: A systematic meta-narrative literature review," *PLoS One*, vol. 17, no. 8, e0272409, Aug., 2022. DOI: 10.1371/journal.pone.0272409.

22 K. Witte, "Social acceptance of carbon capture and storage (CCS) from industrial applications," *Sustainability*, vol. 13, no. 21, p. 12278, Nov., 2021. DOI: 10.3390/su132112278.

23 S. Low, L. Fritz, C. M. Baum, and B. K. Sovacool, "Public perceptions on carbon removal from focus groups in 22 countries," *Nature Communications*, vol. 15, p. 3453, Apr., 2024. DOI: 10.1038/s41467-024-47853-w.

24 T. Mikunda, L. Brunner, E. Skylogianni, J. Monteiro, L. Rycroft, and J. Kemper, "Carbon capture and storage and the sustainable development goals," *International Journal of Greenhouse Gas Control*, vol. 108, 103318, June, 2021. DOI: 10.1016/j.ijggc.2021.103318.

25 M. Lahsen and J. Ribot, "Politics of attributing extreme events and disasters to climate change," *Wiley Interdisciplinary Reviews: Climate Change*, vol. 13, no. 1, e750, Jan./Feb., 2022. DOI: 10.1002/wcc.750.

26 D. Song, T. Jiang, and C. Rao, "Review of policy framework for the development of carbon capture, utilization and storage in China," *International Journal of Environmental Research and Public Health*, vol. 19, no. 24, 16853, Dec., 2022. DOI: 10.3390/ijerph192416853.

27 K. Jiang and P. Ashworth, "The development of carbon capture utilization and storage (CCUS) research in China: A bibliometric perspective," *Renewable and Sustainable Energy Reviews*, vol. 138, 110521, Mar., 2021. DOI: 10.1016/j.rser.2020.110521.

28 L. Rosa, D. L. Sanchez, G. Realmonte, D. Baldocchi, and P. D'Odorico, "The water footprint of carbon capture and storage technologies," *Renewable and Sustainable Energy Reviews*, vol. 138, 110511, Mar., 2021. DOI: 10.1016/j.rser.2020.110511.

29 N. Thonemann, L. Zacharopoulos, F. Fromme, and J. Nühlen, "Environmental impacts of carbon capture and utilization by mineral carbonation: A systematic literature review and meta life cycle assessment," *Journal of Cleaner Production*, vol. 332, 130067, Jan., 2022. DOI: 10.1016/j.jclepro.2021.130067.

30 D. Scott, "Chapter 14: Ethics of climate engineering: Chemical capture of carbon dioxide from air," in *Ethics of Chemistry*, J. Schummer and T. Børsen, Eds. Singapore: World Scientific, 2021, pp. 375–401, DOI: 10.1142/9789811233548_0014.

31 F. Mulyasari, A. K. Harahap, A. O. Rio, R. Sule, and W. G. A. Kadir, "Potentials of the public engagement strategy for public acceptance and social license to operate: Case study of carbon capture, utilisation, and storage *Gundih* pilot project in Indonesia," *International Journal of Greenhouse Gas Control*, vol. 108, 103312, June, 2021. DOI: 10.1016/j.ijggc.2021.103312.

32 P. Perreault, N. B. Kummamuru, A. G. Quiroga, and S. Lenaerts, "CO_2 capture initiatives: Are governments, society, industry and the financial sector ready?," *Current Opinion in Chemical Engineering*, vol. 38, p. 100874, Dec., 2022. DOI: 10.1016/j.coche.2022.100874.

2

Climate Change, Global Emissions, and the Need for Carbon Capture

2.1 Overview of Global Climate Target and Role of CCS

Climate change refers to the long-term alteration of temperature and weather patterns on Earth caused primarily by human activity since the Industrial Revolution [1]. While natural factors such as volcanic eruptions, solar radiation variations, and oceanic cycles can influence the climate, the dominant driver of the current warming trend is the accumulation of greenhouse gases (GHGs) in the atmosphere, principally carbon dioxide (CO_2), methane (CH_4), and nitrous oxide (N_2O). Due to the greenhouse effect, GHGs trap heat and prevent infrared radiation from escaping into space. This leads to an increase in the temperature at the planet's surface [2].

The period from 1850 to 1900 is often used as the baseline for "preindustrial levels," with an estimated world mean surface temperature of about 13.7°C. Under international efforts like the Paris Agreement, countries have agreed to keep the rise in global average temperature well below 2°C above preindustrial levels. Since the average temperature between 1850 and 1900 was about 13.7°C, this means the global temperature should stay below roughly 15.7°C. The 2°C limit, however, came from political negotiations, not scientific safety [3]. However, scientists now recognize that even this limit is too high to be considered safe. With global temperatures already about 1.2°C warmer, the world is experiencing severe impacts, such as stronger heatwaves, melting ice, and rising sea levels.

Figure 2.1 presents a powerful visual representation of the Earth under stress as a result of human industrial activity. The image depicts factories emitting dark smoke into the atmosphere, symbolizing the GHG emissions that contribute to global warming. There appears to be a layer of pollution and fire covering the upper part of the planet, reflecting rising temperatures, wildfires, and ecological destruction. The lower part of the Earth represents the fragile balance of the natural environment, which is now under threat. As a result of this contrast, it is evident that there is an urgent need to reduce carbon emissions and transition toward cleaner energy sources in order to prevent further damage to the environment.

There are a number of consequences associated with this warming. The melting of glaciers and the thermal expansion of the ocean contribute to rising sea levels that threaten coastal cities. There has been an increase in the frequency and severity of extreme weather events, such as heatwaves, droughts, wildfires, and heavy rainfall. Ecosystems are

Figure 2.1 Visual representation of global warming and industrial pollution, showing the Earth overheated and covered by emissions from factories, symbolizing the impact of human activities on the planet's climate. (Courtesy of Shutterstock.com).

Figure 2.2 Coastal road erosion and damage caused by rising sea levels and stronger waves due to the local impacts of global warming and ocean expansion. (Courtesy of Shutterstock.com).

disrupted as species struggle to adapt to shifting temperature zones. Ocean acidification, driven by higher CO_2 absorption, endangers marine biodiversity and food chains [4].

As can be seen in Figure 2.2, rising sea levels and coastal erosion are two of the most significant consequences of global warming. Increasing temperatures cause glaciers and ice sheets to melt and oceans to expand as a result of melting ice and warming seawater. By raising sea levels, coastlines are gradually eroded, and nearby infrastructure such as roads, buildings, and farmland is damaged. The collapsed section of the coastal road illustrates how powerful waves and continuous erosion can damage the land, resulting in costly and dangerous consequences. As climate change accelerates, coastal communities around the world face an increasing level of risk.

There is an equal vulnerability to human systems. Agricultural productivity is threatened by droughts and shifting rainfall patterns, while water scarcity is intensifying in dry and desert regions [5]. Around the world, climate change is making life more difficult for many people. As a result of hot weather, floods, and droughts, people are forced to leave their homes and seek shelter in safer areas. As well as health issues such as heat stress (when the body becomes too hot and cannot cool down), mosquito-borne diseases such as malaria are prevalent in warmer climates. Hunger, illness, and conflict can result from these problems in different parts of the world. Poor countries cause the least pollution but suffer the most from climate change, making this a serious fairness issue [6].

The Sixth Assessment Report of the Intergovernmental Panel on Climate Change (IPCC) confirms that human influence is definitely responsible for global warming. According to the report, the world is likely to experience warming greater than 1.5°C by 2050 without rapid, deep, and sustained reductions in GHG emissions [7]. This level of warming is very serious because it can damage nature forever. Coral reefs may die if the ocean water becomes too warm and acidic. Ice sheets in places like Greenland can melt faster, which raises sea levels and cannot be reversed. Warmer oceans can also change ocean currents, affecting the weather and marine life around the world.

To address this crisis, international efforts like the Paris Agreement aim to limit global temperature rise below 2°C, ideally 1.5°C, by transitioning to low-carbon energy, adopting renewable energy plants, enhancing efficiency, and advancing carbon capture, utilization, and storage (CCUS) technologies [3]. Climate change is not only an environmental issue but also a systemic global challenge. It connects with energy policy, economics, public health, and social equity. The understanding of its mechanisms and consequences is the prerequisite for the exploration of mitigation strategies, particularly carbon capture, to be discussed in subsequent sections.

2.2 Sources and Trends of Greenhouse Gas Emissions

The majority of GHG emissions (GHGs) originate from human activities that have intensified since the Industrial Revolution. Energy production, industrial processes, agriculture, land-use, and waste production are the most significant contributors. Nearly 75% of total anthropogenic GHG emissions come from carbon dioxide (CO_2), followed by methane (CH_4) at about 17%, and nitrous oxide (N_2O) at 6%. Other pollutants include fluorinated gases (F-gases) emitted from industrial and refrigeration processes, which, despite being emitted in small quantities, have extremely high global warming potentials (GWPs) [8].

2.2.1 Major Sources of Emission

1) **Energy and Power Generation**: The combustion of fossil fuels, such as coal, oil, and natural gas, for electricity and heat production remains the largest single source of CO_2 emissions globally. The CO_2 emissions from coal-fired power plants are substantial per

unit of energy generated. Although renewable energy is expanding rapidly, global electricity generation still depends heavily on fossil fuels [9].

2) **Transportation**: The combined emissions of road vehicles, aviation, and maritime transport account for approximately 25% of global GHG emissions. The continued use of internal combustion engines, along with the growing need for transportation and travel around the world (called global mobility), especially in developing countries, has reduced the benefits gained from efficiency improvements, such as making engines use less fuel, and from the adoption of electric vehicles [10].
3) **Industry and Manufacturing**: About 20% of global CO_2 emissions are attributed to industrial activities such as cement, steel, chemicals, and petrochemical production. Cement production alone contributes around 7–8% to emissions, due to both fossil-fuel combustion and CO_2 release during limestone or calcium carbonate ($CaCO_3$) heating for chemical changes. Steel and the chemical sectors also emit large quantities of CO_2 and methane during production and process heating [11].
4) **Agriculture, Forestry, and Land Use (AFOLU)**: Farming is a big source of methane (CH_4) and nitrous oxide (N_2O). Cows and other animals produce methane when digesting food, and rice fields release methane when dead plant material breaks down underwater. Fertilizers that contain nitrogen also produce nitrous oxide, a gas that traps 300 times more heat than carbon dioxide (CO_2). Cutting down forests worsens things because it removes trees that naturally absorb and store CO_2 from the atmosphere [12]. This means proper land management is very critical to reducing emissions. Proper land management means using farming and land practices that protect soil, reduce emissions, and keep carbon stored in the ground and vegetation. An example includes using less chemical fertilizer or applying it more efficiently [13].
5) **Waste Products**: A landfill or wastewater treatment facility produces methane when waste breaks down without oxygen, similar to how dead plants decompose in rice fields. Landfills are large sites where household and industrial solid waste is buried. In these sites, food scraps, paper, and other organic materials slowly decompose underground, releasing methane gas [14]. The purpose of wastewater treatment plants is to clean dirty water from homes, industries, and sewage systems before it is released into the environment. In this process, organic matter in the water, such as human waste, food particles, oils, and plant materials, is broken down by microorganisms. When this happens in low-oxygen or oxygen-free (anaerobic) conditions, methane is produced as a byproduct [15]. Although these emissions are smaller than those from farming or energy production, they are still significant in urban areas, and reducing them is important for climate protection.

2.2.2 Global Emission Trends

The trend of global GHG emissions has been rising over the past century, driven by industrialization, population growth, and increasing energy demand. Between 1990 and 2020, global CO_2 emissions grew by more than 60%, reaching around 37 billion tonnes per year by 2022. The Asia-Pacific region, led by China and India, now accounts for over 50% of global CO_2 emissions, reflecting rapid economic and industrial growth [16].

By contrast, emissions in Europe and North America have stabilized or declined due to improvements in energy efficiency, fuel switching, and policy-driven actions to decrease carbon emissions [16]. Fuel switching means replacing high-carbon fuels (such as coal or oil) with a lower-carbon fuel such as natural gas or renewable alternatives [17]. For example, switching from coal to natural gas in power plants can reduce carbon dioxide emissions by about half per unit of energy produced, since natural gas is a lower-carbon fuel. Increasing the use of renewable energy sources such as solar, wind, and hydropower also contributes to reducing the use of fossil fuels and supporting cleaner energy systems.

Despite global efforts to reduce emissions, the average concentration of CO_2 in the atmosphere has risen sharply from about 280 parts per million (ppm) during preindustrial times (1850–1900) to over 420 ppm in 2024, according to NOAA data. (NOAA stands for the National Oceanic and Atmospheric Administration, a U.S. government agency that monitors the oceans, the atmosphere, weather, and climate.) [18]

Figure 2.3 illustrates the close relationship between global temperature and carbon dioxide (CO_2) emissions over time. The graph shows a steady rise in both CO_2 levels and average global temperature from 1880 to 2020. As industrial activities expanded and fossil fuel use increased, atmospheric CO_2 rose from about 280 ppm in 1880 to over 420 ppm in 2020, while global temperatures also climbed significantly. The clear upward trend in both curves demonstrates the strong connection between CO_2 and GHG emissions and global warming, emphasizing the impact of human activities on the Earth's climate system.

This is the highest CO_2 level recorded in 800,000 years. This indicates that human activities contribute to the addition of carbon dioxide to the atmosphere at a faster rate than nature is able to remove it. Earth's natural carbon sinks, primarily forests and oceans, absorb and store CO_2 from the atmosphere [18]. The term carbon sink refers to anything that absorbs more carbon dioxide from the atmosphere than it releases. Since people burn a large amount of fossil fuels and cut down forests, natural systems cannot keep up with the pace of human activity. This results in an increased amount of CO_2 remaining in the atmosphere, which traps heat and causes the planet to warm faster.

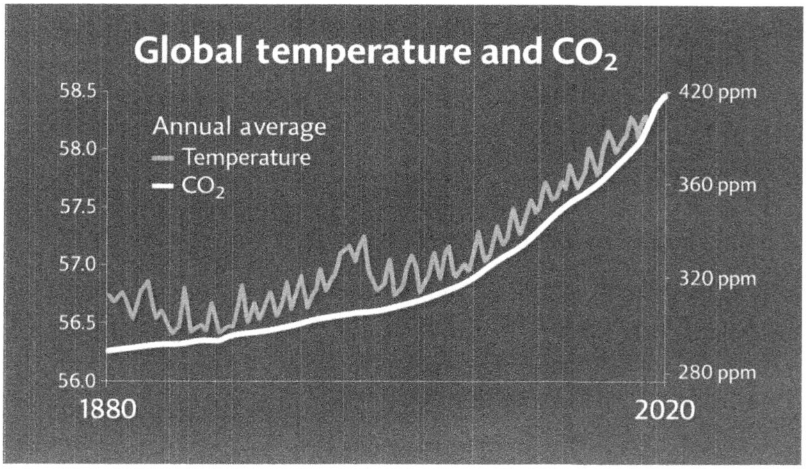

Figure 2.3 Relationship between global temperature and atmospheric CO_2 concentration from 1880 to 2020, showing a steady rise in both indicators. The graph highlights how increasing CO_2 levels from about 280 ppm to over 420 ppm correlate with long-term warming of the planet.

As a result, developed countries are historically responsible for the highest cumulative emissions, with the United States and Europe accounting for more than 40% of all CO_2 emissions since 1850. Today, developing countries are the biggest producers of GHG emissions. This raises questions about fairness regarding who should take more action on climate change. The amount of CO_2 released per person also varies significantly. People in the poorest countries produce less than 1 tonne per year, while those in the richer countries emit over 15 tonnes per person [19].

Without significant systemic changes, current emissions patterns are incompatible with limiting global warming to below 2°C above the preindustrial level. According to the International Energy Agency (IEA), even with existing policies, global CO_2 emissions may remain above 30 billion tonnes annually by 2050, which would contribute to the continuation of global warming [20]. The transition to renewable energy, energy efficiency enhancement, electrification of transportation, reduction of industrial emissions through low-carbon technologies, and deployment of carbon capture and storage (CCS) are all necessary to achieve carbon neutrality. Carbon neutrality means equalizing the amount of carbon dioxide (CO_2) emitted into the atmosphere with an equivalent amount removed or offset, so that net carbon emissions are zero.

It is evident from the persistent increase in GHG emissions that a global transformation of the energy and industrial sectors is urgently required. GHGs come from many sources, but the main challenge is how to grow our economy without increasing carbon emissions. To do this, we need new and advanced technologies that reduce or capture emissions, such as carbon capture and utilization. These technologies will be discussed in the next section.

2.3 Role of CO_2 and Its Global Impact

Carbon dioxide (CO_2) is the most significant human-made GHG and the primary cause of modern climate change. While naturally present in the atmosphere and essential for regulating Earth's temperature, human activities—especially since industrialization, have disrupted its natural balance [21]. CO_2 concentrations, which remained relatively stable at around 280 parts per million (ppm) during the preindustrial era (1850–1900), have risen sharply to over 420 ppm by 2024, according to NOAA data. This increase represents a 50% rise in atmospheric CO_2 within just two centuries, a rate of change unprecedented in geological history [18].

2.3.1 The Role of CO_2 in the Greenhouse Effect

Without carbon dioxide (CO_2) and other GHGs, the Earth would be a much colder and less livable planet. The Sun sends energy to Earth as sunlight. Some of this sunlight warms the surface and is used by plants, and then the Earth releases that energy back into space as heat (infrared radiation). If there were no GHGs like CO_2, methane, and water vapor in the atmosphere, almost all of that heat would escape into space. Nothing would trap it near the surface [22].

CO_2 is a gas that traps heat in the atmosphere of the Earth. Solar radiation reaches the planet's surface, where part of it is absorbed and reemitted as infrared radiation. The GHGs,

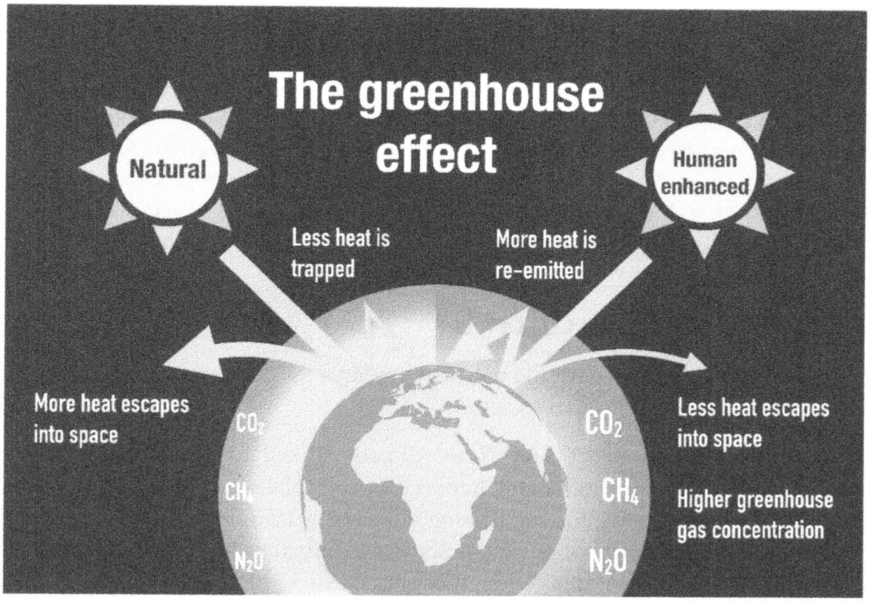

Figure 2.4 There are two types of greenhouse effects: natural and human enhanced. Naturally occurring greenhouse gases like CO_2, CH_4, and N_2O trap some of the Sun's heat and keep the Earth warm enough for life. However, human activities have increased these gases, trapping more heat in the atmosphere and causing global warming. (Courtesy of Shutterstock.com).

such as CO_2, absorb some of this outgoing infrared energy and reradiate it in all directions, thereby effectively insulating the Earth. Due to this natural process, Earth's average surface temperature remains at approximately 15°C instead of the frigid—18°C it would be without GHGs. However, excessive CO_2 concentrations amplify this effect and lead to global warming [22].

Figure 2.4 illustrates the natural and human-enhanced greenhouse effects. Several naturally occurring GHGs, such as CO_2, CH_4, and N_2O, trap enough heat to keep the Earth's temperature at a comfortable level for life. It should be noted, however, that human activities have caused an increase in these gases, contributing to the accumulation of heat and causing global warming.

CO_2 remains in the atmosphere for hundreds to thousands of years, unlike short-lived gases like methane, which only last about a decade. This means that even if we cut emissions drastically today, the CO_2 already in the atmosphere will continue to impact global temperatures for centuries [23]. That's why it's crucial to focus not just on reducing emissions but also on finding ways to remove CO_2 from the atmosphere, using both natural and technological solutions.

The main sources of carbon dioxide (CO_2) emissions caused by human activities come from a few key areas. The biggest contributor is burning fossil fuels like coal, oil, and natural gas for electricity, industrial heating, and transportation [24]. Industrial processes, such as making cement and steel, also release significant CO_2, both from burning fuel and the chemical reactions involved in their production. Deforestation and changes in land use add to the problem by reducing forests' ability to absorb CO_2 and releasing carbon stored in trees and soil [25].

When trees are cut down, the soil that once stored carbon begins to release it. Without tree roots and shade, the soil becomes warmer and exposed to the air, causing microbes to break down organic matter faster and release carbon dioxide (CO_2) [25]. Together, these activities account for about 75% of global GHG emissions, with cement production alone responsible for 7–8%. This shows how challenging it is to cut emissions in sectors like construction and heavy industries, such as cement and steel production.

2.3.2 Global Impacts of Elevated CO_2 Levels

There are many effects associated with the increase in atmospheric CO_2 on the Earth's climate system, environment, and human societies.

1) **Temperature Rise**: Elevated CO_2 levels are directly linked to global temperature increases. The IPCC's Sixth Assessment Report confirms that CO_2 is responsible for about 66% of the total radiative forcing driving observed warming. The average global temperature has already risen to about 1.2°C above preindustrial levels. Several studies suggest that without significant mitigation, it could exceed 2.5–3°C by 2100 [21].
2) **Ocean Acidification**: The world's oceans absorb about 25–30% of the CO_2 we release, which mixes with seawater to create carbonic acid. This lowers the ocean's pH, making it more acidic. This process, known as ocean acidification, poses a serious threat to marine life, especially coral reefs and shellfish. It also disrupts food chains and fisheries, impacting coastal economies and food security [26]. Figure 2.5 shows how atmospheric

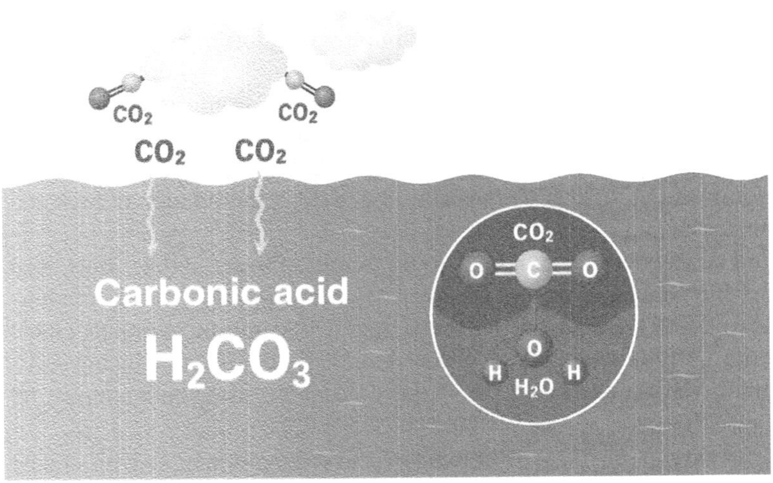

Figure 2.5 The formation of carbonic acid in ocean water. Atmospheric carbon dioxide (CO_2) dissolves in the ocean and reacts with water (H_2O) to form carbonic acid (H_2CO_3). In addition to contributing to ocean acidification, this chemical process is an important part of ocean carbon capture.

carbon dioxide (CO_2) interacts with ocean water to form carbonic acid (H_2CO_3). CO_2 from the atmosphere dissolves into the surface layer of the ocean, where it reacts chemically with water molecules (H_2O). This reaction produces carbonic acid, a weak acid that lowers the pH of seawater and contributes to ocean acidification.

3) **Melting of Ice and Sea-Level Rise**: CO_2-induced warming speeds up the melting of glaciers and ice, causing global sea levels to rise. Rising sea levels endanger coastal infrastructure, roads, and millions of people in those areas [21].
4) **Ecosystem and Biodiversity Disruption**: Changes in temperature and rainfall affect where plants and animals live, forcing many to move or disappear. Forests, wetlands, and the tundra are especially at risk. The tundra is a cold, treeless region found near the Arctic, where the ground stays frozen for most of the year. Even small temperature increases can melt frozen soil and disturb plants and animals. Higher CO_2 levels can make some plants grow faster, but they also reduce the amount of protein, iron, and zinc in many food crops like rice and wheat. This means that even if crops grow more, they may be less nutritious for people and animals [27].
5) **Socioeconomic Consequences**: CO_2-induced warming contributes to extreme weather events, droughts, floods, and crop failures. These consequences lead to economic losses, food insecurity, and health challenges, particularly in developing countries. When extreme weather damages farms, roads, and homes, economic losses occur because people and governments must spend more money rebuilding. Food insecurity happens when droughts, floods, or changing rainfall patterns reduce crop yields, making it hard for people to get enough food. Health problems also increase: heatwaves cause heat stress, changing climates spread diseases like malaria, and a lack of clean water can lead to illness. These impacts make life harder for communities that already have fewer resources to adjust to. Climate-related migration is expected to increase as regions become uninhabitable due to rising temperatures and other issues discussed in this paragraph [28].

2.3.3 Carbon Cycle and Management

The carbon cycle is the natural process that moves carbon between the air, oceans, plants, animals, and Earth's soil and rock. Carbon moves through this cycle in several ways. For example, plants absorb CO_2 from the air during photosynthesis, animals release CO_2 when they breathe, oceans absorb and store carbon, and carbon from dead plants and animals can become trapped in the soil or rocks for millions of years. This cycle helps keep the level of carbon dioxide (CO_2) in the atmosphere balanced. However, human activities, mainly burning fossil fuels and cutting down forests, have added too much CO_2, upsetting this natural balance. For example, as the planet warms, warmer oceans can hold less CO_2. These changes create feedback loops, where warming leads to even more emissions, which in turn warm the Earth faster.

Because carbon dioxide (CO_2) stays in the air for a long time and warms the planet, we need to both reduce new emissions and remove CO_2 that is already there. Technologies such as CCUS trap CO_2 from factories and store it deep underground. Direct air capture (DAC) works like a giant air filter as it pulls CO_2 directly from the air and stores or reuses it [29]. Bioenergy with carbon capture (BECCS) means using plants to make energy and capturing the CO_2 that is released when the plants are burned [30].

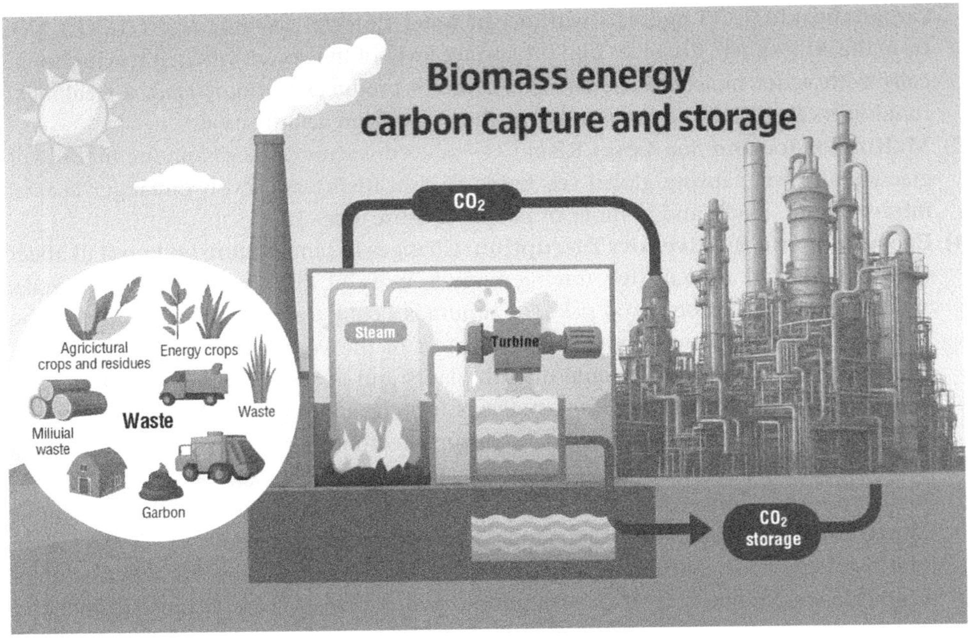

Figure 2.6 Illustration of Bioenergy with Carbon Capture and Storage (BECCS), showing how biomass is used for power generation while a carbon capture plant on the right captures and stores CO_2 to reduce atmospheric emissions.

Figure 2.6 illustrates the BECCS system. Electricity is generated by burning biomass from sources such as forest waste, crops, and organic residues. Instead of discharging carbon dioxide into the atmosphere, this process and the combined carbon capture plant capture, compress, and transport carbon dioxide underground for permanent storage. The process not only provides renewable energy from biomass but also reduces CO_2 emissions, making it a key technology for achieving net-negative emissions.

Natural methods also help. Storing carbon in the soil means farming practices like planting cover crops or reducing tilling. Cover crops are plants like grass that farmers grow not to harvest, but to protect and improve the soil. Reducing tilling means disturbing the soil as little as possible when planting. These practices keep carbon locked in the ground instead of escaping as CO_2 [31].

Ocean-based solutions include improving the ocean's natural ability to absorb CO_2 from the atmosphere. For example, the ocean absorbs about one-fourth of human-made CO_2. Scientists are studying ways to increase this natural capacity safely, such as by encouraging plankton growth (tiny plants that take up CO_2) or by increasing ocean alkalinity (making the water slightly less acidic so it can hold more CO_2) [32]. By combining these technological and eco-friendly approaches, we can move closer to net-zero emissions and help restore Earth's climate balance.

In conclusion, carbon dioxide is not merely one of many GHGs; it is the foundation of global warming [21]. Its long atmospheric lifetime, vast emission sources, and long-term effects on climate, ecosystems, and human systems make CO_2 management a central part of any effective climate strategy. Understanding its nature provides scientific approaches to developing and deploying capture technologies. These technologies form the basis of mitigation pathways discussed in subsequent sections.

2.4 The Scientific Basis: IPCC Reports and Climate Models

Climate change is a result of decades of scientific research and data synthesis, largely coordinated by the IPCC, established in 1988 by the World Meteorological Organization (WMO) and United Nations Environment Program (UNEP). Its reports, contributed to by thousands of scientists, provide the most comprehensive global agreement on climate change [33].

Climate reports are released by the IPCC every six to seven years, with each report providing stronger evidence that human activity contributes to global warming. It is stated in the Sixth Assessment Report (AR6), which was published between 2021 and 2023, that humans have contributed to the warming of the atmosphere, oceans, and land [33]. According to this study, Earth's surface temperature has increased by approximately 1.2°C since preindustrial times (1850–1900), and many of these changes have not occurred for thousands of years [34]. As a result, every slight increase in temperature will have a greater effect on human health, the environment, and infrastructure, so reducing emissions as soon as possible will be essential to maintain a stable climate.

A key aspect of the IPCC's work is the use of climate models, which are powerful computer programs that simulate how the Earth's climate system operates. Scientists can use these models to predict future climate changes under different levels of GHG emissions based on how the atmosphere, oceans, land, and ice interact with each other. The models are designed based on physical laws, such as the movement of heat, air, and radiation. They are tested with actual climate data and satellite observations. Scientists are confident that their predictions for the future will be accurate because these models can match past and current climate trends accurately [33].

The Coupled Model Intercomparison Project (CMIP) brings scientists together to study climate change. CMIP6, the latest version, was used in the IPCC's recent report. It examines future scenarios shaped by energy use and human lifestyles, impacting GHG emissions. SSP1–1.9, for example, is an example of a sustainable future in which people adopt low-emission lifestyles by using renewable energy, public transportation, and environmentally friendly consumption habits. The economy grows responsibly, and global warming is kept to a maximum of 1.5°C in this world. On the other hand, SSP5–8.5 predicts a future with a high level of carbon emissions, a high rate of energy demand, and a consumption-oriented lifestyle. Economic growth continues without strong environmental controls, which will result in a global temperature increase of over 4°C by 2100 and a severe climate change impact [35].

The IPCC's climate models illustrate what could happen if we fail to take action to combat climate change. Increasing GHG emissions will result in an increase in extreme weather events such as heatwaves, floods, and droughts. The sea level could rise by over one meter by 2100, dislocating millions of people and causing lasting damage to coastal ecosystems. However, strong climate action taken within the next 20 years could still limit global warming to below 2°C, preventing the worst impacts of global warming. It is clear that while there is still time for action, the window of opportunity is rapidly closing [36].

Another key part of the IPCC's research is the study of climate feedbacks, which are processes that can either amplify or reduce the effects of global warming. In climate science, "positive" feedback does not mean beneficial. It refers to processes that reinforce the initial warming and make it stronger. A common example is ice melting: as reflective ice disappears, darker ocean or land surfaces absorb more solar radiation, leading to additional warming. "Negative" feedback does not mean harmful. Instead, it refers to processes that counteract warming and help stabilize the climate system. For example, higher CO_2 concentrations may stimulate plant growth, allowing more CO_2 to be absorbed from the atmosphere, although this effect is limited. Together, these feedback mechanisms strongly influence the long-term stability of the Earth's climate system [36, 38].

The IPCC's scientific foundation is based on its multidisciplinary integration of atmospheric science, oceanography, ecology, and economics in order to understand the causes and effects of climate change [36]. In order to ensure accuracy, its reports are carefully reviewed by a team of experts to assist governments in making informed decisions. Not only does the IPCC explain how the climate is changing, but it also provides examples of how renewable energy, carbon capture, and carbon pricing can be used to combat climate change. The overall conclusion of these reports is that climate change is real, largely caused by humans, and is becoming more severe.

2.5 The Paris Agreement and Global Carbon Targets

The Paris Agreement, adopted in December 2015 at the COP21 meeting of the United Nations Framework Convention on Climate Change (UNFCCC), is the world's most influential global plan to fight climate change. It unites 195 countries on a shared goal to limit global warming and move toward cleaner, lower-carbon development. As opposed to the older Kyoto Protocol, which set strict emission targets for some countries, the Paris Agreement allows countries to set their own climate targets [37].

The Paris Agreement aims to keep global warming below 2°C above preindustrial levels and limit it to 1.5°C. This target is based on strong scientific evidence from the IPCC's reports. These reports show that if temperatures rise more than 2°C, the world could face serious and irreversible damage to both nature and human life. The 1.5°C goal is especially significant because it would reduce extreme weather risks, rising sea levels, and loss of ecosystems [38]. The future of the planet can be dramatically affected by even a small change of half a degree.

To achieve its goals, the Paris Agreement created a system called Nationally Determined Contributions (NDCs). Each country prepares its own climate action plans, explaining how it will reduce emissions, adapt to climate change, and support clean development. In contrast to earlier agreements that required all countries to abide by the same rules, the NDCs allow countries to set their own goals in accordance with their resources, technology, and development needs [39].

The Paris Agreement has also created rules to ensure countries are transparent about their progress regarding climate change and global warming. Each country must report regularly on how much GHG it produces and how well it meets its climate goals in NDCs' systems. Experts can review and compare reports from all countries. Additionally, every five years, the Global Stocktake (GST) is conducted to assess how well the world is progressing toward its climate goals. In 2023, the first GST revealed that current plans are not sufficient to keep global warming below 1.5°C, which means countries must take stronger and faster measures to reduce emissions [38, 39].

A key component of the Paris Agreement is the provision of financial assistance and technology sharing, particularly for developing countries and countries that are vulnerable. It has been agreed that rich nations will provide a large amount of money each year to assist poorer countries in reducing emissions [40]. Among the projects funded by this money are renewable energy projects, such as the construction of solar panels and wind farms, and stronger infrastructure projects, such as the construction of roads and bridges capable of withstanding floods and storms.

Despite the fact that the Paris Agreement represents a significant step forward, it still faces many challenges. It is possible that, even if all countries meet their current climate plans (NDCs), global temperatures will rise by about 2.5–2.7°C by 2100, which is much higher than the 1.5°C goal. CO_2 emissions also increase [41]. There are many countries that have difficulty putting climate policies into action due to problems relating to politics, economics, or technology. Due to the nature of the Paris Agreement, which is based on voluntary commitments, it lacks strict penalties. Therefore, the success of the Agreement depends heavily on each country's political will and on international cooperation to take stronger measures [38, 39].

Although the Paris Agreement faces many challenges, it remains an important step in the global fight against climate change. In order to reduce carbon emissions, it provides a flexible yet structured system that allows countries, cities, industries, and communities to work together [38]. In addition to its impact on government policies, it has led to the development of clean technologies, the growth of renewable energy, and improved carbon management. There are now several large economies that have set net-zero targets, such as the European Union by 2050, China by 2060, and India by 2070, demonstrating widespread support for the goals of the Paris Agreement [42].

In summary, the Paris Agreement creates a strong global plan to fight climate change. This was done by setting clear temperature goals, holding countries responsible, and supporting them through finance and technology. Despite the fact that there is still a gap between current promises and actual action, the agreement can help the transition of the world toward long-term sustainability [38]. Reaching its goals will require not only cutting emissions but also using CCUS technologies to remove remaining emissions and achieve net-zero carbon levels. This leads to the next section, which looks at the gaps in current mitigation efforts. It also looks at the significant role carbon capture technologies play in reaching global climate goals.

2.6 Limits of Current Climate Action and the Need for Carbon Capture

Global GHG emissions continue to rise despite numerous scientific warnings and international agreements. As a result, there appears to be a growing gap between what countries promise to do and what actually takes place. IPCC's Sixth Assessment Report (AR6) warns that current emissions levels do not meet the goals of the Paris Agreement. Despite all the current climate plans (NDCs), the world could still warm by about 2.5°C to 2.7°C above the preindustrial level by the end of this century. This is well in excess of the 1.5°C limit needed to avoid climate change's worst effects [7, 33]. As a result of this gap, it can be concluded that current efforts will not suffice and that additional solutions, such as CCUS, are essential for reducing emissions and achieving climate goals.

One of the main reasons why the world is still not cutting enough emissions is due to the heavy reliance on fossil fuels in the global economy. Approximately 80% of the world's energy comes from coal, oil, and natural gas, especially in developing countries that require more energy to run factories and grow [9, 24]. In spite of the rapid growth in renewable energy sources, such as solar, wind, and hydro, they cannot yet completely replace fossil fuels. . This is mainly due to their intermittent nature, since solar and wind generation depend on weather conditions and are subject to variability, as well as current limitations in large-scale energy storage and grid infrastructure.The cement, steel, chemicals, and aviation industries are considered "hard-to-abate" industries because they produce CO_2 during chemical processes, burn fossil fuel, or require very high temperatures that are difficult to achieve solely through heat from solar energy [43].

As another reason for slow progress, there is a disconnect between climate policies and real-world action. Several countries have committed to reaching net-zero emissions, but only a few have made these goals legally binding or established clear plans for achieving them. Many government policies are aimed at meeting today's energy needs rather than achieving long-term clean energy goals. It is also challenging for developing countries to take strong action on climate change due to differences in technology, funding, and government capacity [44]. Therefore, the world's response is uneven. Some regions are reducing emissions, while others are still increasing.

It is also difficult to reduce emissions due to economic and social factors. As fossil fuels remain cheap and widely used, it is difficult to switch to cleaner alternatives. There has been a significant increase in carbon pricing systems, which make polluters pay for their emissions. However, these systems are still too small and too weak to produce substantial changes. Carbon taxes and energy reforms are also opposed by many people and politicians because they raise costs. Carbon budget refers to the amount of carbon dioxide that can still be released before the Earth warms by more than 1.5°C. Due to the fact that we continue to produce more emissions every year, this budget is becoming smaller [45]. If we don't cut emissions soon, we could use up the remaining carbon budget for 10 years.

As a result of these challenges, carbon capture technologies have become an integral part of the fight against climate change. As the name implies, CCUS is the process of capturing CO_2 from factories or power plants, transporting it, and either using it in industrial processes or storing it deep underground to prevent it from entering the atmosphere.

Carbon capture eliminates emissions at their source, unlike renewable energy and energy efficiency measures. It is possible to remove CO_2 that is already in the air when used in conjunction with bioenergy (BECCS) [30] or DAC. As a result, CCUS can help the world reach net-zero emission levels. Figure 2.7 shows a DAC facility located in an arid region, equipped with large axial fans and modular air contactor units designed to capture CO_2 directly from ambient air. The extracted CO_2 is then processed and stored in cylindrical tanks for compression or utilization.

Both the IPCC and the IEA agree that global climate goals cannot be met without carbon capture (CCUS). According to the IEA's Net Zero by 2050 plan, CCUS is expected to remove

Figure 2.7 A large-scale Direct Air Capture (DAC) plant featuring modular air contactors and CO_2 processing units for atmospheric carbon removal.

approximately 6–8 billion tonnes of CO_2 per year by the year 2050. This represents approximately 15% of the total reduction in emissions required. However, progress remains slow [46]. In 2024, there will be approximately 50 large CCUS projects operating around the world. It is estimated that these projects capture fewer than 50 million tonnes of CO_2 per year, which is far less than the required amount. The main challenges are high costs, weak government support, and public concerns about underground CO_2 storage [47].

It will be necessary to strengthen policies and provide financial support to accelerate carbon capture and utilization (CCUS) to reduce emissions [47]. Governments can help by setting carbon prices, offering tax incentives, and establishing clear rules for storing CO_2 safely for an extended period of time. In addition, international cooperation can facilitate the sharing of technology, the provision of funding, and the training of experts in developing countries where industries are growing rapidly. A combination of CCUS with blue hydrogen resources, biomass, and bioenergy could also make the technology more affordable and efficient by sharing infrastructure and costs [48].

The reduction of emissions requires more than just technological innovation. It requires us to make substantial changes in the way we produce products, such as cement and steel, and in how we consume energy [11]. We must improve energy efficiency, utilize circular economy ideas (recycling and reusing materials instead of wasting them) [49], and participate in nature-based activities, such as planting trees and storing carbon in the soil. However, despite these efforts, some emissions will remain from heavy industry, transportation, and agriculture for some time to come. That is why carbon capture is an essential component of achieving global climate change goals.

In conclusion, today's global emissions trend indicates that current climate measures such as renewable energy, energy efficiency improvements, and the promotion of low-carbon lifestyles are insufficient to meet the Paris Agreement's goals. The world still relies heavily on fossil fuels as a source of energy, and progress toward cleaner energy systems has been slow [9, 24]. Climate targets are often set by governments, but the policies are often weak or not fully enforced, and some countries lack the funds or technology to effectively implement them. As a result of these problems, there has been a widening gap between what nations have promised and what they have actually accomplished. In order to close this gap, we must employ a combination of strong solutions, such as CCUS, which plays an essential role. As the following chapters will explain, the use of large-scale carbon capture technologies is essential for the reduction of industrial emissions and for creating a path to net-zero emissions and a stable climate.

2.7 Key Highlights (Chapter Summary)

◇ Chapter 2 – Key Highlights:

1) It is undeniable that global warming is a result of human activities, primarily as a result of the burning of fossil fuels and industrial processes that have increased atmospheric CO_2 levels from 280 parts per million to over 420 parts per million in 2024.
2) GHG emissions continue to rise, with CO_2 responsible for approximately 75% of total GHG emissions, followed by methane (CH_4) and nitrous oxide (N_2O).
3) The global average temperature has already increased by around 1.2°C above pre-industrial levels (1850–1900), leading to intensified heatwaves, droughts, floods, and sea-level rise.
4) Over half of the global CO_2 emissions are attributed to fossil fuel-based power generation, transportation, heavy industry (cement, steel, chemicals), and deforestation.
5) IPCC climate models and assessments confirm that current emission trends are not in accordance with Paris Agreement goals. They warn that continued warming will lead to catastrophic climate system changes.
6) Nearly all countries have committed to limiting global temperature rise below 2°C and striving for 1.5°C as part of the Paris Agreement (2015).
7) Estimates suggest that global warming may reach 2.5–2.7°C by 2100, which is far greater than the Paris targets.
8) Because of fossil fuel reliance and limited technological alternatives in hard-to-abate sectors, the commitments to reduce GHG emissions remain difficult to achieve.
9) In industries and sectors where direct decarbonization is difficult, carbon capture, utilization, and storage (CCUS) is becoming increasingly important as an essential technology for addressing residual emissions.

10) Achieving net-zero emissions requires a combination of several strategies, including the expansion of renewable energy, energy efficiency, circular economy practices, and large-scale deployment of carbon capture technologies.

Keywords: Climate Change; Global Warming; Greenhouse Gas Emissions; Carbon Dioxide (CO_2); IPCC Climate Models; Paris Agreement; Nationally Determined Contributions (NDCs); Carbon Neutrality; Fossil-Fuel Combustion; Carbon Capture Utilization and Storage (CCUS); Mitigation Strategies; Global Temperature Rise; Climate Policy; Renewable Energy Transition; Net-Zero Emissions.

References

1 D. S.-K. Ting and A. Vasel-Be-Hagh, *Mitigating Climate Change: Proceedings of the Mitigating Climate Change 2021 Symposium and Industry Summit (MCC 2021)*. Cham, Switzerland: Springer, 2022, DOI: 10.1007/978-3-030-92148-4.

2 H. Ritchie, P. Rosado, and M. Roser, "CO_2 and greenhouse gas emissions," Our World in Data, 2023. [Online]. Available: https://ourworldindata.org/co2-and-greenhouse-gas-emissions

3 M. Joshi, E. Hawkins, R. Sutton, *et al.*, "Projections of when temperature change will exceed 2 °C above pre-industrial levels," *Nature Climate Change*, vol. 1, pp. 407–412, 2011. DOI: 10.1038/nclimate1261.

4 A. Gulzar, M. A. Mehmood, S. A. Ganie, and I. Showqi, "A brief review on global warming and climate change: Consequences and mitigation strategies," *International Journal of Advance Research in Science and Engineering (IJARSE)*, vol. 7, no. 4, pp. 1–11, Mar., 2018.

5 C. A. Nobre, "Tropical agriculture and global warming: Impacts and mitigation options," *Scientia Agricola*, vol. 64, no. 1, pp. 83–89, Feb., 2007. DOI: 10.1590/S0103-90162007000100013.

6 W. N. Rom and K. E. Pinkerton, "Introduction: Consequences of global warming to planetary and human health," in *Climate Change and Global Public Health*, W. N. Rom and K. E. Pinkerton, Eds. Cham, Switzerland: Springer, 2020, pp. 1–33, DOI: 10.1007/978-3-030-54746-2_1.

7 H. Lee and J. Romero, Eds., *Climate Change 2023: Synthesis Report, Summary For Policymakers. Contribution Of Working Groups I, II And III To The Sixth Assessment Report Of The Intergovernmental Panel On Climate Change*. Geneva, Switzerland: Intergovernmental Panel on Climate Change (IPCC), 2023, DOI: 10.59327/IPCC/AR6-9789291691647.001.

8 L. Jeffry, M. Y. Ong, S. Nomanbhay, M. Mofijur, M. Mubashir, and P. L. Show, "Greenhouse gases utilization: A review," *Fuel*, vol. 301, p. 121017, Oct., 2021. DOI: 10.1016/j.fuel.2021.121017.

9 M. Höök and X. Tang, "Depletion of fossil fuels and anthropogenic climate change—a review," *Energy Policy*, vol. 52, pp. 797–809, Jan., 2013. DOI: 10.1016/j.enpol.2012.10.046.

10 C. Wang, J. Wood, Y. Wang, X. Geng, and X. Long, "CO_2 emission in transportation sector across 51 countries along the Belt and Road from 2000 to 2014," *Journal of Cleaner Production*, vol. 266, p. 122000, Sept., 2020. DOI: 10.1016/j.jclepro.2020.122000.

11 W. Shen *et al.*, "Quantifying CO_2 emissions from China's cement industry," *Renewable and Sustainable Energy Reviews*, vol. 50, pp. 1004–1012, Oct., 2015. DOI: 10.1016/j.rser.2015.05.031.

12 Q. Yue, X. Xu, J. Hillier, K. Cheng, and G. Pan, "Mitigating greenhouse gas emissions in agriculture: From farm production to food consumption," *Journal of Cleaner Production*, vol. 149, pp. 1011–1019, Apr., 2017. DOI: 10.1016/j.jclepro.2017.02.172.

13 X. Chuai, X. Huang, W. Wang, R. Zhao, M. Zhang, and C. Wu, "Land use, total carbon emissions change and low carbon land management in coastal Jiangsu, China," *Journal of Cleaner Production*, vol. 103, pp. 77–86, Sept., 2015. DOI: 10.1016/j.jclepro.2014.03.046.

14 C. Zhang, T. Xu, H. Feng, and S. Chen, "Greenhouse gas emissions from landfills: A review and bibliometric analysis," *Sustainability*, vol. 11, no. 8, p. 2282, 2019. DOI: 10.3390/su11082282.

15 C. Song, J. Zhu, J. L. Willis, D. P. Moore, M. A. Zondlo, and Z. J. Ren, "Methane emissions from municipal wastewater collection and treatment systems," *Environmental Science & Technology*, vol. 57, no. 6, pp. 2248–2261, Feb., 2023. DOI: 10.1021/acs.est.2c04388.

16 M. Crippa *et al.*, CO_2 Emissions of All World Countries: 2022 Report, JRC/IEA/PBL, European Commission, 2022.

17 F. Michelet, "How fuel switching impacts the environmental value of renewable energy," SSRN, 32 pages, Jul. 8, 2025. [Online]. Available: https://papers.ssrn.com/sol3/papers.cfm?abstract_id=5344538

18 M. Vollmer and W. Eberhardt, "A simple model for the prediction of CO_2 concentrations in the atmosphere, depending on global CO_2 emissions," *European Journal of Physics*, vol. 45, no. 2, p. 025803, Feb., 2024. DOI: 10.1088/1361-6404/ad230d.

19 M. Khan, A. T. Rana, and W. Ghardallou, "FDI and CO_2 emissions in developing countries: The role of human capital," *Natural Hazards*, vol. 117, pp. 1125–1155, 2023. DOI: 10.1007/s11069-023-05949-4.

20 International Energy Agency, *Net Zero by 2050: A Roadmap for the Global Energy Sector.* IEA, 2021.

21 M. Filonchyk, M. P. Peterson, L. Zhang, V. Hurynovich, and Y. He, "Greenhouse gases emissions and global climate change: Examining the influence of CO_2, CH_4, and N_2O," *Science of the Total Environment*, vol. 935, p. 173359, Jul., 2024. DOI: 10.1016/j.scitotenv.2024.173359.

22 D. Coe, W. Fabinski, and G. Wiegleb, "The impact of CO_2, H_2O and other 'greenhouse gases' on equilibrium Earth temperatures," *International Journal of Atmospheric and Oceanic Sciences*, vol. 5, no. 2, pp. 29–40, Aug., 2021. DOI: 10.11648/j.ijaos.20210502.12.

23 J. W. B. Rae, Y. G. Zhang, X. Liu, G. L. Foster, H. M. Stoll, and R. D. M. Whiteford, "Atmospheric CO_2 over the past 66 million years from marine archives," *Annual Review of Earth and Planetary Sciences*, vol. 49, pp. 609–641, May, 2021. DOI: 10.1146/annurev-earth-082420-063026.

24 B. Li and N. Haneklaus, "The role of renewable energy, fossil fuel consumption, urbanization and economic growth on CO_2 emissions in China," *Energy Reports*, vol. 7, no. Suppl. 7, pp. 783–791, Nov., 2021. DOI: 10.1016/j.egyr.2021.09.194.

25 I. P. Obani, Z. I. Obani, F. C. Anaeto, T. A. Ojewe, and C. S. Nwachukwu, "Impact of deforestation on carbon emissions: A data-driven study of the Amazon and Southeast Asia," *European Journal of Life Safety and Stability*, vol. 50, pp. 17–36, 2025. [Online]. Available: https://ejlss.es/index.php/ejlss/article/view/80

26 H. S. Findlay and C. Turley, "Chapter 13—ocean acidification and climate change," in *Climate Change (Third Edition): Observed Impacts on Planet Earth*, D. A. Vallero, Ed. Amsterdam, Netherlands: Elsevier, 2021, pp. 251–279, DOI: 10.1016/B978-0-12-821575-3.00013-X.

27 K. R. Shivanna, "Climate change and its impact on biodiversity and human welfare," *Proceedings of the Indian National Science Academy*, vol. 88, pp. 160–171, 2022. DOI: 10.1007/s43538-022-00073-6.

28 N. Karacasulu and İ. A. Karakir, "Climate-related migration: Formulating a contemporary framework," *Journal of International and Area Studies*, vol. 28, no. 1, pp. 99–118, 2021. [Online]. Available: https://www.jstor.org/stable/27210937

29 P. Roy, A. K. Mohanty, and M. Misra, "Prospects of carbon capture, utilization and storage for mitigating climate change," *Environmental Science: Advances*, vol. 2, pp. 409–423, 2023. DOI: 10.1039/D2VA00236A.

30 B. H. M. Silveira, H. K. M. Costa, and E. M. Santos, "Bioenergy with carbon capture and storage (BECCS) in Brazil: A review," *Energies*, vol. 16, no. 4, 2023. DOI: 10.3390/en16042021.

31 C. Paul *et al.*, "Carbon farming: Are soil carbon certificates a suitable tool for climate change mitigation?," *Journal of Environmental Management*, vol. 330, 117142, Mar., 2023. DOI: 10.1016/j.jenvman.2022.117142.

32 A. Abreu *et al.*, Plankton-Based Solutions to Support Life On Our Planet: CO_2 Sink Carbon And Mitigate Climate Change, Doctoral dissertation, UN Compact, 2024.

33 R. P. Allan *et al.*, "Summary for policymakers," in Climate Change 2021: The Physical Science Basis.," in *Contribution of Working Group I to the Sixth Assessment Report of the Intergovernmental Panel on Climate Change*, P. A. Arias *et al.*, Eds. Cambridge, UK: Cambridge Univ. Press, 2023, pp. 3–32.

34 T. Noor, I. Nazeer, Z. Attique, M. Shahzad, and A. Baqi, "Global temperature variations since pre-industrial era," *International Journal of Innovative Science and Technology*, vol. 3, pp. 67–74, 2021.

35 D. F. B. Usta, M. Teymouri, and U. Chatterjee, "Assessment of temperature changes over Iran during the twenty-first century using CMIP6 models under SSP1-2.6, SSP2-4.5, and SSP5-8.5 scenarios," *Arabian Journal of Geosciences*, vol. 15, p. 416, 2022. DOI: 10.1007/s12517-022-09709-9.

36 G. A. Meehl, "The role of the IPCC in climate science," *Oxford Research Encyclopedia of Climate Science*, vol. 19, Apr., 2023. DOI: 10.1093/acrefore/9780190228620.013.933.

37 N. Höhne *et al.*, "The Paris agreement: Resolving the inconsistency between global goals and national contributions," in *Climate Policy After the 2015 Paris Climate Conference*, J. Depledge, J. E. Viñuales, E. Lees, and D. M. Reiner, Eds. London, UK: Routledge, Dec., 2021, pp. 30–46.

38 R. Warren, O. Andrews, S. Brown, *et al.*, "Quantifying risks avoided by limiting global warming to 1.5 or 2 °C above pre-industrial levels," *Climatic Change*, vol. 172, p. 39, 2022. DOI: 10.1007/s10584-021-03277-9.

39 T. Leiter, "Nationally determined contributions (NDCs) as a governance instrument—accounting for politics, negotiation progress, and related mechanisms under the Paris agreement," *Environmental Politics*, vol. 33, no. 3, pp. 552–557, 2023. DOI: 10.1080/09644016.2023.2252312.

40 G. I. Iacobuta, C. Brandi, A. Dzebo, and S. D. Elizalde Duron, "Aligning climate and sustainable development finance through an SDG lens: The role of development assistance in implementing the Paris agreement," *Global Environmental Change*, vol. 74, p. 102509, May, 2022. DOI: 10.1016/j.gloenvcha.2022.102509.

41 M. Meinshausen, J. Lewis, C. McGlade, *et al.*, "Realization of Paris agreement pledges may limit warming just below 2 °C," *Nature*, vol. 604, pp. 304–309, 2022. DOI: 10.1038/s41586-022-04553-z.

42 H. L. van Soest, M. G. J. den Elzen, and D. P. van Vuuren, "Net-zero emission targets for major emitting countries consistent with the Paris agreement," *Nature Communications*, vol. 12, p. 2140, 2021. DOI: 10.1038/s41467-021-22294-x.

43 S. Paltsev, J. Morris, H. Kheshgi, and H. Herzog, "Hard-to-Abate sectors: The role of industrial carbon capture and storage (CCS) in emission mitigation," *Applied Energy*, vol. 300, p. 117322, Oct., 2021. DOI: 10.1016/j.apenergy.2021.117322.

44 W. Peng *et al.*, "Climate policy models need to get real about people—here's how," *Nature*, vol. 594, no. 7862, pp. 174–176, Jun., 2021. DOI: 10.1038/d41586-021-01500-2.

45 W. N. Digitemie and I. O. Ekemezie, "Assessing the role of carbon pricing in global climate change mitigation strategies," *Magna Scientia Advanced Research and Reviews*, vol. 10, no. 2, pp. 22–31, 2024. DOI: 10.30574/msarr.2024.10.2.0040.

46 H. C. Lau and S. C. Tsai, "Net zero by 2050: A tale of three economies," *Energy & Fuels*, vol. 38, no. 21, pp. 21079–21094, Oct., 2024. DOI: 10.1021/acs.energyfuels.4c04212.

47 G. Leonzio, "Recent advancements and challenges in carbon capture, utilization and storage," Università di Cagliari Institutional Research Information System (IRIS), Jan. 2024. [Online]. Available: https://hdl.handle.net/11584/393983

48 S. M. Alizadeh, Y. Khalili, and M. Ahmadi, "Comprehensive review of carbon capture and storage integration in hydrogen production: Opportunities, challenges, and future perspectives," *Energies*, vol. 17, no. 21, p. 5330, Oct., 2024. DOI: 10.3390/en17215330.

49 N. Wang, Y. Bai, Z. Guo, Y. Fan, and F. Meng, "Synergies between the circular economy and carbon emission reduction," *Science of the Total Environment*, vol. 951, 175603, Nov., 2024. DOI: 10.1016/j.scitotenv.2024.175603.

3

Environmental Justice and Carbon Management Policies

3.1 What is Environmental Justice?

Environmental justice means making sure that everyone no matter their race, income, or background has the same right to live in a clean and healthy environment. It's about fairness in how environmental benefits and harms are shared [1]. For example, if a city builds a park or plants trees, everyone should enjoy the clean air and green space. But if a factory pollutes the air or water, no group should suffer more than others just because they are poor or belong to a minority.

In many parts of the world, history shows that this fairness hasn't always happened [2, 3]. Often, low-income neighborhoods or communities of color are placed near highways, disposal sites, power plants, or chemical factories. These areas experience more pollution, higher health problems like asthma or cancer, and less green space. People living in these places usually have less political power, so their voices are ignored when decisions are made [4].

Environmental justice also involves giving everyone a voice in the development of their own environment [5]. This includes involving communities in the planning of projects such as power plants, waste treatment centers, or carbon capture facilities. It is not sufficient to build something and then inform people about it afterward. As part of true justice, people must be consulted first, listened to their concerns, and included in the decision-making process [6].

One example comes from the small town in North Carolina called Warren County, where a movement began in the 1980s to protest the siting of toxic waste dumps near Black and Latino communities. In addition to raising awareness about environmental racism, this movement helped bring to light the idea that some communities suffer from greater environmental harm simply because of their racial background [7]. During the past few decades, environmental justice has become an important component of discussions regarding energy, climate change, and urban development.

In the case of carbon capture and storage (CCS), environmental justice asks important questions: Where are these projects being built? Who gets the benefits like jobs or cleaner air? And who takes on the risks, such as living near a CO_2 pipeline or injection site? As a result, if the answers to these questions reveal unfair patterns, then it is possible that CCS may provide environmental benefits while still imposing social risks on nearby communities, and that is not fair.

Figure 3.1 visually captures the core dimensions of environmental justice, which go beyond pollution control and address fairness, voice, and participation in environmental decision-making. The top left panel represents the positive goal: inclusive access to clean air, green space, and healthy environments for all communities. In contrast, the top right panel illustrates the harsh reality many communities face such as living near polluting industries. The bottom left panel highlights community resistance and activism, where affected groups demand justice and equity through public protest. Finally, the bottom right panel emphasizes the importance of participation and consultation, showing community members engaged in planning and decision-making processes.

Figure 3.1 Illustration of environmental justice in four key dimensions: (top left) equal access to clean and healthy environments, (top right) disproportionate exposure to pollution in communities, (bottom left) public protest for justice, and (bottom right) meaningful participation in environmental planning and decision-making.

In short, environmental justice is not just about pollution or CO_2. It's about people, fairness, and inclusion. Any climate solution, including CCS, must respect these values if it wants to be accepted and successful in the long run.

3.2 Historical Examples of Environmental Inequities

The impact of environmental problems has not been equally distributed among all communities. Compared to other groups, low-income families and communities of color have experienced more pollution, more health problems, and fewer environmental protections [8]. The unfair situations described above are known as environmental inequities, and they have occurred in many countries in recent decades.

One well-known example is from the United States. In the 1980s, a small town in North Carolina called Warren County became a national symbol for environmental injustice [7]. The state decided to place a toxic waste landfill in a poor, mostly Black community. People in the area protested, saying it was unfair to place dangerous waste in their neighborhood just because they had less power and money. Even though the protest didn't stop the landfill, it helped start the environmental justice movement in the United States [7].

Another example is from South Africa, where, during apartheid, Black communities were often forced to live near industrial zones and mining sites with high levels of pollution. These areas had little access to clean air, water, or public services, while wealthier white neighborhoods were built far from these risks. The effects of those decisions are still felt today, with some communities still facing higher rates of illness [9].

Another example can be found in India and other parts of South Asia, where slum areas are located near industrial plants. People living nearby are unable to demand cleaner conditions from factories that release harmful chemicals into the air and water [10]. Most of these individuals work in informal jobs, have no legal claim to their homes, and are not often invited to participate in planning and environmental decisions.

Even in wealthier countries, unfair patterns exist. In the United Kingdom, studies have shown that landfills and polluting industries are more often located in poorer areas [11]. Residents in these places have fewer political connections and may not be heard when they complain about the smell, noise, or health impacts.

These examples show a clear pattern: environmental burdens often fall on those who are least able to resist or protect themselves. This history is important because when new projects like carbon capture are proposed, some people worry that the same unfair treatment will happen again [12]. Understanding these past examples helps us ask better questions today: Are we repeating the same mistakes? Are we listening to the right voices? And are we making decisions that are truly fair for everyone?

3.3 Justice in the Context of Carbon Capture

As CCS projects become more common, questions about fairness and justice are getting more attention. CCS is often seen as a helpful technology for fighting climate change. But if these projects are not designed and managed fairly, they can create new inequalities and repeat old mistakes from past industrial projects [13].

The location of CCS projects is one of the most important issues [14]. In some cases, the land is inexpensive, or the community has less power to oppose development. Most often, these areas are rural, economically challenged, or house a large minority population. The placement of a CO_2 pipeline or storage site in a community that has not been properly consulted can create the impression that residents are being used as a dumping ground for environmental problems [15]. This does not reflect climate justice; rather, it reproduces environmental injustice.

Another important concern in CCS projects is who gets the benefits and who carries the risks. Large companies involved in CCS may earn money through carbon credits, government incentives, or new business opportunities. But the people who live near the project site may face different realities such as the fear of leaks, noise from operations, construction traffic, or changes to the local environment and landscape. If these local communities do not receive clear benefits, in return like well-paying jobs, better healthcare, improved infrastructure, or investment in schools and public services, the project can feel unfair [16].

By doing so, a power imbalance is created where companies and officials make the decisions and receive the rewards, while local residents are responsible for dealing with risks and problems. This can lead to a sense of mistrust, especially if community members were not involved from the start of the planning and decision-making process. Occasionally, residents may feel ignored or even exploited, particularly in areas that have experienced environmental injustice in the past. In order to gain public support for CCS, developers and policymakers must ensure that local residents are not just observers, but active participants, and that the benefits of the project are shared in fair, visible, and lasting ways [6, 17].

This vision of fairness in CCS projects reflects the broader concept of climate justice, which lies at the intersection of social justice and climate change as shown in Figure 3.2.

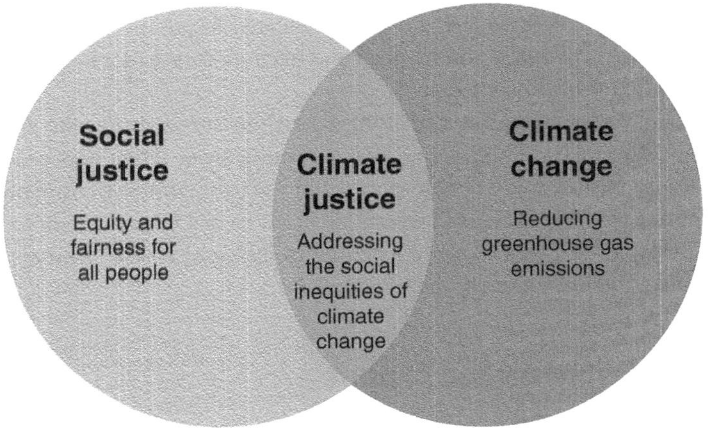

Figure 3.2 The concept of climate justice as the overlap of social justice and climate change. Climate justice emphasizes that climate solutions like CCS must not only address emissions but also be fair, inclusive, and responsive to social inequalities.

To make CCS projects more just, developers and policymakers must ensure that local people are part of the decision-making process [17]. That means asking questions like:

- Did the community agree to the project?
- Were they involved early on?
- Do they have access to clear, honest information?
- Are their voices taken seriously?

It's also important to provide local benefits such as clean jobs, infrastructure, health services, or investment in schools and other communities [18, 19]. When people feel that they are gaining something, and not just bearing the risks, they are more likely to support the project [6].

For example, in a proposed CCS project near a semirural town, the local government partnered with the project developer to ensure the community received tangible benefits beyond emissions reductions. As part of the agreement, a portion of the revenue from carbon storage leases was directed toward improving local schools, especially providing modern lab facilities, and scholarships for students pursuing environmental science and engineering [18, 19]. In addition, the project supported the expansion of a local health clinic. These visible, direct investments helped shift public perception. Rather than seeing the CCS facility as a risk or imposition, residents began to view it as a long-term opportunity for economic and social development. This case illustrates how aligning carbon capture development with community priorities such as education, health, and public infrastructure can significantly enhance social acceptance and trust [6].

In short, justice in CCS deployment means more than just following safety rules. It means treating people with respect, sharing benefits, and making sure no group is left behind. Climate solutions must work for everyone, not just for those with money or power.

3.4 Policy Frameworks Supporting Just Carbon Management

3.4.1 Policy Tools for a Just Transition

In order for carbon capture and other climate solutions to be accepted by the public, strong and thoughtful policies are required [20]. As with rulebooks, policy frameworks establish the standards for how projects should be carried out, who should be involved, and what protections should be in place. By designing policies with justice in mind, we can ensure that no group is left out or harmed unfairly.

Some governments have started using what's called "just transition" policies. As we move away from polluting industries such as coal or oil and begin to use cleaner energy sources such as renewable energy or carbon capture technology, these plans and programs will help to ensure that people and communities are not left behind [21].The idea is to protect workers, families, and towns that depend on those old industries for their jobs and income. For example, if a coal power plant shuts down and is replaced by a CCS facility, a just transition plan might include job training for coal workers so they can learn how to work in the new CCS plant or in renewable energy [22]. It could also include investment in schools, healthcare, or small businesses in the local area to help the whole community adjust [23]. One example comes from Spain, where the government signed an agreement with coal unions to provide retirement plans, retraining programs, and economic support for coal regions when several mines were closed. The goal of a just transition is to make the shift to clean energy fair so we don't create a situation where some people benefit, while others lose their jobs, homes, or opportunities [24].

Another important tool for making CCS projects fairer is the use of Community Benefit Agreements (CBAs). These are official agreements or contracts between the companies building the carbon capture project and the people who live nearby. A CBA clearly says what benefits the community will receive in return for hosting something like a CO_2 pipeline or storage site. This could include things like local job opportunities, support for schools, healthcare programs, or ongoing environmental monitoring to make sure the area stays safe. These agreements are important because they build trust. They show that the project is not just happening "to" the community, but with the community's involvement [25].

In some countries, equity scoring or environmental justice screening tools are now being used to help make better decisions about where to build CCS projects and how to protect people. These tools use data such as income levels, race or ethnicity, health conditions, pollution levels, and access to services to identify which communities have been most affected by environmental harm in the past or are most at risk today [26].

For example, a government might use a screening tool to find neighborhoods that have high health problems, low incomes, and are located near factories or highways. These areas would then be marked as "high priority" for protection or environmental cleaning investment. By using these tools, policymakers and companies can make sure that new projects like CCS facilities don't add more stress to already vulnerable communities [26]. Instead, they can focus on giving more support such as funding for clean energy, health clinics, or environmental monitoring. These screening tools help bring fairness into decision-making and make it easier to reduce environmental inequalities.

Figure 3.3 illustrates four key policy tools that support a fair and inclusive approach to carbon capture deployment. It shows how just transition frameworks protect workers and communities during energy shifts, how CBAs ensure that local populations receive tangible support, how environmental justice screening tools help identify vulnerable areas for special attention, and how regulatory laws and technical safeguards like inspection and monitoring provide long-term oversight and protection.

3.4.2 Participation, Inclusion, and Global Perspectives

International organizations, such as the United Nations (UN), emphasize that climate policies should focus not only on reducing emissions but also on fairness, inclusion, and social protection [27]. As part of its carbon reduction plan, the UN encourages countries to consider how the plan will affect different populations, including low-income families, Indigenous peoples, rural communities, and those who are already affected by extreme weather conditions or pollution. Under the Paris Agreement, countries are also required to report on these impacts as part of their national climate commitments [28].

Figure 3.4 summarizes the three core principles that international organizations, including the United Nations, emphasize for shaping just and effective climate policy: fairness,

Figure 3.3 Policy tools to support fair and inclusive carbon capture deployment. Just transition plans, community agreements, equity tools, and safeguards are key to building public trust and ensuring climate solutions benefit everyone, not just the powerful.

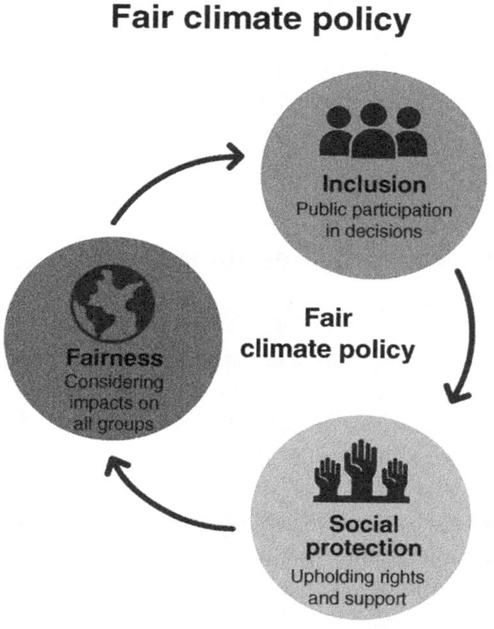

Figure 3.4 Three pillars of fair climate policy: fairness ensures that climate impacts and solutions are distributed equitably; inclusion guarantees public participation in decisions; and social protection upholds the rights and well-being of vulnerable communities.

inclusion, and social protection. These principles are foundational for ensuring that climate action benefits all segments of society.

Good climate policies don't just give orders to companies. They also create ways for the public to get involved and have a voice in CCS projects. This means people from all parts of the community, not just experts or politicians, can take part in decisions that affect their lives. For example, a government might hold public meetings where local people can ask questions and share concerns about a CCS project. In some places, community representatives are invited to join planning groups or review reports before big decisions are made. These steps help make sure that everyone, not just powerful groups, can share their opinions, give feedback, and help shape the final outcome [23].

But there are still important gaps in how CCS projects are managed. In many places, the laws and regulations mostly focus on technical safety and engineering standards but say very little about social inclusion or fairness [29]. As a result, some communities only find out about a new CCS project after it's already been approved, leaving them feeling excluded and even betrayed. This lack of early communication can lead to frustration, protests, or long-term distrust, even if the project is safe and well-designed. That's why it's so important to build justice and fairness into the policy from the very beginning, not as something added later. True environmental justice means giving people a chance to be involved before decisions are made, and making sure their voices are part of shaping the project from the start [30].

In summary, fair and effective carbon management requires clear policies that support inclusion, equity, and community empowerment. When communities are given a real voice and their needs are considered from the start, climate solutions like CCS can serve not only the environment but also the people most affected by change. With the right frameworks in place, CCS can become a tool for both climate protection and social justice; not one at the expense of the other [31].

3.5 Environmental Racism and Community Resistance

It is believed that environmental racism occurs when some communities, especially those made up of people of color or Indigenous groups, live in close proximity to polluting industries or harmful infrastructure. Even though it is not always on purpose, unfair systems and decisions have, over time, resulted in the same groups experiencing more environmental harm than others. It is possible to observe these patterns in both rich and poor countries [2].

For example, in the United States, studies have shown that Black and Latino neighborhoods are more likely to be close to highways, oil refineries, or industrial waste sites [2]. In Brazil, Afro-Brazilian and Indigenous communities often live near mining projects or hydroelectric dams that damage their land and water [32]. In parts of Africa and Asia, poorer villages are chosen for waste dumping or coal ash disposal because they are seen as less powerful. These examples show that environmental risks are not shared equally.

When new technologies like CCS are introduced, people in these communities may be suspicious or resistant, and for good reason. They may have experienced broken promises in the past. Maybe they were promised jobs that never came, or safety that wasn't delivered.

Figure 3.5 A visual of environmental injustice and the intentions of climate justice. This illustration shows the contrast between a polluted, marginalized world and a hopeful, green future. Climate justice is not only about outcomes, but about the pathway that determines which side communities end up on through fairness, participation, and policy transformation.

So even if a carbon capture project is clean and well-managed, people might still say "no" because they don't trust the people behind it [33].

Figure 3.5 presents a powerful visual contrast between environmental injustice and the promise of a just, inclusive future. It does not depict climate justice itself as a fixed outcome, but rather the diverging paths to a clean environment and climate justice shaped by policy, inclusion, and social responsibility.

This is where community resistance becomes important. Resistance doesn't always mean public protests or loud opposition. Often, it takes the form of people asking difficult but important questions, requesting more detailed information about potential risks, or seeking legal protections for their land, health, or safety. In some cases, residents may organize town hall meetings to discuss the project, launch objections to ask for changes, or even demand that a project be paused until there is greater transparency and fairness [34]. For example, imagine a company wants to build a carbon capture pipeline near a small town. The local community may not be against clean energy, but they might be concerned about safety, land use, or environmental impact. In response, they may organize public meetings, ask the company for detailed safety reports, and request emergency response plans [35].

It's important to understand that resistance is not ignorance. It's often based on real experience and deep knowledge of local history and harm. Instead of treating resistance as a problem, developers and policymakers should see it as a chance to learn and improve [34]. Listening to community concerns can lead to better projects that are safer, more inclusive, and more likely to succeed [31]. In short, environmental racism is a serious issue that must be considered in any climate project, including carbon capture.

And community resistance is not something to fight; it's something to respect, understand, and work with [2].

3.6 Ensuring Equitable Access and Participation

For carbon capture and other climate technologies to be successful and fair, everyone should have the chance to participate and benefit, especially communities that are most affected by pollution and climate change. This means giving people not just a voice, but also real power in how decisions are made [36].

Equitable access means more than just allowing people into a meeting. It means giving them the tools, knowledge, and support they need to understand what's happening and shape the outcomes. For example, if a carbon capture project is being planned near a rural or minority community, information should be shared in the local language, using simple and clear explanations, not full of technical terms that only experts understand [6].

Participation should also begin early in the project, not after everything has already been decided. People need time to ask questions, talk to their neighbors, and raise any concerns they might have. In many successful projects, community members are part of advisory groups, planning teams, or oversight committees. This kind of shared decision-making helps build trust and leads to better outcomes [17].

One way to support fairness is by making sure meetings are open and accessible. That could mean offering childcare during town halls, holding meetings at times that work for working people, or providing transportation support. Without these things, many voices will be left out even if the invitation is open.

Governments and companies can also support fairness by creating participation funds, small budgets set aside to help local communities take part in the planning and decision-making process [27]. These funds can be used to hire independent experts who advise the community, print educational materials, or organize public workshops and meetings. Many communities do not have the money or technical knowledge to fully understand complex carbon capture projects or to speak up confidently. Participation funds help bridge that gap [23].

It's also important to support long-term community involvement, not just one-time consultations at the beginning of a project. True participation means that local people are included throughout the entire life of the project, from early planning and construction to daily operations and even after the site is closed or decommissioned. Keeping communities informed and involved over time shows respect, builds trust, and helps prevent misunderstandings or conflicts later on. When people feel they are truly part of the process, not just temporarily consulted, they are more likely to support the project and contribute to its success [23].

Finally, ensuring equity means recognizing and valuing different types of knowledge. While community members may not be engineers or scientists, they often have a deep understanding of their local environment such as the history of the land, patterns of water flow, or cultural practices that affect how a project is experienced. Their insights are not only valid but essential. By listening to these voices, carbon capture projects become more grounded in real-life conditions and community values, rather than being based only on technical models [23].

In summary, equitable access and participation are not extras; they are core parts of justice in carbon management. When people are informed, respected, and included from the beginning, projects are more likely to succeed and more likely to be fair.

3.7 Key Highlights (Chapter Summary)

◇ **Chapter 3—Key Highlights:**

1) Environmental justice means everyone deserves equal protection and a healthy environment, regardless of income, race, or background.
2) Past decisions have often placed polluting projects near marginalized or low-income communities.
3) Carbon capture projects must avoid repeating historical environmental injustices.
4) Community trust depends on early and honest involvement in decision-making process.
5) Choosing fair project locations, sharing the benefits, and making sure risks aren't placed on just one group are all necessary for CCS projects to succeed.
6) Policy tools like Just Transition programs and Community Benefit Agreements can improve equity.
7) Environmental racism is real and must be addressed in climate infrastructure projects such as carbon capture.
8) Community resistance is often based on past experiences and deserves respect and attention.
9) Public participants must be active, supported, and accessible not just symbolic or last-minute.
10) A justice-based approach to carbon management strengthens both social trust and climate outcomes.

Keywords: Environmental Justice, Carbon Capture, Environmental Racism, Policy Equity, Community Participation, Just Transition, Public Inclusion, Risk Distribution, Climate Governance, Vulnerable Communities

References

1 F.-A. Hoover, S. Meerow, Z. J. Grabowski, and T. McPhearson, "Environmental justice implications of siting criteria in urban green infrastructure planning," *Journal of Environmental Policy and Planning*, vol. 23, no. 5, pp. 665–682, Jun., 2021. DOI: 10.1080/1523908X.2021.1945916.

2 S. Henderson and R. Wells, "Environmental racism and the contamination of black lives: A literature review," *Journal of African American Studies*, vol. 25, pp. 134–151, 2021. DOI: 10.1007/s12111-020-09511-5.

3 N. Donley *et al.*, "Pesticides and environmental injustice in the USA: Root causes, current regulatory reinforcement and a path forward," *BMC Public Health*, vol. 22, p. 708, 2022. DOI: 10.1186/s12889-022-13057-4.

4 A. M. Azul *et al.*, "Unhealthy lifestyles, environment, well-being and health capability in rural neighbourhoods: A community-based cross-sectional study," *BMC Public Health*, vol. 21, p. 1628, 2021. DOI: 10.1186/s12889-021-11661-4.

5 L. Álvarez and B. Coolsaet, "Decolonizing environmental justice studies: A Latin American perspective," *Capitalism Nature Socialism*, vol. 31, no. 2, pp. 50–69, 2020. DOI: 10.1080/10455752.2018.1558272.

6 F. Mulyasari, A. K. Harahap, A. O. Rio, R. Sule, and W. G. A. Kadir, "Potentials of the public engagement strategy for public acceptance and social license to operate: Case study of carbon capture, utilisation, and storage *Gundih* pilot project in Indonesia," *Int. J. Greenh. Gas Control*, vol. 108, p. 103312, Jun., 2021. DOI: 10.1016/j.ijggc.2021.103312.

7 R. S. Newman, "*North Carolina: Protesting unwelcome toxic dumps*," in *Environmental Racism in the United States and Canada: Seeking Justice and Sustainability*, B. E. Johansen, Ed, 1st ed., 2020, p. 172.

8 A. R. Pearson, J. P. Schuldt, R. Romero-Canyas, and D. Larson-Konar, "Diverse segments of the US public underestimate the environmental concerns of minority and low-income Americans," *Proceedings of the National Academy of Sciences*, vol. 115, no. 49, pp. 12429–12434, Oct., 2018. DOI: 10.1073/pnas.1804698115.

9 V. Stull, M. M. Bell, and M. Ncwadi, "Environmental apartheid: Eco-health and rural marginalization in South Africa," *Journal of Rural Studies*, vol. 47, pp. 369–380, Oct., 2016. DOI: 10.1016/j.jrurstud.2016.04.004.

10 G. N. T. Hasnat, M. A. Kabir, and M. A. Hossain, "Major environmental issues and problems of South Asia, particularly Bangladesh," in *Handbook of Environmental Materials Management*, C. M. Hussain, Ed. Cham: Springer, 2019, pp. 109–148, DOI: 10.1007/978-3-319-73645-7_7.

11 A. Nigam and M. Parikh, "Comparative analysis of climate justice framework in USA, UK, Japan, China, Germany and France," *Journal of Humanities and Social Sciences Studies*, vol. 2, no. 5, pp. 49–61, 2020. DOI: 10.32996/jhsss.

12 S. Tyree and M. Greenleaf, "The environmental injustice of 'clean coal': Expanding the national conversation on carbon capture and storage technology to include an analysis of potential environmental justice impacts," *Environmental Justice*, vol. 2, no. 4, pp. 167–171, Dec., 2009. DOI: 10.1089/env.2009.0040.

13 K. Bickerstaff, H. Bulkeley, and G. Walker, *Energy Justice in a Changing Climate: Social Equity and Low-carbon Energy*. London: Bloomsbury Publishing, 2013.

14 Y. Lam, J. Ventrella, A. I. Baptista, and J. D. Rodriguez, "Analysis of proposed carbon capture projects in the US power sector and co-location with environmental justice communities," *PLoS One*, May 16,, vol. 20, p. e0323817, 2025. DOI: 10.1371/JOURNAL.PONE.0323817.

15 B. K. Sovacool, "Who are the victims of low-carbon transitions? Towards a political ecology of climate change mitigation," *Energy Research & Social Science*, vol. 73, 101916, 2021. DOI: 10.1016/J.ERSS.2021.101916.

16 S. K. Mahjour and S. A. Faroughi, "Risks and uncertainties in carbon capture, transport, and storage projects: A comprehensive review," *Gas Science and Engineering*, vol. 119, no. Part A, 205117, 2023. DOI: 10.1016/J.JGSCE.2023.205117.

17 L. Fritz, C. M. Baum, S. Low, Fritz L, Baum CM, Low S, Sovacool BK. "Public engagement for inclusive and sustainable governance of climate interventions," *Nature Communications*, vol. 15, art. no. 4168, 2024. 10.1038/S41467-024-48510-Y

18 G. Siciliano, L. Wallbott, F. Urban, A. N. Dang, and M. Lederer, "Low-carbon energy, sustainable development, and justice: Towards a just energy transition for the society and the environment," *Sustainable Development*, vol. 29, no. 6, pp. 1049–1061, 2021. DOI: 10.1002/SD.2193.

19 A. Banerjee and G. Schuitema, "How just are just transition plans? Perceptions of decarbonisation and low-carbon energy transitions among peat workers in Ireland," *Energy Research & Social Science*, vol. 88, p. 102616, Jun., 2022. DOI: 10.1016/J.ERSS.2022.102616.

20 M. Fairbrother, "Public opinion about climate policies: A review and call for more studies of what people want," *PLOS Climate*, May 2, vol. 1, 2022. DOI: 10.1371/JOURNAL.PCLM.0000030.

21 X. Wang and K. Lo, "Just transition: A conceptual review," *Energy Research & Social Science*, vol. 82 , Dec., 102291, 2021. DOI: 10.1016/J.ERSS.2021.102291.

22 B. Blankenship, M. Aklin, J. Urpelainen, and V. Nandan, "Jobs for a just transition: Evidence on coal job preferences from India," *Energy Policy*, vol. 165, Jun., 112910, 2022. DOI: 10.1016/J.ENPOL.2022.112910.

23 R. Leonhardt, B. Noble, G. Poelzer, P. Fitzpatrick, K. Belcher, and G. Holdmann, "Advancing local energy transitions: A global review of government instruments supporting community energy," *Energy Research & Social Science*, vol. 83, Jan., 102350, 2022. DOI: 10.1016/J.ERSS.2021.102350.

24 D. Bolet, F. Green, and M. González-Eguino, "How to get coal country to vote for climate policy: The effect of a 'just transition agreement' on Spanish election results," *American Political Science Review*, vol. 118, no. 3, Dec., 2023. DOI: 10.1017/S0003055423001067.

25 C. Gunton and S. Markey, "The role of community benefit agreements in natural resource governance and community development: Issues and prospects," *Resources Policy*, vol. 73, Oct., 102152, 2021. DOI: 10.1016/j.resourpol.2021.102152.

26 N. C. Osakwe, A. A. Motsinger-Reif, and D. M. Reif, "Environmental health and justice screening tools: A critical examination and path forward," *Frontiers in Environmental Health*, vol. 3, 12 Sep., 2024. DOI: 10.3389/FENVH.2024.1427495.

27 M. D. Davidson, "How fairness principles in the climate debate relate to theories of distributive justice," *Sustainability*, vol. 13, no. 13, p. 7302, Jun., 2021. DOI: 10.3390/SU13137302.

28 L. Perugini *et al.*, "Emerging reporting and verification needs under the Paris agreement: How can the research community effectively contribute?," *Environmental Science & Policy*, vol. 122, pp. 116–126, Aug., 2021. DOI: 10.1016/J.ENVSCI.2021.04.012.

29 M. Batres *et al.*, "Environmental and climate justice and technological carbon removal," *The Electricity Journal*, vol. 34, no. 7, p. 107002, Aug.–Sep., 2021. DOI: 10.1016/J.TEJ.2021.107002.

30 M. Meenar, M. Heckert, and D. Adlakha, "'Green enough ain't good enough:' Public perceptions and emotions related to green infrastructure in environmental justice communities," *International Journal of Environmental Research and Public Health*, vol. 19, no. 3, p. 1448, 2022. DOI: 10.3390/IJERPH19031448.

31 F. S. Khatibi, A. Dedekorkut-Howes, M. Howes, and E. Torabi, "Can public awareness, knowledge and engagement improve climate change adaptation policies?," *Discover Sustainability*, vol. 2, no. 18, 2021. DOI: 10.1007/S43621-021-00024-Z.

32 M. Rougeon, C. Mota, and L. Trad, "Environmental racism and environmental justice: Decolonial inflections and new agendas in Latin America and Brazil," *Journal of Political Ecology*, vol. 30, no. 1, 2023. DOI: 10.2458/V30I1.24748.

33 S. Pianta, A. Rinscheid, and E. U. Weber, "Carbon capture and storage in the United States: Perceptions, preferences, and lessons for policy," *Energy Policy*, vol. 151, 112149, 2021. DOI: 10.1016/j.enpol.2021.112149.

34 M. Thomas, B. DeCillia, J. B. Santos, and L. Thorlakson, "Great expectations: Public opinion about energy transition," *Energy Policy*, vol. 162, 112777, 2022. DOI: 10.1016/j.enpol.2022.112777.

35 E. van de Grift and E. Cuppen, "Beyond the public in controversies: A systematic review on social opposition and renewable energy actors," *Energy Research & Social Science*, vol. 91, 102749, 2022. DOI: 10.1016/j.erss.2022.102749.

36 E. ter Mors and E. van Leeuwen, "It matters to be heard: Increasing the citizen acceptance of low-carbon technologies in the Netherlands and United Kingdom," *Energy Research & Social Science*, vol. 105, 103103, 2024. DOI: 10.1016/j.erss.2023.103103.

4

Principles and Types of Carbon Capture Technologies

4.1 Overview of Carbon Capture Mechanisms

Carbon capture is a set of technologies designed to prevent carbon dioxide (CO_2), the main greenhouse gas, from entering or remaining in the atmosphere. The idea is simple: capture CO_2 from industrial emissions or directly from the air, then either store it safely underground or reuse it in industrial applications [1]. Although the principles of chemistry and engineering are complex, the goal of carbon capture regarding climate change and global warming remains simple.

Carbon capture works in three main stages: capture, compression and transport, and storage or utilization. In the capture stage, carbon dioxide (CO_2) is separated from other gases produced during combustion or industrial processes. This can take place in power plants, cement factories, steel plants, or through direct air capture (DAC) systems that remove CO_2 directly from the atmosphere. Next, the captured CO_2 is compressed into a dense fluid and transported, usually through pipelines, to a suitable storage site or facility for reuse. In the final stage, the CO_2 is either stored permanently underground, for example, in depleted oil and gas reservoirs or deep saline aquifers, or reused in industry, where it can be converted into valuable products such as synthetic fuels, building materials, or carbonated beverages [2].

Carbon capture methods are grouped by where and how CO_2 is captured. Post-combustion removes CO_2 after fossil fuels are burned. Precombustion separates carbon before combustion. Oxy-fuel combustion burns fuels in pure oxygen, producing more pure CO_2 and water vapor that are easy to separate [3]. DAC pulls CO_2 straight from the atmosphere, helping address past emissions [4]. Carbon capture can be combined with bioenergy systems to create Bioenergy with Carbon Capture and Storage (BECCS). This process uses biomass, like forestry residues or crops, to generate renewable energy while removing CO_2 from the atmosphere. Since biomass carbon is taken from atmospheric CO_2 through biomass that is used for photosynthesis, capturing and storing it after combustion results in net-negative emissions [5].

Despite the differences in design and application of carbon capture technologies, all share the goal of removing carbon dioxide (CO_2) emissions from energy production and industrial activities. In other words, they allow us to continue producing energy and industrial goods without emitting large amounts of carbon dioxide into the atmosphere [1, 2].

A significant challenge exists in expanding this technology to be used globally in an economical, safe, and effective manner. Developing and maintaining large capture plants, pipelines, and storage facilities requires significant investment, infrastructure, and technical expertise. Furthermore, the process must be managed carefully to avoid leaks, particularly during transportation and storage [6]. As a result, reliability is ensured, and energy consumption is minimized, since the capture and compression of CO_2 also consumes energy.

Despite these challenges, carbon capture is seen as a crucial tool for tackling climate change. As countries around the world commit to net-zero emission targets, carbon capture can help reduce emissions from hard-to-decarbonize sectors such as cement, steel, chemicals, and power generation. Therefore, it is essential to understand how different carbon capture mechanisms work in theory and in practice [1, 2]. Through the use of these carbon capture plants, policymakers, engineers, and investors can assess their true ability to reduce greenhouse gas emissions. This will enable them to contribute to the transition to a low-carbon future globally.

4.2 Post-combustion Capture

Post-combustion carbon capture is the most common and well-developed method of carbon capture. The technology is primarily used in existing power plants and industrial facilities. In this process, carbon dioxide (CO_2) is separated from flue gases produced after burning fuels such as coal, oil, or natural gas [7].

When fuels are burned in the air, they produce exhaust gases, called flue gases. These gases mainly contain nitrogen, water vapor, carbon dioxide (CO_2), and small amounts of other gases. In post-combustion carbon capture, CO_2 is removed from these flue gases before they are released into the atmosphere. The most common method uses chemical solvents, usually liquids called amines such as monoethanolamine (MEA), which absorb CO_2 [7].

There are two main steps in the process. First, the flue gas is sent through a tall absorber tower, where the solvent comes into contact with the gas and captures the CO_2. Then, the CO_2-rich solvent is transferred to another unit called a regenerator or stripper, where it is heated to release pure CO_2. Once the CO_2 is removed, the solvent is cooled and reused, creating a continuous cycle of carbon capture and regeneration. Purified CO_2 is then compressed into a state between gas and liquid, known as supercritical or dense. It is prepared for transport and storage underground in geological formations or for reuse in industrial applications, such as producing synthetic fuels or carbonated drink products [7].

Figure 4.1 shows a typical post-combustion carbon capture facility integrated with an industrial plant.

In this process, flue gas containing CO_2 from combustion is directed through an absorber column or tower where a chemical solvent, usually an amine-based solution, selectively captures the CO_2. The CO_2-rich solvent is then regenerated in a separate stripper column by applying heat, releasing high-purity CO_2 for compression, transport, and storage.

Post-combustion capture's main advantage is its compatibility with existing facilities. The technology can be applied to coal-fired and natural gas power plants, as well as to industries

Figure 4.1 A typical post-combustion carbon capture system showing absorber and regenerator, heat exchangers, and associated process piping. In this setup, CO_2 is separated from flue gas after combustion and recovered in concentrated form for subsequent compression and storage. (Courtesy of Shutterstock.com).

such as cement, steel, and chemical production. Upgrading existing plants by adding carbon capture systems, known as retrofitting, is usually cheaper and more practical than building entirely new facilities designed for carbon capture from the beginning [8].

This approach allows industries to reduce emissions while continuing to use their current infrastructure, saving time and investment costs. Although building a carbon capture and storage (CCS) system involves significant new investment, retrofitting existing plants can still be more economical and faster than constructing entirely new facilities with built-in capture systems [8]. In a retrofit, the carbon capture unit is new, but it makes use of the plant's existing infrastructure, including boilers, flue gas ducts, power supply, and utility systems, rather than replacing them.

Even though the construction of a carbon capture and storage system (CCS) involves substantial new investments, retrofitting (upgrading) existing plants with CCS can be more economical and faster than building a new industrial facility incorporating a carbon capture system, such as a cement plant. In a retrofit, the carbon capture unit is new, but it utilizes the plant's existing infrastructure, such as boilers, turbines, flue gas ducts, power supplies, and utility systems, rather than replacing them. Through the reuse of major facilities and components, it is possible to avoid the rebuilding cost of the entire production process from scratch. Moreover, the existing plant already occupies an approved industrial site with environmental and safety permits in place. This helps reduce regulatory approval time and complexity [8].

Additionally, the post-combustion capture technology has been developed, tested, and proven reliable over many years of research and practical application. As a result, it is no longer an experimental concept, but a technology with proven methods, suppliers, and operational experience. The effectiveness of this system has been demonstrated in several pilot and full-scale projects around the world [7, 8]. Among the most notable examples are the Boundary Dam project in Canada and the Petra Nova facility in the

United States, both of which demonstrate that carbon capture systems can operate safely and efficiently on a large scale [9].

However, post-combustion capture faces significant technical and economic challenges. The process requires a large amount of energy, mainly in the form of heat, to release the absorbed CO_2 from the solvent during regeneration. This additional energy demand, often called the "energy penalty," can increase the total energy consumption of a power plant by about 10–15%, depending on the system design. In addition, the chemical solvents (typically amines) used in the process can degrade over time and cause equipment corrosion, which further lowers efficiency and raises maintenance and operating costs [8]. In order to deal with these issues, research is currently focused on developing advanced capture materials, such as improved amine blends and solid sorbents, that require less energy for separation from CO_2 (regeneration) and enhance the efficiency of carbon capture systems as a whole.

Optimized post-combustion capture systems reduce CO_2 emissions by up to 90%, making them an important technology in the global transition to a low-carbon economy. With the incorporation of carbon storage (CCS) and carbon utilization (CCU), existing fossil fuel infrastructure can become near-zero-emission systems. The technology is particularly useful for countries heavily dependent on coal or natural gas, as it offers a strong tool for emission reduction while renewable energy capacity continues to expand. In summary, post-combustion capture offers a practical solution for mitigating emissions from current energy and industrial sources. While not without challenges, ongoing innovation and policy support are likely to enhance its performance and cost-effectiveness [7, 8].

4.3 Precombustion Capture

Precombustion capture is a carbon capture method that removes carbon dioxide (CO_2) before the fuel is burned, rather than afterward. It is commonly used in industrial processes such as hydrogen production, which depend on gasification or reforming technologies. In simple terms, the process converts a carbon-based fuel, such as coal, natural gas, or biomass, into a mixture of hydrogen (H_2) and carbon monoxide (CO) called syngas (synthetic gas). The carbon monoxide then reacts with steam in the water–gas shift reaction, producing additional hydrogen and CO_2. The resulting CO_2 is captured, compressed, and transported for storage or reused. The hydrogen-rich syngas that remains can be used as a clean fuel, offering a significant reduction in carbon emissions compared to conventional fossil fuels [10].

Figure 4.2 illustrates the precombustion CO_2 capture process, where fuel is first gasified to produce a syngas containing carbon monoxide (CO) and hydrogen (H_2). The syngas then enters a shift reactor, where CO reacts with steam to form CO_2 and additional hydrogen. The CO_2 is subsequently separated and directed to storage, while the purified hydrogen is collected for use or storage as a low-carbon energy carrier.

Precombustion capture is generally more efficient than post-combustion capture because of the process conditions under which CO_2 is separated. In precombustion systems, CO_2 is removed from a high-pressure gas mixture with a high concentration of CO_2. These conditions

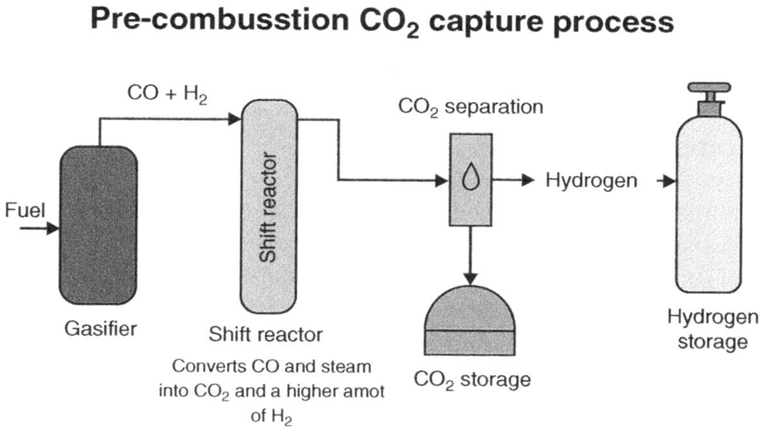

Figure 4.2 Schematic of the precombustion CO_2 capture process, showing the sequence of gasification, shift reaction, CO_2 separation, and separate storage of CO_2 and hydrogen.

make the separation process easier and require less energy, since the CO_2 is already in concentrated and pressurized form [11]. In contrast, post-combustion capture deals with flue gas, which contains CO_2 in low concentrations (usually less than 15%) and at atmospheric pressure. Post-combustion flue gas typically contains carbon dioxide (CO_2), nitrogen (N_2), water vapor (H_2O), oxygen (O_2), and small amounts of other gases such as nitrogen oxides (NO_x), sulfur dioxide (SO_2), carbon monoxide (CO), and particulates. This means a much lower volume of CO_2 must be separated, and more energy is needed to separate a small volume of CO_2.

Precombustion capture can achieve CO_2 removal rates of up to 90–95%, depending on the configuration and technology used. Upgrading an existing power plant to use precombustion capture is difficult because the entire way the fuel is used and burned must change. In a normal power plant, fuel is burned directly to produce heat and generate electricity. But in a pre-combustion system, the fuel must first pass through gasification or reforming, which converts it into hydrogen and carbon monoxide before burning happens. As an example, if an industrial company attempted to convert a coal-fired power station into a gasification-based hydrogen plant, it would have to install new boilers, gasifiers, reactors, and redesign the entire process layout. The process would be almost as expensive and complex as the construction of a new facility from scratch. That's why precombustion capture is usually used in newly designed plants from the beginning, not as an upgrade to old ones [11].

From a climate perspective, precombustion capture plays a dual role: it prevents CO_2 emissions and also produces a hydrogen-based energy economy [12]. This process produces hydrogen, which can be used for transportation, heating, and industrial uses, supporting global decarbonization goals [13]. Countries such as Japan, the United Kingdom, and the United States are actively investing in hydrogen and precombustion technologies as part of their long-term net-zero carbon dioxide emissions strategies.

To conclude, this method of removing CO_2 is highly efficient, but requires extensive facilities. It is generally not practical or cost-effective to retrofit existing plants with precombustion carbon capture systems . While its upfront costs are high, its combination of carbon reduction and clean hydrogen fuel generation is highly advantageous [11, 12]. A diversified and flexible approach to achieving deep decarbonization is possible through precombustion capture, in addition to post-combustion capture and other emerging methods.

4.4 Oxy-Fuel Combustion

Oxy-fuel combustion is a carbon capture method that burns fuels in pure oxygen instead of air. This results in a cleaner exhaust stream, which makes carbon dioxide (CO_2) easier to capture. Under normal conditions, combustion air contains about 78% nitrogen and only 21% oxygen. When fossil fuels are burned in the air, nitrogen mixes with the exhaust gases. This lowers the CO_2 concentration in the flue gas and increases the difficulty and cost of carbon dioxide separation. By contrast, oxy-fuel combustion eliminates nitrogen from the process, producing a flue gas that consists of CO_2 and water vapor. This simple change in the combustion environment makes CO_2 capture far more efficient and straightforward [14].

Figure 4.3 illustrates the oxy-fuel combustion process, where fuel is burned in a nearly pure oxygen environment instead of air. This eliminates nitrogen from the combustion process and produces a flue gas composed mainly of CO_2 and water vapor. The water is condensed and removed, while the CO_2 stream is directed to a separation unit for purification and storage.

In an oxy-fuel combustion system, oxygen is first produced by an air separation unit (ASU), which extracts pure oxygen from the atmosphere [15]. Fuels such as coal, natural gas, or biomass are burned together with this oxygen to produce a high-temperature flame. In spite of this, burning in pure oxygen can cause the temperature to rise above the level necessary for safe operation. To control this, part of the CO_2-rich exhaust gas is recycled back into the boiler and mixed with the hot combustion gases. Because carbon dioxide has a high heat capacity, it can absorb and distribute heat, thereby maintaining a stable flame temperature [16]. It is then necessary to cool the flue gas, which is composed primarily of CO_2 and water vapor, to condense and remove water. A high-purity stream of CO_2 is left behind after this process, which is ready for compression, transportation, and storage.

The main advantage of oxy-fuel combustion is that it produces an exhaust stream made up almost entirely of pure CO_2 and water vapor, which makes carbon capture much

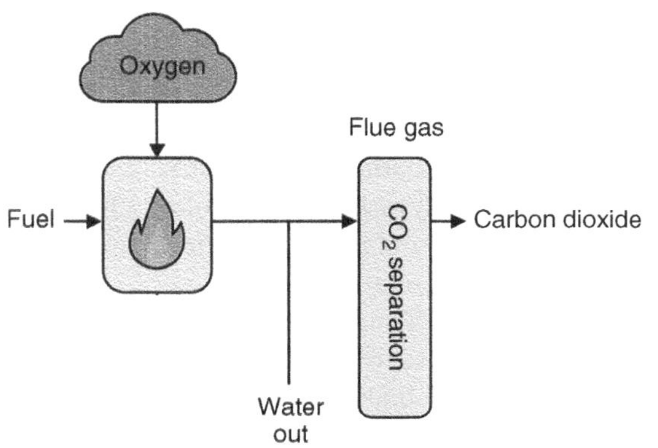

Figure 4.3 Simplified schematic of the oxy-fuel combustion process. It shows oxygen and fuel combustion, generation of CO_2-rich flue gas, and downstream separation of CO_2 for capture and storage.

simpler. This reduces or even eliminates the need for energy-intensive chemical solvents and complex separation or regeneration systems used in other capture methods. In addition, oxy-fuel combustion can achieve very high CO_2 capture efficiencies, up to 95% or more. It can be applied to a wide range of plants and fuels, including biomass, making it a flexible option for low-carbon energy production [16].

In spite of these challenges, oxy-fuel combustion has been successfully demonstrated on a pilot and semicommercial scale, suggesting that it has the potential to be deployed at a large scale in the future. Oxy-fuel technology is being tested in several projects in Europe, Japan, and Australia for capturing high rates of CO_2 from both coal and natural gas plants. Continuous research is underway to improve the production of oxygen from the air [17].

In summary, oxy-fuel combustion is an efficient and promising technology for capturing CO_2 emissions from combustion processes. Its primary advantage lies in the high purity of the CO_2 stream produced, as the absence of nitrogen during combustion results in flue gases consisting mainly of carbon dioxide and water vapor. This leads to highly concentrated CO_2 output, simplifying the capture process [14]. However, the high energy demands associated with oxygen production remain a significant challenge [17]. However, continuous advancements are being made to enhance oxygen separation efficiency.

4.5 Direct Air Capture (DAC)

As the name suggests, DAC is an advanced carbon capture technology that removes carbon dioxide (CO_2) directly from the atmosphere, rather than from specific industrial sources. By comparison to other methods of CO_2 capture, DAC targets a very low concentration of CO_2 in ambient air. The concentration of CO_2 in ambient air is approximately 0.04% (420 ppm). Because large amounts of air must be flowed to produce a significant amount of carbon dioxide, this process poses a technical challenge and is energy-intensive [18]. It is possible to deploy DAC plants anywhere, including deserts, cold climates, and more, which makes it a very flexible and globally applicable technology for reducing atmospheric carbon dioxide emissions [19].

The concept behind DAC is relatively straightforward: ambient air is drawn into a system that chemically extracts CO_2. The captured carbon dioxide is then transported for permanent storage or utilized in industrial processes to produce various products. In practice, DAC facilities use large fans to move air through the contact surfaces of chemical solvents or sorbents that selectively bind CO_2 molecules [18].

There are two main technological approaches: liquid-solvent based and solid-sorbent based. In the liquid approach, air passes through a chemical solution, often amine-based, similar to those used in post-combustion carbon capture, that reacts with CO_2 and separates it from the air. In the solid approach, CO_2 molecules adhere to the surfaces of solid materials such as metal–organic frameworks or functionalized porous solids. Once the sorbent becomes saturated, the CO_2 is released through heating or pressure reduction [20]. It is collected in pure form, compressed, and transported for long-term geological storage. Alternatively, recovered CO_2 can be used to produce synthetic fuels, building materials, or carbonated products.

Figure 4.4 The Direct Air Capture (DAC) plant shows large fans located in front of air contactor units that pull atmospheric air into the system for CO_2 capture. The facility is built from modular, repeatable units, meaning each unit can be independently maintained, replaced, or upgraded with minimal disruption. This modular design also allows the plant to be easily scaled up or down by adding or removing additional units depending on capacity needs.

Scalability and flexibility are the key advantages of DAC. Scalability refers to its modular design, which consists of small units that can be assembled as building blocks. This allows for gradual expansion, starting with a smaller pilot project and scaling up to a larger operation if it proves to be successful [21]. Flexibility means DAC can operate in diverse environments and climates, including deserts, offshore areas, and cold regions.

Figure 4.4 illustrates a large-scale DAC installation designed to remove CO_2 directly from ambient air using industrial fans. There are several fan units that draw atmospheric air inside and pass it through sorbent-based chemicals that selectively capture CO_2 molecules. Modular, repeatable units have been used to construct the plant, allowing capacity to be easily increased or decreased over time by adding or removing identical blocks. Due to their flexibility, DAC systems can be installed in a wide range of areas, including remote deserts, coastal areas, and near CO_2 storage and utilization facilities. After the CO_2 is captured, it can be compressed for permanent geological storage or reused for the production of synthetic fuels, building materials, or industrial applications.

Unlike other methods, DAC facilities can be located in regions ideal for storage, such as near deep saline aquifers or depleted oil and gas reservoirs. This reduces the need for costly pipeline transport from capture sites to storage locations [22]. Moreover, DAC can be powered by renewable energy sources like solar or wind. This ensures that the carbon capture process itself does not contribute any additional carbon to the environment. When captured CO_2 is permanently stored underground, DAC achieves "negative emissions," provided no fossil fuels are used for DAC facility operations such as compressors or regenerators. This makes DAC a crucial technology for compensating emissions from hard-to-decarbonize sectors like aviation, shipping, and agriculture [23].

However, DAC remains an emerging and therefore costly technology. Ongoing advances in material science and process optimization are gradually reducing costs, yet the energy demand continues to be a major limitation. The process requires significant amounts of heat and electricity, particularly for solvent or sorbent regeneration and CO_2 compression [21]. As a result, large-scale deployment will depend on the availability of affordable renewable energy to operate the plants efficiently [23]. Furthermore, government support will be crucial through policies such as carbon pricing, taxation incentives, and dedicated funding programs. This will encourage investment and accelerate the development of DAC projects [24].

There have already been a number of pioneering companies that have demonstrated the feasibility of DAC. The largest DAC plant in the world is operated by Climeworks (Switzerland). The Orca plant in Iceland is a facility that injects captured CO_2 underground and mineralizes it. DAC systems are being developed in the United States by Carbon Engineering and Global Thermostat. Their focus is on integrating DAC with synthetic fuel production that recycles CO_2 captured into usable energy. By implementing these and other projects, DAC will be able to demonstrate the growing role it is playing not only as a tool for climate mitigation but also as a driver for carbon circularity, so that captured carbon is used as a resource rather than disposed of [25]. Carbon circularity in simple terms means keeping carbon in use instead of releasing it into the atmosphere.

In summary, DAC represents one of the most promising technologies for removing existing carbon emissions in the atmosphere. Although currently challenged by costs and high energy requirements, its potential for global expansion and long-term carbon removal from the air makes it an essential component of future net-negative emissions strategies [21]. As renewable energy expands and technological efficiencies regarding capturing and compressing carbon dioxide improve, DAC is expected to become a vital part of the climate solution.

4.6 Bioenergy with Carbon Capture and Storage (BECCS)

The BECCS technology eliminates carbon dioxide from the atmosphere as well as prevents CO_2 emissions. Therefore, it is a highly effective method for achieving net-negative emissions. This process begins with biomass, such as forestry residues, agricultural waste, crops, or organic materials. These materials naturally absorb CO_2 from the atmosphere during photosynthesis. Biologically, photosynthesis is the process by which green plants, microorganisms, and some bacteria convert sunlight into sugars (glucose). It is possible to produce electricity, heat, or fuel from this biomass by burning it. Unlike traditional biomass plants, BECCS facilities are equipped with carbon capture systems. These systems capture the CO_2 released during the combustion of biomass, such as agricultural waste or crops, rather than allowing it to enter the atmosphere [26].

Once captured, the CO_2 is compressed and transported, normally through pipelines, to a certified geological storage site, such as a depleted oil or gas reservoir or a deep saline aquifer, where it is permanently stored underground [27]. In the case of BECCS, the carbon originally comes from the atmosphere and is absorbed by biomass (plants or organic matter). When that biomass is burned for energy, the released CO_2 is captured again and stored instead of being released back into the air. This double cycle means BECCS can be net-negative, actively removing more CO_2 from the atmosphere than it emits [28].

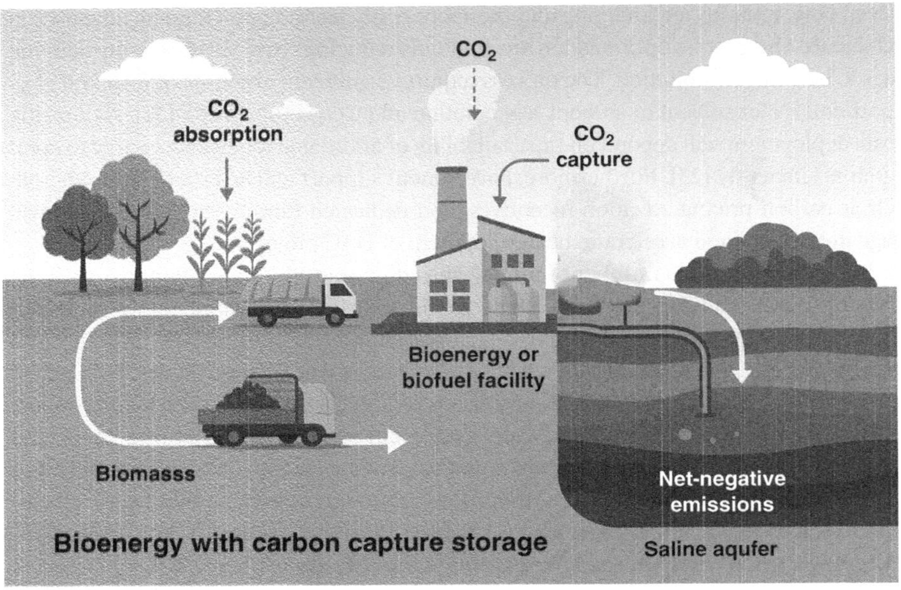

Figure 4.5 Illustration of the Bioenergy with Carbon Capture and Storage (BECCS) process. It shows CO_2 absorption by trees during growth, energy generation from biomass, capture of resulting CO_2 emissions, and permanent geological storage to achieve net-negative emissions.

Figure 4.5 illustrates the complete process of combining BECCS. It starts with the natural absorption of CO_2 from the atmosphere by biomass such as trees during growth. The biomass is then collected and transported to a facility that produces bioenergy or biofuel. As a result of burning or heating biomass, electricity, heat, or fuel is generated [28]. Energy production involves capturing CO_2 emissions rather than releasing them into the atmosphere. Once captured, the CO_2 is compressed and transported, usually via pipeline or ship, to geological storage sites such as deep saline aquifers shown on the bottom left of the picture or depleted oil and gas reservoirs where it is permanently stored underground [27].

BECCS offers a dual benefit: it generates renewable energy, such as heat, electricity, or hydrogen, while simultaneously removing CO_2 from the atmosphere. This makes it highly attractive for countries aiming not only to reach net-zero but also to achieve net-negative emissions by actively reducing atmospheric carbon levels [29].

BECCS is particularly relevant in sectors that already rely on biomass, such as power generation, biofuels, and hydrogen production. In power generation, biomass can be burned to produce electricity or heat. Integrating CO_2 capture ensures emissions are permanently removed rather than released [30]. In biofuel production by anaerobic bacterial fermentation of biomass, organic materials such as crops, agricultural residues, and forestry waste are processed into fuels like bioethanol and biodiesel. Biofuels are produced by converting biomass into fuel through biological or chemical processes [31].

During the production of hydrogen, biomass is gasified, which means that it is heated at high temperatures with limited oxygen in order to produce synthesis gas (syngas), a mixture of hydrogen, carbon dioxide, and oxygen. Following gasification, the hydrogen can be separated and used, while the CO_2 released during the process can be captured and stored [32].

Hydrogen can be a low-carbon energy carrier when produced with carbon capture or from renewable sources, making it an attractive solution for clean energy in the industries.

BECCS is also increasingly considered for hard-to-abate industries like steel and cement, where integrating biomass as part of the fuel production combined with CO_2 capture and permanent storage could provide one of the few viable pathways to deeply reduce or even neutralize emissions [33].

However, BECCS also faces practical challenges. It relies on a large environmentally sustainable biomass supply, which must be managed carefully to avoid environmental harm, for example, the unsustainable removal of forest wood. Proper land-use is essential to prevent interference with agricultural land and protect food production. Technology also requires substantial infrastructure for biomass such as storage and handling as well as CO_2 transport and long-term storage [34]. Costs remain high, and successful large-scale deployment will depend heavily on strong policy support, effective carbon pricing, and investment incentives.

Figure 4.6 shows a modern biomass bioenergy facility processes agricultural residues, organic waste, and crops into biogas or bioenergy. Typically, such biomass facilities are

Figure 4.6 Illustration of a biomass-based facility located in an agricultural region. It shows both the opportunity for renewable energy production and the potential land-use challenges related to threatening farmland and agriculture.

located near farmland, as shown in the image, to reduce transportation distances and ensure a reliable supply of biomass. However, biomass reliance creates challenges. If not managed environmentally, biomass consumption can threaten agricultural land, potentially reducing land availability for food production. In this way, if farmers start growing energy crops like corn or switchgrass for bioenergy instead of food crops, food shortages may occur. Additionally, cutting trees for logs can lead to deforestation [34]. Significant infrastructure is also required for biomass collection, transport, storage, and handling, which increases costs and logistical complexity. Ensuring that biomass is sourced sustainably and without harming ecosystems or food security is therefore critical for the responsible deployment of BECCS and bioenergy systems.

Despite these challenges, BECCS is widely considered by the Intergovernmental Panel on Climate Change (IPCC) and the International Energy Agency (IEA) as one of the most essential technologies for achieving net-negative global emissions. BECCS is particularly important in scenarios that aim to limit global warming to 1.5°C above the preindustrial period temperature level. Essentially, BECCS is a solution for generating renewable energy and removing carbon dioxide from the atmosphere. As a result, it can be used as an effective tool in the transition to a climate-stable, low-carbon future [35].

4.7 Comparative Evaluation—Efficiency, Cost, and Feasibility

Different carbon capture technologies, such as post-combustion, precombustion, oxy-fuel combustion, DAC, and BECCS, offer distinct advantages and limitations when evaluated in terms of efficiency, cost, scalability, and practicality. An understanding of these costs and benefits is essential for determining where, how, and to what extent each technology can be implemented most effectively.

Post-combustion capture is currently the most developed and widely implemented carbon capture technology. Its main advantage is that it can be easily added to existing fossil-fuel power plants and industrial facilities with minimal disturbance to the core process. But it requires a lot of heat and electricity, especially for solvent regeneration and CO_2 separation, so it's quite energy intensive. The cost of post-combustion capture typically ranges from \$50 to \$100 per tonne of CO_2 captured, depending on the size of the plant, the type of fuel, and the configuration of the process. On average, it can be slightly more expensive than precombustion and oxy-fuel capture methods [7, 8].

It is generally more efficient to capture CO_2 before combustion than after combustion due to the fact that the CO_2 is separated under high pressure, making it easier to capture. Using this method, fossil fuels or biomass are first converted into syngas, a mixture of hydrogen (H_2) and carbon dioxide (CO_2). The high CO_2 concentration makes CO_2 extraction simpler. Hydrogen production facilities are the most common application of this technology. Precombustion capture typically costs around \$40–80 per tonne of CO_2, making it slightly cheaper than post-combustion in many cases. It is only feasible for new-build plants, as upgrading existing plants with precombustion facilities requires extensive process redesign and high costs [10, 11].

Oxy-fuel combustion produces very high CO_2 purity and simplified separation by removing nitrogen from the combustion process. However, the energy required to produce pure oxygen through air separation remains a major challenge. It is attractive for new plants but difficult to apply to existing infrastructure without major modifications. The cost of oxy-fuel capture typically ranges from $40 to 90 per tonne of CO_2, placing it between pre-combustion and post-combustion, generally cheaper than post-combustion, but slightly more expensive or comparable to precombustion [14, 16].

DAC technology is the only technology capable of removing CO_2 from the atmosphere. This makes it extremely important in compensating for emissions from hard-to-abate industries, such as steel and cement. Approximately 60% of the CO_2 emitted by cement production comes from the chemical reaction itself (when limestone is heated). Even if clean energy is used in the production of cement, process emissions cannot be avoided. DAC is therefore used to remove an equivalent amount of CO_2 from the atmosphere to compensate for or offset the cement plant's unavoidable emissions. However, it is currently the most expensive option, costing $400–600 per tonne, and requiring a substantial amount of clean energy [18, 19].

BECCS and DAC are the only carbon capture technologies with true net-negative carbon emission potential, as it combines biomass-based renewable energy generation with carbon removal. It is strategically important for achieving both net-zero and carbon-negative national targets, but its large-scale deployment faces challenges related to sustainable biomass availability, land-use conflict with agriculture, and the need for extensive infrastructure for biomass handling, storage, and CO_2 storage. In terms of cost, BECCS is typically estimated at $100–200 per tonne of CO_2 removed, depending on biomass sourcing, transport, and storage logistics [26].

In summary:

- Post-combustion is most suitable for upgrading existing fossil fuel plants and short-term implementation.
- Precombustion and oxy-fuel are suited as new plants, while precombustion is integrated with hydrogen systems.
- DAC is essential for global atmospheric CO_2 removal and is the most expensive.
- BECCS is critical for achieving net-negative emissions, the same as DAC, but relies on sustainable biomass resources.

No single carbon capture technology is suitable for every situation. Instead, different solutions should be selected based on the specific project conditions—such as sector, cost, location, and local environmental or policy factors.

4.8 Key Highlights (Chapter Summary)

◇ Chapter 4—Key Highlights:

1) Carbon capture technologies aim to prevent CO_2 from entering the atmosphere or remove it directly from the air.
2) The main stages in carbon capture are capture, compression/transport, and storage or utilization.
3) Post-combustion capture is the most widely used and ideal for retrofitting (upgrading) existing power plants and industries.

4) Precombustion capture is highly efficient but mainly suitable for new facilities, especially hydrogen and gasification-based systems.
5) Oxy-fuel combustion burns fuel in pure oxygen after removing nitrogen from air, producing a CO_2-rich combustion exhaust that is easier to capture.
6) Direct Air Capture (DAC) removes CO_2 directly from the ambient atmosphere and is crucial for negative emission strategies.
7) BECCS combines biomass and bioenergy production with CO_2 capture, offering net-negative emissions.
8) Each technology varies in cost, scalability, energy, and infrastructure requirements.
9) No single technology is suitable for all cases; instead, a combination is required to meet global decarbonization targets.
10) Successful deployment of carbon capture technologies depends on government policies, international cooperation, and long-term financial support.

Keywords: Carbon Capture, Post-Combustion, Precombustion, Oxy-Fuel Combustion, Direct Air Capture, BECCS, Negative Emissions, Climate Mitigation, Carbon Neutrality, Net-Zero Transition, Carbon Utilization, Industrial Decarbonization, Renewable Energy Integration, Energy Transition, Clean Hydrogen

References

1 P. Roy, A. K. Mohanty, and M. Misra, "Prospects of carbon capture, utilization and storage for mitigating climate change," *Environmental Science: Advances*, vol. 2, pp. 409–423, 2023. DOI: 10.1039/D2VA00236A.

2 T. Wilberforce, A. Baroutaji, B. Soudan, A. H. Al-Alami, and A. G. Olabi, "Outlook of carbon capture technology and challenges," *Science of the Total Environment*, vol. 657, pp. 56–72, Mar., 2019. DOI: 10.1016/j.scitotenv.2018.11.424.

3 Z. Zhang, T. N. G. Borhani, and M. H. El-Naas, "Chapter 4.5—carbon capture," in *Exergetic, Energetic and Environmental Dimensions*, I. Dincer and C. O. Colpan, Eds. Academic Press, 2018, pp. 997–1016, DOI: 10.1016/B978-0-12-813734-5.00056-1.

4 M. Ozkan, S. P. Nayak, A. D. Ruiz, and W. Jiang, "Current status and pillars of direct air capture technologies," *iScience*, vol. 25, no. 4, 103990, Apr., 2022. DOI: 10.1016/j.isci.2022.103990.

5 M. Bui, M. Fajardy, and N. Mac Dowell, "Bio-energy with carbon capture and storage (BECCS): Opportunities for performance improvement," *Fuel*, vol. 213, pp. 164–175, Feb., 2018. DOI: 10.1016/j.fuel.2017.10.100.

6 R. Shirmohammadi, A. Aslani, and R. Ghasempour, "Challenges of carbon capture technologies deployment in developing countries," *Sustainable Energy Technologies and Assessments*, vol. 42, 100837, Dec., 2020. DOI: 10.1016/j.seta.2020.100837.

7 C. Chao, Y. Deng, R. Dewil, J. Baeyens, and X. Fan, "Post-combustion carbon capture," *Renewable and Sustainable Energy Reviews*, vol. 138, 110490, Mar., 2021. DOI: 10.1016/j.rser.2020.110490.

8 F. Raganati and P. Ammendola, "CO_2 post-combustion capture: A critical review of current technologies and future directions," *Energy & Fuels*, vol. 38, no. 15, pp. 13858–13905, Jul., 2024. DOI: 10.1021/acs.energyfuels.4c02513.

9 H. C. Mantripragada, H. Zhai, and E. S. Rubin, "Boundary Dam or Petra Nova—which is a better model for CCS energy supply?," *International Journal of Greenhouse Gas Control*, vol. 82, pp. 59–68 , Mar., 2019. DOI: 10.1016/j.ijggc.2019.01.004.

10 D. Jansen, M. Gazzani, G. Manzolini, E. van Dijk, and M. Carbo, "Pre-combustion CO_2 capture," *International Journal of Greenhouse Gas Control*, vol. 40, pp. 167–187, Sept., 2015. DOI: 10.1016/j.ijggc.2015.05.028.

11 Z. Chen, "A review of pre-combustion carbon capture technology," in Proc. 2022 7th Int. Conf. on Social Sciences and Economic Development (ICSSED 2022), vol. 656, pp. 142–146, Apr. 2022, DOI: 10.2991/aebmr.k.220405.029

12 F. Petrakopoulou and G. Tsatsaronis, "Production of hydrogen-rich fuels for pre-combustion carbon capture in power plants: A thermodynamic assessment," *International Journal of Hydrogen Energy*, vol. 37, no. 9, pp. 7554–7564, May, 2012. DOI: 10.1016/j.ijhydene.2012.01.147.

13 S. Sharma and S. K. Ghoshal, "Hydrogen the future transportation fuel: From production to applications," *Renewable and Sustainable Energy Reviews*, vol. 43, pp. 1151–1158, Mar., 2015. DOI: 10.1016/j.rser.2014.11.093.

14 S. Yadav and S. S. Mondal, "A review on the progress and prospects of oxy-fuel carbon capture and sequestration (CCS) technology," *Fuel*, vol. 308, 122057, Jan., 2022. DOI: 10.1016/j.fuel.2021.122057.

15 A. Darde, R. Prabhakar, J.-P. Tranier, and N. Perrin, "Air separation and flue gas compression and purification units for oxy-coal combustion systems," *Energy Procedia*, vol. 1, no. 1, pp. 527–534, Feb., 2009. DOI: 10.1016/j.egypro.2009.01.070.

16 T. Wall *et al.*, "An overview on oxyfuel coal combustion—state of the art research and technology development," *Chemical Engineering Research and Design*, vol. 87, no. 8, pp. 1003–1016, Aug., 2009. DOI: 10.1016/j.cherd.2009.02.005.

17 M. A. Nemittallah *et al.*, "Oxy-fuel combustion technology: Current status, applications, and trends," *International Journal of Energy Research*, vol. 41, no. 13, pp. 1670–1708, Feb., 2017. DOI: 10.1002/er.3722.

18 C. Breyer, M. Fasihi, C. Bajamundi, and F. Creutzig, "Direct air capture of CO_2: A key technology for ambitious climate change mitigation," *Joule*, vol. 3, no. 9, pp. 2053–2057, Sep., 2019. DOI: 10.1016/j.joule.2019.08.010.

19 A. B. Hernández, Direct Air Capture by Large Scale Reforestation of Sahara Desert, Master's Thesis, Politecnico di Milano, Milan, Italy, 2019/2020. [Online]. Available: https://hdl.handle.net/10589/164425

20 G. Li and J. Yao, "Direct air capture (DAC) for achieving net-zero CO_2 emissions: Advances, applications, and challenges," *Eng*, vol. 5, no. 3, pp. 1298–1336, Jul., 2024. DOI: 10.3390/eng5030069.

21 N. McQueen, K. V. Gomes, C. McCormick, K. Blumanthal, M. Pisciotta, and J. Wilcox, "A review of direct air capture (DAC): Scaling up commercial technologies and innovating for the future," *Progress in Energy*, vol. 3, no. 3, 032001, Apr., 2021. DOI: 10.1088/2516-1083/abf1ce.

22 T. Terlouw, K. Treyer, C. Bauer, and M. Mazzotti, "Life cycle assessment of direct air carbon capture and storage with low-carbon energy sources," *Environmental Science & Technology*, vol. 55, no. 16, pp. 11397–11411, Aug., 2021. DOI: 10.1021/acs.est.1c03263.

23 M. Mostafa, C. Antonicelli, C. Varela, D. Barletta, and E. Zondervan, "Capturing CO_2 from the atmosphere: Design and analysis of a large-scale DAC facility," *Carbon Capture Science & Technology*, vol. 4, 100060, Sept., 2022. DOI: 10.1016/j.ccst.2022.100060.

24 M. Erans, E. S. Sanz-Pérez, D. P. Hanak, Z. Clulow, D. M. Reiner, and G. A. Mutch, "Direct air capture: Process technology, techno-economic and socio-political challenges," *Energy & Environmental Science*, vol. 15, pp. 1360–1405, 2022. DOI: 10.1039/D1EE03523A.

25 S. O'Neill, "Direct air carbon capture takes baby steps—Giant strides are needed," *Engineering*, vol. 8, pp. 3–5, 2022, https://www.sciencedirect.com/journal/engineering.

26 S. V. Hanssen, V. Daioglou, Z. J. N. Steinmann, *et al.*, "The climate change mitigation potential of bioenergy with carbon capture and storage," *Nature Climate Change*, vol. 10, pp. 1023–1029, 2020. DOI: 10.1038/s41558-020-0885-y.

27 M. S. A. Joarder *et al.*, "A comprehensive review of carbon dioxide capture, transportation, utilization, and storage: A source of future energy," *Environmental Science and Pollution Research*, vol. 32, pp. 9299–9332, 2025. DOI: 10.1007/s11356-025-36284-9.

28 A. I. Osman, N. Mehta, A. M. Elgarahy, *et al.*, "Conversion of biomass to biofuels and life cycle assessment: A review," *Environmental Chemistry Letters*, vol. 19, pp. 4075–4118, 2021. DOI: 10.1007/s10311-021-01273-0.

29 A. C. Köberle, "The value of BECCS in IAMs: A review," *Current Sustainable/Renewable Energy Reports*, vol. 6, pp. 107–115, 2019. DOI: 10.1007/s40518-019-00142-3.

30 J. Full, S. Merseburg, R. Miehe, and A. Sauer, "A new perspective for climate change mitigation—introducing carbon-negative hydrogen production from biomass with carbon capture and storage (HyBECCS)," *Sustainability*, vol. 13, no. 7, 4026, Apr., 2021. DOI: 10.3390/su13074026.

31 P. Gupta and A. Gupta, "Biogas production from coal via anaerobic fermentation," *Fuel*, vol. 118, pp. 238–242, Feb., 2014. DOI: 10.1016/j.fuel.2013.10.075.

32 W. X. Peng, L. S. Wang, M. Mirzaee, H. Ahmadi, M. J. Esfahani, and S. Fremaux, "Hydrogen and syngas production by catalytic biomass gasification," *Energy Conversion and Management*, vol. 135, pp. 270–273, Mar., 2017. DOI: 10.1016/j.enconman.2016.12.056.

33 S. E. Tanzer, K. Blok, and A. Ramírez, "Decarbonising industry via BECCS: Promising sectors, challenges, and techno-economic limits of negative emissions," *Current Sustainable/Renewable Energy Reports*, vol. 8, pp. 253–262, 2021. DOI: 10.1007/s40518-021-00195-3.

34 H. J. Buck, "Challenges and opportunities of bioenergy with carbon capture and storage (BECCS) for communities," *Current Sustainable/Renewable Energy Reports*, vol. 6, pp. 124–130, 2019. DOI: 10.1007/s40518-019-00139-y.

35 A. Hansson, J. Anshelm, M. Fridahl, and S. Haikola, "Boundary work and interpretations in the IPCC review process of the role of bioenergy with carbon capture and storage (BECCS) in limiting global warming to 1.5 °C," *Frontiers in Climate*, vol. 3, Apr., 2021. DOI: 10.3389/fclim.2021.643224.

5

Social Acceptance and Public Perception of Carbon Capture Projects

5.1 Why Public Perception Matters?

When we talk about carbon capture and storage (CCS), we usually focus on technology, engineering, and environmental benefits. However, there is another equally important aspect: public perception. Public perception can make or break a CCS project. Even if the science is solid and the design is safe, if the local community doesn't trust the project or feels left out, the project might never move forward if the local community doesn't trust it or feels excluded from the process [1].

Public perception matters because large energy projects are often built in or near communities. People who live close to CO_2 pipelines, compressors, or storage sites may worry about safety, health, or changes to their environment. Some may fear leaks or accidents, even if the risks are low. Others might worry about how the project will affect land value, agriculture, or their daily life [2]. These concerns are real and valid, even if they aren't always based on technical facts.

Governments and companies have learned from past experience that ignoring public opinion can lead to serious problems such as project delays, legal challenges, or even complete project cancellation [3]. For example, in some areas, wind farms and waste treatment plants were stopped because communities were not involved early or didn't feel that their concerns were taken seriously. CCS projects face the same risk. Without trust and good communication, they could be delayed, blocked, or cancelled in the same way [4].

On the other hand, when communities are informed, involved, and treated with respect, they are more likely to support new projects, even those that are complex or unfamiliar, like CCS [1]. Complex projects are those that involve new or unfamiliar technologies, have potential environmental or safety risks, or require large infrastructure. People want to be part of the conversation, not just receive decisions made by others. They want transparency about risks, benefits, and how their voices will be included in the process.

In short, public perception isn't just a public relations issue, and it's a critical part of project success. Technical teams need to work alongside communication experts, sociologists, and local leaders to build bridges with the community. In the end, building trust is just as important as building pipelines.

5.2 Historical Cases of Public Opposition

To understand how people might react to carbon capture projects, it helps to look at past examples of public opposition to large-scale energy and industrial developments. History shows us that even well-designed projects can face strong pushback when local communities feel excluded, uninformed, or unsafe. These cases offer important lessons for the CCS sector today.

One clear example comes from the nuclear energy industry. In many countries during the 1970s and 1980s, governments tried to build nuclear power plants without much input from the public. While the technology was advanced and promised cleaner energy, the risks, especially after accidents like Three Mile Island in the United States or Chernobyl in the Soviet Union, caused fear and distrust [5, 6]. Protests grew large, and in several countries like Germany and Italy, nuclear development was slowed down or completely stopped due to public pressure. Even today, the public memory of those events still influences how people view new energy technologies.

Another example is the construction of high-voltage power lines and wind farms in rural areas. Although these projects support renewable energy, they often face opposition from local residents. People often raise concerns related to renewable projects, , noise, or potential health effects, even if the science shows minimal risk. In some regions, the term "Not In My Backyard" or NIMBY has become common [7]. It refers to individuals who generally support clean energy but oppose having the infrastructure built near their homes. This mindset has been intensified by poor communication between the project team and the local community.

Figure 5.1 highlights three key lessons from past infrastructure projects that are relevant to CCS: how public memory of past accidents influences current attitudes, how NIMBY

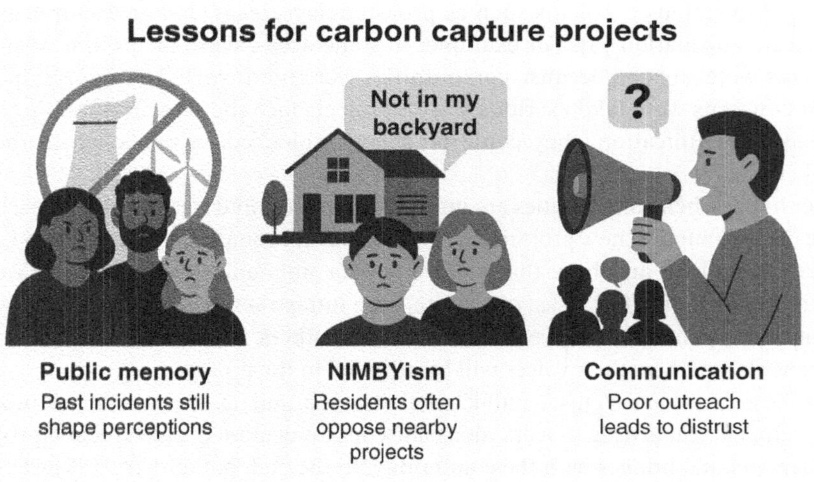

Figure 5.1 Common barriers to public acceptance of CCS projects: public memory of past failures, local resistance (NIMBYism), and inadequate communication efforts.

("Not In My Backyard") concerns drive local opposition, and how poor communication can lead to mistrust and resistance.

A more recent and relevant case is the public reaction to CO_2 storage sites in Europe. In the Netherlands, a project aimed at storing CO_2 in an empty gas field near the town of Barendrecht faced strong opposition. The local community was not properly consulted in the early stages, and people feared gas leaks or explosions. Despite technical studies showing the site was safe, the project was cancelled in 2010 due to public protests and political pressure [8]. This case became a turning point in how CCS projects are approached in Europe. It showed that technical safety alone is not enough, and people need to feel informed and respected.

In the United States, some proposed CO_2 pipelines have also sparked protests, especially in areas where people are already concerned about land rights, water safety, or past environmental damage. For example, when companies don't clearly explain the route, risks, or emergency plans, communities may assume the wrong thing. In several Midwestern states, public hearings about CCS pipelines are filled with emotional statements from farmers, community representatives, and local officials. These statements express fears about leaks, property damage, or lack of compensation.

These historical cases make one thing clear: public opposition is often not about the technology itself, but about how the project is introduced, explained, and managed. People are more likely to resist when decisions are made without them, or when their concerns are dismissed. In contrast, when developers engage early, provide clear information, and invite community input, projects tend to move forward more smoothly even when risks are involved [9].

In short, CCS developers today should avoid repeating past mistakes. Instead, they should study these cases, learn from them, and prioritize building strong relationships with communities, not just delivering technical presentations. Trust, transparency, and open communication are essential. Without these, opposition is likely even if the science is solid.

5.3 Risk Communication and Transparency

5.3.1 Understanding and Communicating Risk Perception

Risk communication in CCS is essential to ensure that communities understand the potential hazards and trust the project. In any carbon capture project, one of the most sensitive issues is risk; not just the technical risk, but how people feel about risk. Even if engineers and scientists say a system is safe, the public may still worry. As shown in Figure 5.2, technical and emotional views of risk often differ, which creates a communication gap between CCS developers and local communities. This is why risk communication is so important [10, 11].It is not only about sharing factual information. It's about building understanding, trust, and openness between the people running a project and the people living near it.

Many communities fear CCS projects because they don't fully understand how the technology works or what could go wrong. For example, they may worry that CO_2 could leak from underground storage or that a pipeline could explode. Even if the actual probability of these events is very low, the perceived risk can still be high, especially if past projects in

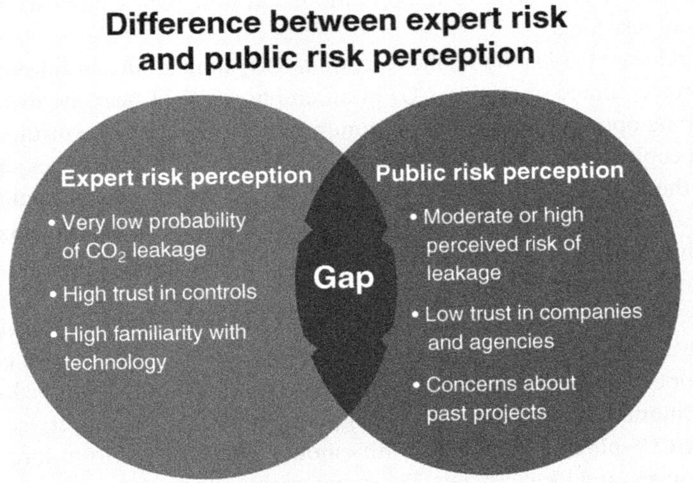

Figure 5.2 Difference between expert-assessed risk and public-perceived risk in CCS projects. Misalignment between technical evaluation and public emotion highlights the need for risk communication.

other sectors have caused harm, or if there is a general lack of trust in companies or government agencies [12].

Good risk communication means talking early and honestly. People want to know not only what the risks are, but how those risks will be managed. Vague answers or overly technical language can create more fear. On the other hand, clear explanations, real data, and concrete safety plans can help reduce anxiety [13]. For example, showing people how leak detection systems work, or how pressure sensors can automatically shut down a pipeline in case of emergency, can turn fear into understanding.

5.3.2 Transparency, Community Engagement, and Trust Building

Another key part of communication is transparency [14]. Communities don't want to feel that information is being hidden or filtered. They want access to the same data that engineers and regulators see [14]. In some successful projects, public monitoring tools have been created so that local residents can check CO_2 levels, pressure readings, or incident reports in real time. This level of transparency enhances institutional credibility within the community It's also imperative to remember that different people respond to risk in different ways. Some care most about the environmental impacts. Others worry about their health, land, or future property values [15]. Risk communication should therefore be tailored to local concerns. One-size-fits-all messages don't work. In fact, simplifying too much like saying " Do not worry, everything is fine" without giving enough explanation can feel dismissive. Dismissive means that people might feel that their concerns are being ignored. The best approach is to listen first, understand what the community is truly concerned about, and then respond with empathy and clarity.

Figure 5.3 CCS community engagement strategies that use honest, transparent, and locally-focused communication to build trust and improve public acceptance.

A good example comes from a CCS pilot project in Alberta, Canada. The project team organized small, in-person community meetings instead of large, formal presentations. They used maps, models, and Q&A sessions to explain risks and safety measures. Importantly, they admitted the uncertainties that still existed, rather than pretending everything was guaranteed. This honesty helped build respect and, over time, support [11]. As Figure 5.3 shows, respectful and open community engagement strategies, like those used in Alberta, are more likely to succeed than top-down approaches.

In contrast, when project developers avoid tough questions or try to minimize concerns, it often results in failure. People feel ignored or tricked, and rumors spread quickly among people. In today's world of social media, a single poorly handled meeting or unclear statement can turn into a major problem. That's why honesty, clarity, and two-way communication are essential parts of any CCS risk handling strategy [11].

In summary, risk communication is not about convincing people that CCS is safe. It's about sharing information in a respectful, transparent way and allowing communities to form their own conclusions based on facts. When people feel included and informed, they are far more likely to support a project even if it carries some risk.

5.4 Influence of Media and Social Networks

5.4.1 Influence of Traditional and Social Media

In today's world, the way people learn about carbon capture projects is often not through science papers or technical reports but through the media and social networks. News articles, online videos, tweets, and Facebook posts can shape public opinion more than any formal presentation or public meeting. This makes media influence a powerful factor in how communities react to CCS [16].

Traditional media such as newspapers, television, and radio still play a significant role. If a major outlet like CNN, BBC, or The New York Times reports that a CO_2 pipeline has leaked somewhere in the world, people in other regions may become concerned, even if the incident is rare. These reports often highlight the most visible aspects of an incident, which can shape public perception and sometimes exaggerate the seriousness of the situation from a technical perspective.

In some cases, media coverage may be inaccurate or misleading [17]. For example, an article might describe CO_2 as "toxic" or compare CCS to nuclear waste, even though the scientific facts are different. When reporters don't understand the technology, or when they want to make dramatic headlines, the public may get a mistaken picture of the real risks and benefits. But media can also play a positive role. Well-informed journalists can help explain complex technologies in clear language. Local media, in particular, can help spread accurate information in a way that feels trustworthy to the community [17]. In some CCS projects, project teams have built relationships with local reporters to ensure local community and public receive updates, and access to reliable data.

Social media is even more influential in shaping opinion, especially among younger audiences. Platforms like Twitter, Facebook, Instagram, and TikTok allow people to share their views quickly and widely. The downside is that misinformation can spread very fast. On the other hand, social media also gives communities a voice. People can ask questions, express concerns, and share personal stories [16, 17]. Project developers can use these platforms not only to share facts, but to listen, respond, and build relationships. Figure 5.4 illustrates how different media sources such as newspapers, TV, social platforms, and local websites shape public awareness and opinions about CCS. In the middle of the image, there

Figure 5.4 The influence of traditional and social media on public perception of CCS projects. News reports, radio, and social media posts (shown in the top clouds) can quickly shape how people in the community (bottom) perceive CO_2 pipelines and storage systems, sometimes spreading useful awareness, and sometimes causing confusion or fear if information is inaccurate or incomplete.

are tanks labeled "CO_2." with connected piping. A warning sign appears above one of the piping, suggesting that there might be a leak or danger even if that's not confirmed. This is an example of how rumors or panic can start, especially if people don't have full information. At the bottom, we see people in the community, three men and a woman, looking at their phone and tablet. They are trying to understand what's happening. But depending on what kind of media they follow, they might receive different messages: some helpful, some false. Thus, the image also highlights how fast misinformation can spread and how project teams can respond through official channels and trusted sources.

5.4.2 Strategies for Media Response and Public Trust

When people hear about a carbon capture project, they often go online to learn more [16]. If the only information they find is from people who are afraid or misinformed, public opinion may turn negative even if the project is safe and well-designed. That's why project teams need to have a strong presence on platforms like news websites, social media (e.g. Twitter/X, Facebook, and YouTube), and local community websites. By posting updates, answering questions, and correcting misinformation quickly, they can help people feel more confident and informed [17].

For example, imagine that a local resident posts a video on social media saying that "the CO_2 pipeline near my house is dangerous and might explode." If this message spreads without any response, many people might believe it. But if the project team shares a clear and respectful video explaining the safety systems in place like pressure sensors, automatic shut-off valves, and emergency plans, people get a more complete picture. This builds public trust.

A useful example comes from a CCS project in Norway. After early media criticism, the project team worked closely with journalists and launched a strong social media campaign using infographics and behind-the-scenes videos. They answered questions in real time and posted regular updates. Over time, public opinion became more balanced, and some initial critics changed their views [18]. The key lesson here is that media and social networks aren't just communication channels; they are battlegrounds for trust. Being proactive, honest, and available online is now a necessary part of gaining public support.

In today's world, managing how a CCS project is talked about in the media and on social networks is not just helpful but essential [16, 17]. This doesn't mean trying to "control the narrative," which is a phrase that means trying to make everyone only hear your version of the story or ignore other opinions. Instead, the goal is to be part of the public conversation to listen, respond, and share trustworthy information with the public. So, in summary, being active in media and social networks helps CCS project teams share facts, calm public fears, and show they care about community concerns. It's not about silencing others. It's about helping people understand what's really going on.

5.5 Behavioral Psychology and Climate Technology

Even when people have access to facts about carbon capture and its benefits, they don't always respond with logic. That's because human decision-making is not only based on facts, but it's also shaped by emotions, habits, and mental shortcuts [19]. Understanding how people actually think and feel is essential when introducing new technologies like CCS. That's where behavioral psychology becomes valuable.

One important concept is loss aversion. People tend to fear losing something they already have more than they value gaining something new [20]. In the case of CCS, a community may worry that a new pipeline or injection well will harm their land, water, or lifestyle, even if it offers long-term climate benefits. The perceived "loss" feels more real and immediate than the environmental "gain," which is often abstract or far in the future.

Another idea from psychology is status quo bias, a natural human tendency to prefer things the way they are. Many people feel uncomfortable with change, especially when it involves unfamiliar technology [21]. If they've never heard of CO_2 storage before, their first reaction might be suspicion or fear. That doesn't mean they are anti-science; it just means that the new idea challenges their comfort zone [22].

Trust is also a big part of behavior. People are more likely to accept new technology if they trust the people behind it. This includes scientists, engineers, companies, and government agencies [2]. If those institutions have a history of mistakes or secrecy, public resistance grows. On the other hand, if people feel respected and heard, they are more open to new ideas, even risky ones.

Framing plays a major role, too. Messages are received differently depending on how they are presented. Saying "this CO_2 pipeline is safe" may seem misleading. But saying "we understand your concerns, and here's how we're addressing them step-by-step" creates a feeling of openness. Behavioral science shows that how you say something often matters more than what you say [23].

Psychologists use the term "social proof" to describe how people often look to others when deciding what to think or do, especially when they are unsure or don't have enough information [24]. In the case of CCS, this means that if people see that trusted local figures like village leaders, community elders, teachers, or respected neighbors support the project, they are more likely to support it too. People feel more confident when they know others like them approve of something.

Another factor is mental distance. Climate change is often seen as a global, long-term issue, not something that affects people's daily lives. CCS can seem too technical, distant, or irrelevant [25]. To overcome this, communication should focus on local impacts: "this project will create jobs here," or "this technology helps protect your children's future."

Finally, many people today face information overload. There is so much news, social media, and technical content that people often feel overwhelmed. When this happens, they may stop paying attention [26]. That's why it's important to communicate CCS in a clear, simple, and memorable way. Instead of long charts or technical terms, project teams should use tools that people connect with like real-life stories, simple visual materials, or short videos. For example, explaining CCS as a "filter for the planet's air" can help people understand it quickly. This approach helps make the message stick and feel relevant to their own lives.

Figure 5.5 shows how behavioral psychology helps explain public reactions to CCS projects. People often focus more on what they might lose like land or safety than on long-term climate benefits (loss aversion). They are also more likely to support a project if trusted local voices support it (social proof). But too much technical information can overwhelm people and reduce their interest (information overload).

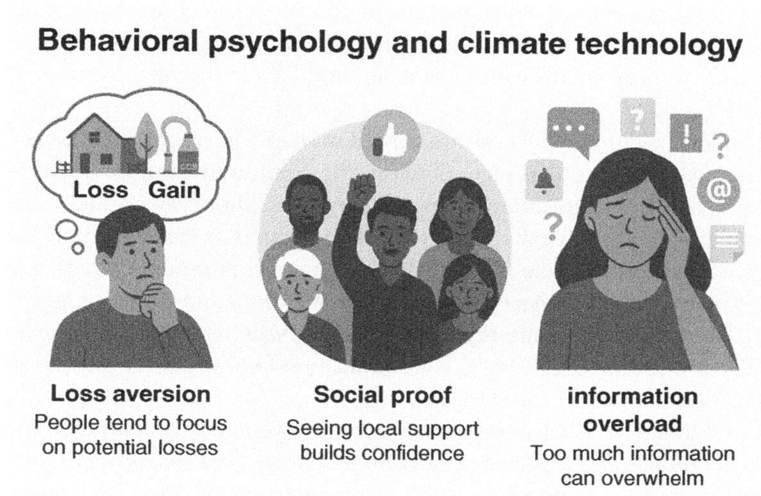

Figure 5.5 Core behavioral psychology principles that influence public response to CCS technologies: loss aversion, social proof, and information overload.

In short, behavioral psychology shows that people don't just follow facts. They also need to feel understood, respected, and informed in a clear way. When CCS project leaders know how people think and feel, they can explain things better and reduce fear. And when people feel included, they are more likely to support new climate technologies.

5.6 Strategies for Building Trust and Acceptance

After understanding why people resist carbon capture projects and how their perceptions are shaped, the next question is: how can we earn their trust and support? The answer is not only improved technology, but also stronger relationships. [27]. In this section, we explore practical strategies that CCS developers, policymakers, and researchers can use to build trust and gain social acceptance.

The first and most important step is early engagement. Don't wait until everything is decided before talking to the community. Invite them into the conversation from the beginning. When people feel included from day one, they are more likely to cooperate. Holding small meetings, listening sessions, or open-house events helps create a sense of partnership rather than top-down planning [10, 28].

Next, be transparent and honest. Share both the benefits and the risks. If something is uncertain, say so. If there are safety systems in place, explain how they work. Avoid overly technical language and speak in plain terms. People respect honesty, and when they see that you're not hiding anything, trust grows [28]. Another powerful strategy is local involvement. Hire people from the community. Partner with local schools or small businesses in a way that benefits both the project and the community. The project team could run workshops or educational programs about climate change and CCS. This helps students learn, and it also builds trust in the community. Show that the project is not something happening "to" the community, but "with" it and "for" it. Economic benefits like jobs, training programs, and infrastructure improvements can make a big difference, especially in rural or underserved areas [28].

It also helps to create local champions. These are trusted individuals, teachers, elders, or environmental activists, who can act as bridges between the community and the project. Another useful tool is visual communication. Maps, infographics, short videos, and even simple animations can explain complex processes like CO_2 injection or leak detection far better than long technical reports. People remember what they see more than what they read. Consistency also matters. If the project team communicates once and then disappears for six months, trust fades. Instead, create regular updates, newsletters, or online dashboards where people can follow progress. Make communication two-way, allow people to ask questions, give feedback, and feel heard [10, 11].

It's also important to respond to criticism respectfully. Not everyone will support the project at first, and this is acceptable and expected in early project stages. The goal is not to convince everyone overnight, but to create space for dialogue and growth. Even opponents deserve a seat at the table. Often, long-term critics become supporters when they feel truly listened to. Finally, work with local governments and institutions. When local leaders feel confident about the project, they can support and help with legal or social challenges [11, 28]. As a result, it is important to build coalitions that include environmental groups, academic institutions, and indigenous organizations in order to enhance the credibility and strength of the process. Ultimately, trust is not built through one speech or one document; rather, it is built through time, honesty, and honest connection. In order for CCS to be successful on a large scale, it must not only be technically sound but also socially acceptable. These strategies can help ensure communities are not just informed, but engaged, respected, and empowered.

5.7 Key Highlights (Chapter Summary)

◇ Chapter 5 – Key Highlights:

1) Public perception can significantly influence the success or failure of CCS projects, regardless of technical safety.
2) Historical examples show that poor communication and lack of engagement often lead to public opposition.
3) Risk communication must be early, honest, and in language that nonexperts can understand.
4) Behavioral psychology helps explain why people resist new technologies, even with clear benefits.
5) Emotional and psychological factors such as fear, distrust, and loss aversion heavily impact public opinion.
6) Traditional and social media can either amplify misinformation or serve as tools for transparent dialogue.

7) Communities are more accepting when they are actively included in planning and decision-making processes.
8) Trust is built through consistent engagement, transparency, and local participation; not just facts.
9) Creating local champions and community partnerships enhances project credibility and acceptance.
10) Successful CCS deployment requires both technical excellence and social legitimacy.

Keywords: Social Acceptance, Public Perception, Carbon Capture, Risk Communication, Community Engagement, Behavioral Psychology, Trust Building, Media Influence, Stakeholder Involvement, Effective Communication

References

1 J. A. E. Nielsen, K. Stavrianakis, and Z. Morrison, "Community acceptance and social impacts of carbon capture, utilization and storage projects: A systematic meta-narrative literature review," *PLoS One*, vol. 17, no. 8 , Aug. 2022, Art. no. e0272409, 10.1371/journal.pone.0272409.

2 K. Broecks, C. Jack, E. ter Mors, C. Boomsma, and S. Shackley, "How do people perceive carbon capture and storage for industrial processes? Examining factors underlying public opinion in the Netherlands and the United Kingdom," *Energy Research and Social Science*, vol. 81, 102236, Nov., 2021. DOI: 10.1016/j.erss.2021.102236.

3 L. Waller, E. Cox, and R. Bellamy, "Carbon removal demonstrations and problems of public perception," *WIREs—Climatic Change*, vol. 15, no. 1, Art. no. e857, Jan./Feb. 2024, 10.1002/wcc.857.

4 T. Evanson, "Public perception's role in carbon removal at scale: The importance of public opinion on barriers to implementation of CDR," Undergraduate Honors Thesis, Univ. of Texas at Austin, Austin, TX, USA, 2023.

5 B. Wynne, "Public engagement as a means of restoring public trust in science—hitting the notes, but missing the music?," *Community Genetics*, vol. 9, no. 3, pp. 211–220, 2006.

6 M. V. Ramana, "Nuclear power and the public," *Bulletin of the Atomic Scientists*, vol. 67, no. 4, pp. 43–51 , Jul., 2011. DOI: 10.1177/0096340211413358.

7 T. Wolsink, "Wind power and the NIMBY-myth: Institutional capacity and the limited significance of public support," *Renewable Energy*, vol. 21, no. 1, pp. 49–64, 2000.

8 C. F. J. Feenstra, T. Mikunda, and S. Brunsting, What happened in Barendrecht? Case study on the planned onshore carbon dioxide storage in Barendrecht, the Netherlands, Netherlands, 2010.

9 J. Sekera and A. Lichtenberger, "Assessing carbon capture: Public policy, science, and societal need," *Biophysical Economics and Sustainability*, vol. 5, no. 14, 2020. DOI: 10.1007/s41247-020-00080-5.

10 K. Arning, A. Linzenich, M. Ziefle, L. Meuleneers, D. Rezo, and N. Von Der Aßen, "Eliciting laypeople's mental models and risk perceptions of direct air carbon capture and storage: Implications for effective risk communication," *Energy Reports*, vol. 12, pp. 1068–1079, Dec., 2024. DOI: 10.1016/j.egyr.2024.06.070.

11 W. Leiss and P. Larkin, "Risk communication and public engagement in CCS projects: The foundations of public acceptability," *International Journal of Risk Assessment and Management*, vol. 22, no. 3–4, pp. 384–403, 2019. DOI: 10.1504/IJRAM.2019.103339.

12 S. L. Seigo, S. Dohle, and M. Siegrist, "Public perception of carbon capture and storage (CCS): A review," *Renewable and Sustainable Energy Reviews*, vol. 38, pp. 848–863, Oct., 2014. DOI: 10.1016/j.rser.2014.07.017.

13 W. Leiss, "Effective risk communication practice," *Toxicology Letters*, vol. 149, no. 1–3, pp. 399–404, Apr., 2004. DOI: 10.1016/j.toxlet.2003.12.050.

14 E. Kurz-Milcke, G. Gigerenzer, and L. Martignon, "Transparency in risk communication: Graphical and analog tools," *Annals of the New York Academy of Sciences*, vol. 1128, no. 1, pp. 18–28 , Apr., 2008. DOI: 10.1196/annals.1399.004.

15 M. Gerrard, F. X. Gibbons, and M. Reis-Bergan, "The effect of risk communication on risk perceptions: The significance of individual differences," *JNCI Monographs*, vol. 1999, no. 25, pp. 94–100, Jan., 1999. DOI: 10.1093/oxfordjournals.jncimonographs.a024217.

16 A. M. Feldpausch-Parker, M. Burnham, M. Melnik, M. L. Callaghan, and T. Selfa, "News media analysis of carbon capture and storage and biomass: Perceptions and possibilities," *Energies*, vol. 8, no. 4, pp. 3058–3074, Apr., 2015. DOI: 10.3390/en8043058.

17 K. A. Swain, "How media framing shapes public understanding of carbon capture," in *Handbook of Climate Change Mitigation and Adaptation*, M. Lackner, B. Sajjadi, and W. Y. Chen, Eds. New York, NY, USA: Springer, 2025, pp. 1–90, DOI: 10.1007/978-1-4614-6431-0_201-1.

18 J. Wettestad, T. H. J. Inderberg, and L. H. Gulbrandsen, "Exploring paths and innovation in Norwegian carbon capture and storage policy," *Environmental Policy and Governance*, vol. 34, no. 2, pp. 125–136, Apr., 2024. DOI: 10.1002/eet.2068.

19 M. Wang, Y. Gong, S. Wang, Y. Li, and Y. Sun, "Promoting support for carbon capture and storage with social norms: Evidence from a randomized controlled trial in China," *Energy Research & Social Science*, vol. 74, 101979, Apr., 2021. DOI: 10.1016/j.erss.2021.101979.

20 A. Ropret Homar, An examination of the effects of loss aversion on pro-environmental decisions, Ph.D. dissertation, Univ. of Ljubljana, School of Economics and Business, Ljubljana, Slovenia, 2023.

21 S. Rabaa, S. Geisendorf, and R. Wilken, "Why change does (not) happen: Understanding and overcoming status quo biases in climate change mitigation," *Zeitschrift Für Umweltpolitik & Umweltrecht*, vol. 45, no. 1, pp. 100–134, 2022.

22 C. Lang, M. Weir, and S. Pearson-Merkowitz, "Status quo bias and public policy: Evidence in the context of carbon mitigation," *Environmental Research Letters*, vol. 16, no. 5, p. 054076, May, 2021. DOI: 10.1088/1748-9326/abeeb0.

23 M. Ferguson and P. Ashworth, "Message framing, environmental behaviour and support for carbon capture and storage in Australia," *Energy Research and Social Science*, vol. 73, p. 101931, Mar., 2021. DOI: 10.1016/j.erss.2021.101931.

24 T. Bowden-Green and M. Vafeas, "Recognising motivation in others: The effectiveness of using social proof to change driving behaviour," *Journal of Social Marketing*, vol. 14, pp. 382–397, 2024. DOI: 10.1108/jsocm-02-2024-0045.

25 P. Shah, J. Z. Yang, and L. Kahlor, "Psychological distance, risk perception, and affect: Texas residents' support for carbon capture and storage," *Journal of Risk Research*, vol. 26, no. 2, pp. 184–198, 2023. DOI: 10.1080/13669877.2022.2116084.

26 M. Arnold, M. Goldschmitt, and T. Rigotti, "Dealing with information overload: A comprehensive review," *Frontiers in Psychology*, vol. 14, 1122200, 2023. DOI: 10.3389/fpsyg.2023.1122200.

27 B. Wang *et al.*, "How to effectively communicate about greenhouse gas emissions with different populations," *Environmental Science & Policy*, vol. 147, pp. 29–43, Sep., 2023. DOI: 10.1016/j.envsci.2023.05.015.

28 F. Mulyasari, A. K. Harahap, A. O. Rio, R. Sule, and W. G. A. Kadir, "Potentials of the public engagement strategy for public acceptance and social license to operate: Case study of carbon capture, utilisation, and storage *Gundih* pilot project in Indonesia," *Int. J. Greenh. Gas Control*, vol. 108, p. 103312, Jun., 2021. DOI: 10.1016/j.ijggc.2021.103312.

6

Health, Safety, and Environmental (HSE) Considerations

6.1 Overview of HSE in Carbon Capture Projects

Health, safety, and environmental (HSE) considerations are critical to the successful development of carbon capture, utilization, and storage (CCUS) projects. Because carbon capture involves handling large volumes of compressed CO_2 at high pressures across multiple facilities and system components, a firm HSE policy is essential to ensure safe operations and maintain public confidence. HSE principles must be applied throughout the entire CCUS lifecycle, from capture facility operation and pipeline transport to well injection and long-term underground storage [1].

It is important to note that in carbon capture installations, there are potential hazards associated with the use of chemical solvents such as amines, high-pressure equipment, and process heat, which may pose a threat to both employees and the surrounding environment if not properly controlled. During transportation, especially via pipeline, supercritical or dense-phase CO_2, which exhibits properties of both liquid and gas phases, may cause asphyxiation and rapid depressurization if a leak occurs. Long-term storage of CO_2 in geological formations is a primary concern at the storage stage. Consequently, the injected CO_2 must remain safely trapped underground and not migrate to the surface or contaminate groundwater [2].

In order to mitigate these risks, CCS projects utilize a combination of strategies such as risk assessment, engineering or process controls, compliance with national and international codes and standards, and advanced monitoring systems [3, 4]. In international standards for HSE protection, the importance of safe design is highlighted [5]. For instance, valves must be designed to maintain integrity during fire exposure to prevent escalation . In addition, continuous monitoring and a strong safety culture are supported by well-trained personnel. The following sections examine each area in detail. They begin with occupational safety risks, followed by environmental impact considerations, the regulatory and permitting system, and finally, emergency preparedness and response strategies.

6.2 Occupational and Community Health Concerns

Carbon capture, transportation, and storage (CCS) projects present specific health risks that must be carefully managed to protect both workers in the plants and the surrounding communities. Generally, CCS does not involve flammable gases like natural gas or

Figure 6.1 Two scenarios of accidental CO_2 release: ground freezing due to CO_2 leakage from a high-pressure CO_2 pipeline rupture; right–dense CO_2 plume spreading at ground level, illustrating its asphyxiation hazard.

hydrogen, except in the case of precombustion capture methods, where hydrogen may be present. However, carbon dioxide (CO_2) poses significant asphyxiation hazards because it is colorless, odorless, and heavier than air. Leaks, particularly in low-lying or enclosed areas, can displace oxygen and lead to unconsciousness or even death without any warning [6]. This risk is especially relevant in capture plant facilities, along pipeline routes, and at injection well sites.

Figure 6.1 shows two real-world examples of CO_2 pipeline release scenarios, highlighting the potential safety hazards associated with CCS infrastructure. The image on the left illustrates a pipeline rupture where high-pressure CO_2 rapidly depressurized, causing extreme cooling that froze the surrounding soil and ground surface. The image on the right shows a dense CO_2 cloud spreading close to the ground. This shows how CO_2, which is heavier than air, can form asphyxiating areas at low levels, posing a serious danger to nearby workers, residents, or wildlife [7].

Additionally, hazardous impurities that are commonly found in captured CO_2 streams, such as hydrogen sulfide (H_2S), sulfur oxides (SO_x), nitrogen oxides (NO_x), or residual solvents, could introduce further toxic or corrosive risks to personnel [8]. Hydrogen sulfide, in particular, is highly toxic and has a characteristic rotten egg smell, which can be lethal to humans even at low concentrations [9]. Sulfur oxides (SO_x) and nitrogen oxides (NO_x) can be highly damaging to the respiratory system. They may cause symptoms such as coughing, lung inflammation, and long-term damage, especially in individuals with asthma or heart conditions. Residual solvents, which are utilized in carbon dioxide capture through chemical absorption processes, can lead to skin and eye irritation, headaches, or neurological

Figure 6.2 H_2S leak from a corroded piping in an industrial plant. It shows toxic gas exposure and severe corrosion risks.

effects, depending on their type and concentration [10, 11]. Additionally, these impurities increase the risk of corrosion in equipment, which can result in leaks or accidental releases if not properly managed through purification, monitoring, and safety controls.

Figure 6.2 shows a realistic scenario of a hydrogen sulfide (H_2S) leak from heavily corroded piping in an industrial facility. A dense yellowish gas cloud of H_2S is seen rapidly escaping from the damaged pipe section. One worker wearing a respirator approaches cautiously [9]. However, another worker, without proper respiratory protection, appears to be choking and showing signs of immediate exposure. This highlights the acute toxicity and suffocation risk of H_2S. The image also illustrates severe external corrosion on the pipeline, emphasizing the role of corrosion and material degradation as a key cause of accidental hydrogen sulfide release impact [12, 13].

Workers are exposed to occupational hazards as a result of a variety of aspects of CCS operations. High-pressure equipment, such as compressors and pipelines, operate at pressures exceeding 100 bar. As a result, any mechanical failure or leak may result in a sudden release of energy, which could result in severe physical injury. Chemical solvents, especially amine-based solvents used in post-combustion CO_2 capture, can be toxic or irritating, posing risks of skin burns, eye damage, or inhalation-related health issues if not properly handled. In addition to involving very high temperatures, heat-containing systems like boilers may also result in thermal burns during maintenance or accidental contact [2].

Personal protective equipment (PPE) is necessary to manage these hazards, including flame-resistant clothing, chemical-resistant gloves, face shields, safety goggles, gas-tight masks, and respiratory protection. Gas detection systems are installed to continuously monitor dangerous levels of CO_2, H_2S, or other toxic gases. These systems trigger alarms and automatic plant or unit shutdowns before concentrations reach unsafe levels. Proper ventilation design ensures that leaked gas does not accumulate in enclosed or low-lying areas, allowing fresh air circulation to prevent asphyxiation hazards. In addition, regular

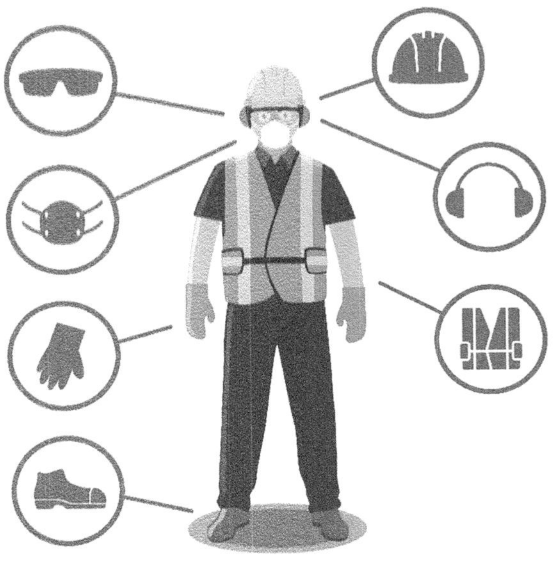

Figure 6.3 Illustration of full industrial Personal Protective Equipment (PPE) used to protect workers from hazards at work, including gas exposure, chemical hazards, high temperatures, noise, and physical injuries in CCS and similar industrial operations. (Courtesy of Shutterstock.com).

safety drills are conducted to train workers on how to react quickly to emergency scenarios such as gas leaks, fires, or equipment failures [14]. These drills simulate real accident conditions and ensure personnel know evacuation routes, alarm responses, and lifesaving procedures.

Figure 6.3 shows a worker equipped with full PPE typically required in CCS, oil and gas, and other industrial environments. The image highlights key protective items, including a safety helmet, safety goggles, respiratory mask, hearing protection, flame-resistant clothing, a high-visibility vest, chemical-resistant gloves, and safety boots. Such PPE is essential for protecting workers from hazards such as toxic gas exposure (e.g. CO_2 or H_2S), chemical splashes, heat, noise, and mechanical injuries [14]. This visual emphasizes the importance of proper PPE selection and correct usage as a first line of defense in minimizing occupational risks.

For nearby communities, the primary concern is accidental CO_2 release from transport pipelines or underground storage sites [15]. In cases where carbon dioxide release is not stopped, sudden high-volume release could generate a CO_2 plume. This is very dangerous because CO_2 is heavier than air and can accumulate in low-lying areas, depleting oxygen. This can cause breathing difficulties, unconsciousness, or even fatalities among residents, as well as a suffocation risk to livestock and wildlife. In addition to health risks, a major leak could also disrupt local ecosystems and damage agriculture or water sources if not quickly contained [16].

To protect the public, continuous monitoring systems are installed along CCS pipelines and storage sites to detect any gas leaks at an early stage. Early-warning systems, such as automated leak detection and alarm signals, allow operators and engineers to respond immediately before the situation escalates [17]. Exclusion zones are often designated around critical CCS infrastructure to ensure that residential areas or public activities are not located too close to potential risk points. Furthermore, effective communication and emergency coordination between project operators, local authorities, and emergency services ensures that appropriate actions, such as stopping the leak, initiating evacuation procedures, and issuing public safety alerts, can be carried out quickly if needed [14]. Understanding and addressing these occupational and community health risks is essential for regulatory and safety compliance. It is also about building public trust and long-term social acceptance of CCS projects.

6.3 Environmental Impacts and Risk Scenarios

Carbon capture and storage (CCS) systems have the potential to significantly reduce greenhouse gas emissions. However, they also introduce environmental risks that need to be carefully evaluated and managed. The primary concern is the accidental release or leakage of CO_2. While CO_2 is not toxic like conventional pollutants, it can displace oxygen and disrupt local ecosystems, particularly in low-lying areas or enclosed spaces [6, 7]. Additionally, if impurities such as hydrogen sulfide (H_2S), nitrogen oxides (NO_x), sulfur oxides (SO_x), or organic solvents are present in the captured CO_2, leaks could result in soil acidification, infrastructure corrosion, and degradation of local air quality [8].

During the storage phase, improper site selection or inadequate sealing of the geological formation can lead to CO_2 migration into unintended subsurface layers [18]. Such leakage has the potential to impact groundwater quality by lowering the pH (making it more acidic) and mobilizing heavy metals such as arsenic, lead, or mercury, which are naturally trapped in rock layers. These metals can dissolve into groundwater under acidic conditions, posing risks to drinking water and ecosystems [19].

Figure 6.4 illustrates the potential environmental risks associated with the CO_2 storage phase. During this phase, improper site selection or inadequate sealing of the geological formation can lead to unwanted migration (leakage) of injected CO_2 into nearby subsurface

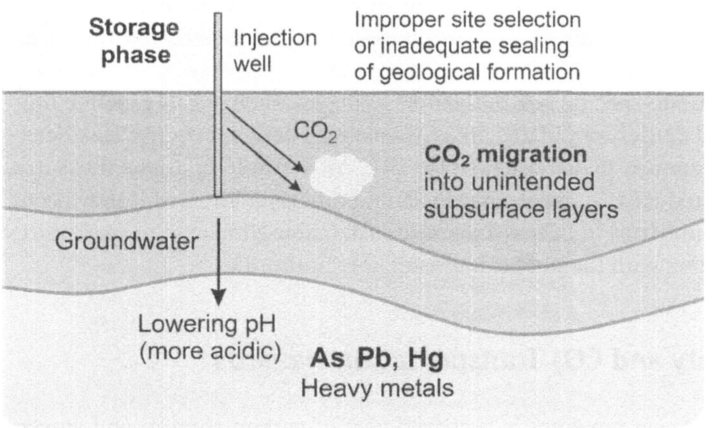

Figure 6.4 CO_2 leakage during geological storage and its impact on groundwater pH and heavy metal mobilization.

layers. When CO_2 reaches groundwater, it dissolves and lowers the pH of the water, making it more [20] acidic. This acidification can mobilize naturally occurring heavy metals found in rocks, such as arsenic (As), lead (Pb), and mercury (Hg), preventing these metals from being trapped in the rocks. As a result, dissolved toxic metals may contaminate groundwater resources, posing a threat to drinking water supplies and surrounding ecosystems.

Although the likelihood of large-scale CO_2 leakage is considered very low when sites are properly evaluated, engineered, and continuously monitored, induced seismicity remains a recognized risk. High-pressure CO_2 injection can increase stress in underground rock layers and trigger microearthquakes. These events are usually too small to cause surface damage or be felt by people, but if not controlled, they could create fractures or pathways that allow CO_2 to escape to the air over time [21]. For this reason, continuous seismic monitoring and controlled injection pressure are essential to maintain the long-term integrity and safety of the storage site.

The pipeline transportation stage poses significant environmental risks, especially when the pipeline passes through sensitive areas such as subsea and coastal regions, agricultural lands, protected ecosystems, or densely populated zones [22]. In the event of a sudden pipeline rupture or leak, the rapid release of high-pressure CO_2 could create a cold, dense gas cloud that settles near the ground. This can temporarily deplete oxygen in the vicinity, damage plants, acidify nearby water sources, and harm wildlife due to asphyxiation or sudden temperature drops [2, 6].

To minimize these risks, CCS projects rely on careful route planning of pipelines to avoid ecologically or socially vulnerable areas whenever possible. Environmental monitoring systems track soil, air, and water quality near pipeline routes constantly [23]. In addition, rapid isolation systems, including automatic shut-off valves and pressure drop, detection systems enable the pipeline to be sealed quickly to prevent extensive CO_2 release and limit the environmental impact [24].

Overall, while CCS provides significant climate benefits by preventing large volumes of CO_2 from entering the atmosphere, it is crucial to ensure that these environmental gains are not adversely affected by unintended ecological harm, mainly due to CO_2 leakage to the environment. This requires a comprehensive Environmental Impact Assessment (EIA) before project approval, specific to the geographical, ecological, and social conditions of each site [25]. Once the project is operational, continuous environmental monitoring of the air, water, soil, and ecosystems is necessary to detect any early signs of negative impacts.

It is also crucial to have site-specific risk mitigation strategies, such as safe pipeline routing to avoid sensitive and densely populated areas, robust well design, effective leak detection, and emergency response plans (ERPs) [14, 24]. As a result of these measures, environmental risks are reduced, actively controlled, and quickly addressed if they occur. Developing and maintaining trust in CCS as a safe and sustainable climate solution requires transparent communication with the public and regulatory authorities.

6.4 Pipeline Safety and CO_2 Transportation Hazards

The transportation of CO_2 via pipelines is a critical part of carbon capture and storage (CCS) infrastructure. It introduces unique safety challenges that are different from those associated with natural gas or oil pipelines in the oil and gas industry. Unlike methane, CO_2

is nonflammable, but it exists in a dense or supercritical phase under high pressure during pipeline transportation [26]. In the event of a pipeline rupture, it can rapidly expand, cool, and form a CO_2 cloud in the low-lying areas. Because CO_2 is heavier than air, it can accumulate near the ground, displacing oxygen, and posing serious asphyxiation risks to humans and animals in the vicinity [7, 22].

In addition to the physical hazards, impurities in the CO_2 stream, such as hydrogen sulfide (H_2S) and sulfur oxides, can increase toxicity and accelerate corrosion within the pipeline, particularly in the presence of water or moisture [8, 9]. This makes it essential to carefully consider pipeline design aspects like material selection, wall thickness calculations, corrosion allowance determinations, and the application of appropriate coatings to prevent both internal and external corrosion [24, 27]. Furthermore, CO_2 pipelines must be engineered to withstand ductile fractures due to the behavior of supercritical carbon dioxide. If not properly managed with fracture arrestors or appropriate pressure management systems, a rupture can propagate rapidly along the pipeline.

Figure 6.5 illustrates a catastrophic longitudinal fracture in a buried high-pressure pipeline caused by supercritical carbon dioxide flowing through the pipeline. A split has occurred in the pipe wall, exposing the internal surface to the atmosphere. Without proper fracture control measures—such as crack arrestors, enhanced material toughness, or pressure management systems—such failures can occur in CO_2 pipelines [28]. The image highlights the critical importance of robust pipeline design and safety engineering to prevent hazardous ruptures in CO_2 transport systems.

Figure 6.5 Example of a longitudinal fracture propagation in a buried high-pressure CO_2 pipeline. This picture shows the severity of rapid rupture in CO_2 transport systems.

To ensure safety, CO_2 transport pipeline systems are equipped with pressure monitoring systems that track operational parameters such as pressure, flow, and temperature continuously and detect any deviations from normal conditions. Many segments of the pipeline are equipped with automatic shut-off valves that can quickly isolate sections of the pipeline during abnormal pressure conditions, which may indicate a CO_2 leak or rupture [24]. Additionally, leak detection technologies, such as fiber-optic sensing and acoustic monitoring, can identify even small leaks within seconds and trigger an immediate response [29].

CO_2 pipelines are carefully routed to avoid densely populated and environmentally sensitive areas, such as wetlands, groundwater reserves, and protected habitats. This strategic routing minimizes the risk of human and environmental harm in case of failure [30]. When rerouting is not possible, additional safety measures, including thicker pipeline walls or deeper burial, are implemented to enhance reliability and reduce potential risks [24].

In most countries, a complete safety and risk assessment is required before a CO_2 transportation pipeline can be constructed. It is therefore imperative that the project developers study and document the risks and consequences that may arise as a result of the project. This process involves the use of consequence modeling, which is a scientific simulation that predicts what would happen if CO_2 were to leak accidentally from the pipeline. In this regard, dispersion modeling is used to determine how far the gas may spread in the air [31]. Authorities define exclusion zones around pipelines, which are areas where people are not permitted to live or build due to possible exposure to the leaked CO_2, in order to ensure the safety of the public [32].

Figure 6.6 illustrates a schematic illustration of a CO_2 pipeline surrounded by clearly defined safety zones. There is an emphasis on the concept of exclusion or controlled zones in order to prevent unauthorized activities close to the pipeline. The exclusion zone is typically a clearly defined part of the controlled area where public access is strictly prohibited, especially under abnormal or emergency conditions, such as a leak or sudden pressure release. The exclusion zone as a part of controlled zone or area represents the highest level of safety restriction and is determined based on dispersion modeling and risk assessment.

It is clearly evident in the image that the right-of-way (ROW) showing the pipeline path is distinct from the controlled area immediately surrounding the buried pipeline, where access is restricted, and consent is required. Beyond that, the buffer area serves as an additional protective safety margin. These zones are established to minimize the risk of accidental damage from third-party (farmers, local people, etc.) activities, such as excavation or heavy machinery operations, ensuring safe operation and environmental protection along the pipeline route.

Regulatory bodies also require a failure scenario analysis, which involves identifying the different types of potential accidents that could lead to pipeline failure, such as a sudden pipe rupture, corrosion-related damage, or external interference from machinery or tools. Each scenario is then assessed in terms of its likelihood of occurrence and the potential consequences or damage it could cause [33].

In addition, a detailed ERP must be developed. This plan outlines exactly how the operator will respond in the event of an incident. It includes procedures for communication with relevant authorities such as the police and hospitals, as well as evacuation protocols and

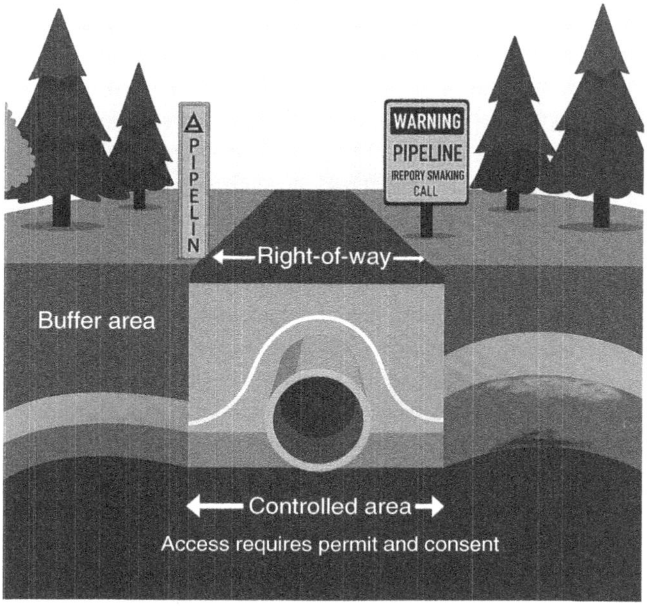

Figure 6.6 Schematic representation of safety zones around a buried CO_2 pipeline, including the right-of-way (ROW), controlled area and exclusion zone, and buffer area, which collectively help prevent unauthorized access and protect against accidental damage or hazards.

rescue actions to protect the public. The plan must be prepared in close coordination with local authorities and emergency organizations, including fire departments, medical teams, and law enforcement agencies [34].

Finally, modern regulations increasingly demand public communication and awareness programs. This means the company must inform nearby communities about the purpose of the pipeline, safety precautions in place, and what people should do in case of an emergency [35]. Globally, CO_2 transport and storage systems (CCS) are expanding, and it is important that all countries adhere to similar safety standards. This includes using real-time monitoring technology to identify any problems, such as leaks, as soon as possible. It is imperative that global guidelines be developed for key areas such as preventing pipeline fractures, setting clear limits for detecting leaks as quickly as possible, protecting monitoring systems from cyberattacks, and maintaining fast, clear communication during emergency situations in order to ensure public safety and build trust.

6.5 Safety Regulations and International Standards

The safe implementation of carbon capture, transportation, and storage (CCS) projects relies heavily on strict adherence to safety regulations and internationally recognized standards. These standards are designed to minimize risks in CCS projects. They cover facility design, CO_2 monitoring, pipeline transport, suitable site selection for carbon dioxide storage, and long-term storage integrity. Regulatory systems vary by region, but there is growing global alignment and harmonization around best practices established by leading organizations such as the International Organization for Standardization (ISO), American Petroleum Institute (API), International Energy Agency (IEA), and the European Union's CCS Directive. For example, API 6A [36], which covers wellhead equipment and components, is technically harmonized with ISO 10423 [37], while API 6D [38], the standard for pipeline valves, is equivalent to ISO 14313 [39]. These equipment standards are so closely aligned that valves, fittings, and wellhead components are frequently dual-certified as API + ISO compliant.

One of the most important international standards for CCS is ISO 27914 [40], which gives detailed guidelines for safely storing CO_2 underground. It covers topics such as choosing the right storage site, analyzing possible risks, modeling how CO_2 might move or leak, and planning proper monitoring systems. Another related standard is ISO 27915 [41], which focuses on the safe pipeline transportation of CO_2. It includes rules for selecting suitable materials, preventing pipeline fractures, detecting leaks, and ensuring safe operation.

In the United States, CO_2 storage sites are regulated by the Environmental Protection Agency (EPA) under special well regulations [42]. These rules require strict monitoring and control of well pressure to prevent leaks. In Europe, the EU CCS Directive (2009/31/EC) requires a full risk assessment for the entire project lifecycle, continuous safety monitoring, and legally binding responsibilities for long-term safe storage [43].

Figure 6.7 presents a structured overview of the major international standards and regulations that govern the safe implementation of carbon capture, transportation, and storage (CCS) projects. The diagram highlights key regulatory bodies, including ISO, API, EPA, and the EU CCS Directive. It illustrates how each entity contributes to various aspects of CCS

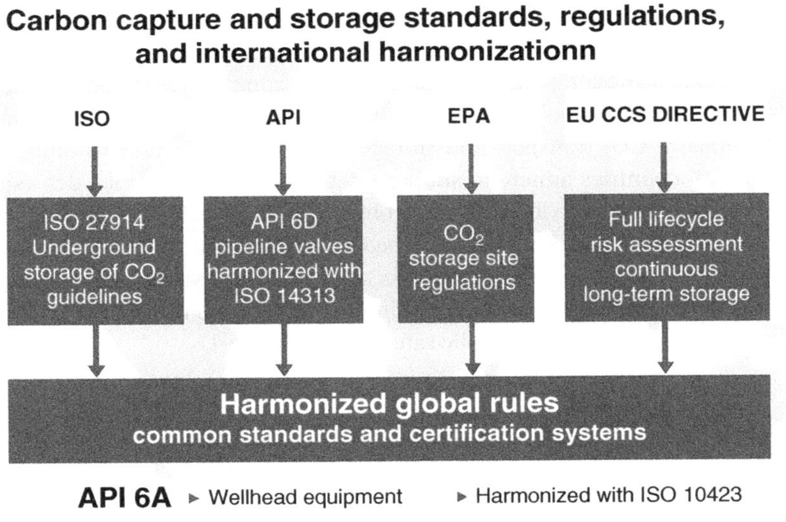

Figure 6.7 International standards and regulatory governing CCS systems, showing how ISO, API, EPA, and the EU CCS Directive align to create harmonized global safety rules for CO_2 storage, transportation, and long-term monitoring.

safety, such as underground CO_2 storage (ISO 27914), pipeline valve integrity (API 6D, harmonized with ISO 14313), site regulations in the United States (EPA), and full lifecycle risk assessment requirements in Europe (EU CCS Directive). The image emphasizes the global movement toward harmonized regulations, where international standards align to create a unified foundation for safety, certification, and regulatory acceptance across borders. Additionally, a note at the bottom refers to API 6A, harmonized with ISO 10423, which pertains to wellhead equipment. This reinforces the strong alignment between the API and ISO frameworks.

As Carbon Capture and Storage (CCS) projects operate in various countries, it is crucial to establish global rules that are harmonized, ensuring they are similar and consistent across borders. This includes faster and clearer permitting procedures from authorities to initiate projects, as well as international certification systems to confirm that a project is safe and properly designed. Having these common rules between standards and regulatory bodies enhances the reliability and safety of CCS projects, facilitating effective international collaboration [44].

It is essential to have strong regulatory management for Carbon Capture and Storage (CCS) projects. This means that governments should carefully monitor and control how these projects are planned and operated. Such management protects people's health, safety, and the environment. Additionally, clear and trustworthy regulations increase confidence in investors, encouraging more funding and facilitating the faster development of CCS projects worldwide. Ultimately, this helps the world achieve its climate goals more quickly and effectively.

6.6 Emergency Response and Contingency Planning

"Emergency response" refers to the measures taken after an accident occurs, such as protecting people, evacuating the area if necessary, and controlling the situation to prevent further damage. By contrast, contingency planning involves planning in advance for an accident by preparing detailed procedures and strategies, as well as training. The purpose of this is to enable the team to respond to an emergency quickly and effectively. In simple terms, contingency planning is about being ready before an incident happens, while emergency response is about taking fast action when an incident actually occurs. In order to ensure the long-term safety and reliability of carbon capture, transportation, and storage (CCS) systems, effective emergency response and contingency planning are critical, especially given the unique hazards associated with high-pressure CO_2 handling [34].

Emergency preparedness starts with risk assessment. This means studying and identifying what kinds of accidents could happen at each stage of the CCS process. These accidents could include leaks, equipment failure, or problems with pressure. Based on this, companies create ERPs. These plans clearly define who is responsible for what action, how people will evacuate safely, and how communication will occur during an emergency [34].

CCS facilities and pipeline operators also use real-time monitoring systems to quickly detect any dangers such as carbon dioxide leakage. These systems may include gas sensors to detect CO_2 leaks and alarms to warn of abnormal pressure. They may also include automatic shutdown systems to stop operations immediately if needed to minimize carbon dioxide leakage. In addition, CO_2 plume modeling predicts how a CO_2 leak might spread in the environment. All of this helps to provide a fast and effective response to any incident [45].

Collaboration with local authorities, emergency services, including fire departments, police, and medical teams, and community representatives is essential for ensuring safety in Carbon Capture and Storage (CCS) projects. A community representative is a trusted member of the local population who serves as a liaison between the community and the project operators. To prepare for emergencies, companies conduct drills and simulations. These exercises involve all stakeholders, including plant operators, firefighters, medical teams, and community representatives, who practice their roles in the event of an emergency [46]. This preparation ensures that everyone understands their responsibilities and can work together effectively to protect people during a crisis.

Figure 6.8 illustrates the key components involved in emergency response and contingency planning for CO_2 leakage incidents at Carbon Capture and Storage (CCS) facilities. The image features a CO_2 processing plant, real-time leak detection alarms, plume

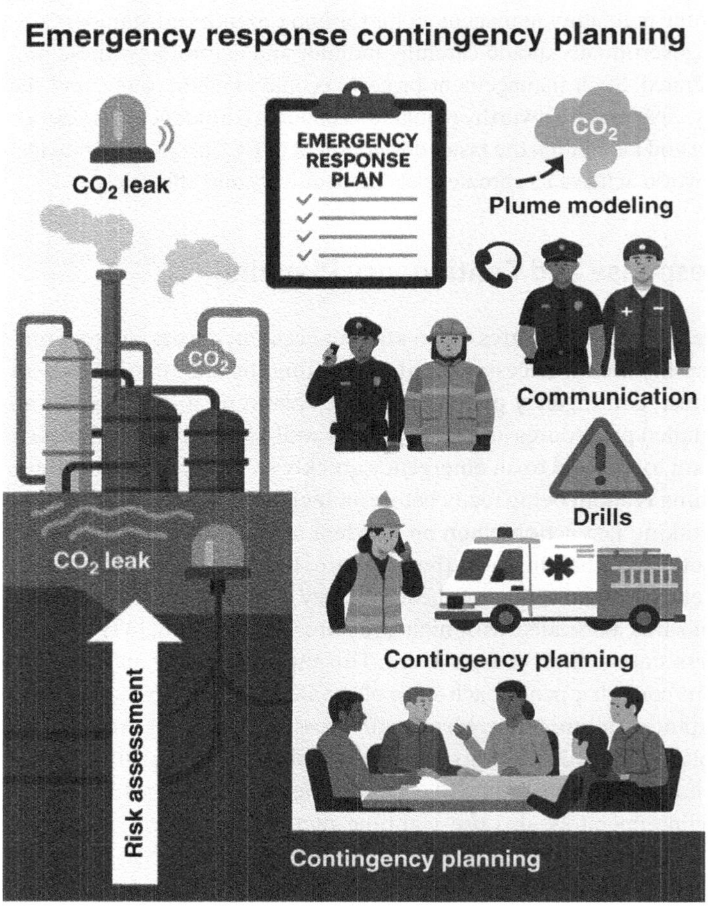

Figure 6.8 Overview of emergency response and contingency planning in CCS operations, illustrating CO_2 leak detection, plume modeling, emergency responders, communication protocols, and stakeholder meetings.

dispersion modeling, and a coordinated emergency response that includes trained personnel such as firefighters, medical teams, and security officers. Additionally, it emphasizes the importance of communication systems, emergency drills conducted with local authorities and communities, and organized contingency planning meetings. These drills are designed to ensure preparedness before any incidents occur.

A contingency plan does not end with the termination of an incident. It also includes actions taken after the incident ends. This can include checking what happened, repairing any damage, and making sure the area is safe again. These actions may involve environmental cleanup, resealing a well if there was a leak, or increasing monitoring to prevent any further issues from arising [47]. International best practices also highlight the importance of building a strong safety culture, being transparent with the public, and earning local communities' trust. People are more likely to support CCS projects when they can clearly see that the company is well-prepared for emergencies, communicates honestly, and manages risks appropriately. In summary, emergency response and contingency planning are not optional extras but essential parts of CCS safety. They ensure that even if an incident occurs, the situation can be quickly controlled, damage can be minimized, and public trust can be maintained.

6.7 Key Highlights (Chapter Summary)

◇ Chapter 6 – Key Highlights:

1) CCS projects involve unique HSE challenges due to the high-pressure handling, transport, and storage of CO_2 across pipelines, injection wells, and underground storage.
2) Worker safety is essential because CO_2 leaks can lead to serious asphyxiation risks, particularly in enclosed or low-lying areas. Furthermore, these leaks may contain dangerous impurities such as H_2S, which is highly toxic.
3) Community safety depends on effective safety distance zones, monitoring systems, and rapid emergency communication protocols to minimize public exposure during accidental releases.
4) Environmental risks include groundwater contamination, water and soil acidification, ecological harm, and, in rare cases, induced earthquakes during high-pressure CO_2 injection.
5) Pipeline safety is a major concern in carbon capture and storage (CCS) projects because a leak can cause rapid decompression of CO_2, which may lead to fracture propagation along the pipeline. In addition, leaked CO_2 can displace oxygen in the air, creating a serious risk of suffocation for people in populated areas.
6) International safety standards such as ISO 27914/27915, API 6A and 6D, EPA, and the EU CCS Directive provide guidance for safe CCS deployment.
7) Strong regulatory compliance, monitoring technologies, and lifecycle risk assessment are essential for ensuring long-term CO_2 containment and operational reliability.

8) Stakeholder engagement, transparency of project data, and public trust play a critical role in the successful acceptance and success of CCS project.
9) In order to minimize harm to people and the environment, emergency response plans should include monitoring of facilities, automated shutdown systems, training, and evacuation strategies.
10) The purpose of contingency planning is to be prepared before an incident occurs, whereas the purpose of emergency response is to take immediate action after an incident has occurred.

Keywords: Health and Safety, CO_2 Exposure, Environmental Risk, Pipeline Safety, Geological Storage Integrity, Regulatory Compliance, Emergency Response, Risk Assessment, Safety Culture, Carbon Capture Infrastructure, Industrial Safety, Asphyxiation Hazard, Leak Detection, Monitoring Systems, Environmental Monitoring, Public Safety, CCS Regulations, Risk Mitigation, Incident Preparedness, Carbon Storage Security

References

1 C. Pawde and R. Parekh, "A discussion of the HSE aspects of carbon dioxide sequestration," Presented at the SPE Asia Pacific Oil and Gas Conference and Exhibition, Jakarta, Indonesia, Oct. 2013, Paper No. SPE-165781-MS, DOI: 10.2118/165781-MS

2 J. Wilday, M. Wardman, M. Johnson, and M. Haines, "Hazards from carbon dioxide capture, transport and storage," *Process Safety and Environmental Protection*, vol. 89, no. 6, pp. 482–491, Nov., 2011. DOI: 10.1016/j.psep.2011.09.002.

3 E. J. Wilson, S. J. Friedmann, and M. F. Pollak, "Research for deployment: Incorporating risk, regulation, and liability for carbon capture and sequestration," *Environmental Science & Technology*, vol. 41, no. 17, pp. 5945–5952, Jul., 2007. DOI: 10.1021/es06227t.

4 A. B. Klass and E. J. Wilson, "Carbon capture and sequestration: Identifying and managing risks," *Issues in Legal Scholarship*, vol. 8, no. 3, 1, 2009.

5 R. Bäßler, "ISO 27913—example for successful standardization activities in the field of CCUS," AMPP Annual Conference and Expo, Paper No. C2024-20592, pp. 1–6 Mar. 2024. DOI: 10.5006/C2024-20592

6 K. Permentier, S. Vercammen, S. Soetaert, and C. Schellemans, "Carbon dioxide poisoning: A literature review of an often forgotten cause of intoxication in the emergency department," *International Journal of Emergency Medicine*, vol. 10, no. 1, p. 14, 2017. DOI: 10.1186/s12245-017-0142-y.

7 R. Ogle, J. Schneider, N. Schulman, and B. Cox, "Risk-based facility siting of CO_2 pipelines," *Process Safety Progress, Early View*, Sep., 2025. DOI: 10.1002/prs.70018.

8 J. Sonke and Y. Zheng, "CO_2 transport and injection, effect of impurities, understanding of reactions and consequences," AMPP Annual Conference and Expo, Paper No. C2023-18756, pp. 1–20, Mar. 2023. DOI: 10.5006/C2023-18756

9 J. Jiang *et al.*, "Hydrogen sulfide—mechanisms of toxicity and development of an antidote," *Scientific Reports*, vol. 6, no. 20831, 2016. DOI: 10.1038/srep20831.

10 S. K. Pandey and J. Singh, "Chapter 22—nitrogen dioxide: Risk assessment, environmental, and health hazard," in *Nitrogen dioxide: Risk assessment, environmental, and health hazard*, P. Kumar and A. Singh, Eds. Elsevier, 2021, pp. 273–288, DOI: 10.1016/B978-0-323-89857-7.00001-3.

11 M. A. Budihardjo, "Risk analysis study of NOx and SOx from transportation (Case study: Main streets of D.I. Jogjakarta)," *TEKNIK*, vol. 28, no. 1, pp. 42–48, Feb., 2012. DOI: 10.14710/jpl.v28y.2047.

12 K. Sotoodeh, *A Practical Guide to Piping and Valves for the Oil and Gas Industry*, 1st ed.. Gulf Professional Publishing, Elsevier, 2021, ISBN: 978–0–12-823796-0

13 K. Sotoodeh, *Piping Engineering: Preventing Fugitive Emission in the Oil and Gas Industry*, 1st ed.. Wiley, 2022. ISBN: 978–1–119–85206-3

14 K. Sotoodeh, *Safety Engineering in the Oil and Gas Industry*, 1st ed.. CRC Press (Taylor & Francis), 2023, ISBN: 978-1-032-47973-6

15 P. Tcvettkov, A. Cherepovitsyn, and S. Fedoseev, "Public perception of carbon capture and storage: A state-of-the-art overview," *Heliyon*, vol. 5, no. 12, p. e02845, Dec., 2019. DOI: 10.1016/j.heliyon.2019.e02845.

16 T. A. Hill, J. E. Fackrell, M. R. Dubal, and S. M. Stiff, "Understanding the consequences of CO_2 leakage downstream of the capture plant," *Energy Procedia*, vol. 4, pp. 2230–2237, 2011. DOI: 10.1016/j.egypro.2011.02.111.

17 I.-K. Waarum *et al.*, "CCS leakage detection technology—industry needs, government regulations, and sensor performance," *Energy Procedia*, vol. 114, pp. 3613–3627, Jul., 2017. DOI: 10.1016/j.egypro.2017.03.1493.

18 Z. Chen, Y. Zhou, and H. Li, "A review of phase behavior mechanisms of CO_2 EOR and storage in subsurface formations," *Industrial & Engineering Chemistry Research*, vol. 61, no. 29, pp. 10298–10318, Apr., 2022. DOI: 10.1021/acs.iecr.2c00204.

19 T. Xiao *et al.*, "Underground sources of drinking water chemistry changes in response to potential CO_2 leakage," *Science of the Total Environment*, vol. 847, 157254, Nov., 2022. DOI: 10.1016/j.scitotenv.2022.157254.

20 S. A. Carroll *et al.*, "Key factors for determining groundwater impacts due to leakage from geologic carbon sequestration reservoirs," *International Journal of Greenhouse Gas Control*, vol. 29, pp. 153–168, Oct., 2014. DOI: 10.1016/j.ijggc.2014.07.007.

21 J. M. Bielicki, M. F. Pollak, J. P. Fitts, C. A. Peters, and E. J. Wilson, "Causes and financial consequences of geologic CO_2 storage reservoir leakage and interference with other subsurface resources," *International Journal of Greenhouse Gas Control*, vol. 20, pp. 272–284, Jan., 2014. DOI: 10.1016/j.ijggc.2013.10.024.

22 M. Vitali, C. Zuliani, F. Corvaro, B. Marchetti, A. Terenzi, and F. Tallone, "Risks and safety of CO_2 transport via pipeline: A review of risk analysis and modeling approaches for accidental releases," *Energies*, vol. 14, no. 15, p. 4601, Jul., 2021. DOI: 10.3390/en14154601.

23 C. van Leeuwen and H. A. J. Meijer, "Detection of CO_2 leaks from carbon capture and storage sites with combined atmospheric CO_2 and O_2 measurements," *International Journal of Greenhouse Gas Control*, vol. 41, pp. 194–209, Oct., 2015. DOI: 10.1016/j.ijggc.2015.07.019.

24 K. Sotoodeh, *ASME Codes and Standards for Piping and Valves: Engineering Applications in Hydrogen, Renewable, and Conventional Energy Systems*, 1st ed.. Elsevier, 2026, ISBN: 978–0443443589

25 K. Yanagi, A. Nakamura, and E. Komatsu, "The importance of designing a comprehensive strategic environmental assessment (SEA) and environmental impact assessment (EIA) for carbon capture and storage in Japan," *International Journal of Greenhouse Gas Control*, vol. 91, 102823, Dec., 2019. DOI: 10.1016/j.ijggc.2019.102823.

26 P. Nikolai, B. Rabiyat, A. Aslan, and A. Ilmutdin, "Supercritical CO_2: Properties and technological applications—a review," *Journal of Thermal Science*, vol. 28, pp. 394–430, 2019. DOI: 10.1007/s11630-019-1118-4.

27 K. Sotoodeh, "Pipeline valves technology, material selection, welding, and stress analysis (a case study of a 30 in class 1500 pipeline ball valve)," *ASME Journal of Pressure Vessel Technology*, vol. 140, no. 4, 044001, 2018. DOI: 10.1115/1.4040139, Paper No. PVT-18-1043

28 L. Chen *et al.*, "Fracture process characteristic study during fracture propagation of a CO_2 transport network distribution pipeline," *Energy*, vol. 283, 129060, Nov., 2023. DOI: 10.1016/j.energy.2023.129060.

29 N. Lalam and R. Wright, "Distributed fiber-optic sensors for pipeline inspection and monitoring," in *Oil and Gas Pipelines: Integrity, Safety, and Security Handbook*, R. W. Revie, Ed, 2nd ed.. Wiley, 2025, , ch. 26, DOI: 10.1002/9781119909668.ch26.

30 N. Compton, "CO_2 dispersion analysis factors and their impacts on pipeline routing," in *Proc. ASME International Pipeline Conference*, Calgary, Alberta, Canada, Sep. 23–27, 2024, Paper No. IPC2024–133701, V005T09A020, 7 pages, Dec. 2024. DOI: 10.1115/IPC2024-133701

31 J. Koornneef, M. Spruijt, M. Molag, A. Ramírez, W. Turkenburg, and A. Faaij, "Quantitative risk assessment of CO_2 transport by pipelines—a review of uncertainties and their impacts," *Journal of Hazardous Materials*, vol. 177, no. 1–3, pp. 12–27, May., 2010. DOI: 10.1016/j.jhazmat.2009.11.068.

32 DNV, *DNV-RP-J202: Design and Operation of CO2 Pipelines*. Høvik, Norway, Apr.: Det Norske Veritas, 2010.

33 M. Pontiggia, T. Vairo, and B. Fabiano, "Risk assessment of buried natural gas pipelines: Critical aspects of event tree analysis," *Chemical Engineering Transactions*, vol. 77, pp. 613–618, 2019. DOI: 10.3303/CET1977103.

34 N. Sacuta, L. Gauvreau, and S. E. Greenberg, "Emergency response planning: An example of international collaboration in CCS community outreach and project development," *Energy Procedia*, vol. 37, pp. 7388–7394, 2013. DOI: 10.1016/j.egypro.2013.06.680.

35 B. W. Terwel, F. Harinck, N. Ellemers, and D. L. D. Daamen, "How organizational motives and communications affect public trust in organizations: The case of carbon dioxide capture and storage," *Journal of Environmental Psychology*, vol. 29, no. 2, pp. 290–299, Jun., 2009. DOI: 10.1016/j.jenvp.2008.11.004.

36 American Petroleum Institute (API), *API Specification 6A: Specification for Wellhead and Christmas Tree Equipment*, 21st ed.. Washington, DC, USA: API Publishing Services, 2018.

37 International Organization for Standardization (ISO), *ISO 10423: Petroleum and Natural Gas Industries—Drilling and Production Equipment—Wellhead and Christmas Tree Equipment*, 5th ed.. Geneva, Switzerland: ISO, 2017.

38 American Petroleum Institute (API), *API Specification 6D: Specification for Pipeline and Piping Valves*, 25th ed.. Washington, DC, USA: API Publishing Services, 2021.

39 International Organization for Standardization (ISO), *ISO 14313: Petroleum and Natural Gas Industries—Pipeline Transportation Systems—Pipeline Valves*, 3rd ed.. Geneva, Switzerland: ISO, 2007.

40 International Organization for Standardization (ISO), *ISO 27914: Carbon Dioxide Capture, Transportation and Geological Storage—Geological Storage*, 1st ed.. Geneva, Switzerland: ISO, 2017.

41 International Organization for Standardization (ISO), *ISO 27915: Carbon Dioxide Capture, Transportation and Geological Storage—Cross-Border CO2 Transport and Storage*, 1st ed.. Geneva, Switzerland: ISO, 2021.

42 W. Harbert, T. M. Daley, G. Bromhal, C. Sullivan, and L. Huang, "Progress in monitoring strategies for risk reduction in geologic CO_2 storage," *International Journal of Greenhouse Gas Control*, vol. 51, pp. 260–275, Aug., 2016. DOI: 10.1016/j.ijggc.2016.05.007.

43 A. Shogenova *et al.*, "Implementation of the EU CCS directive in Europe: Results and development in 2013," *Energy Procedia*, vol. 63, pp. 6662–6670, 2014. DOI: 10.1016/j.egypro.2014.11.700.

44 Q. Qerimi and B. S. Sergi, "The case for global regulation of carbon capture and storage and artificial intelligence for climate change," *International Journal of Greenhouse Gas Control*, vol. 120, p. 103757, Oct., 2022. DOI: 10.1016/j.ijggc.2022.103757.

45 W.-T. Joun, J. Rossabi, W.-J. Shin, and K.-K. Lee, "Real-time multi-level CO_2 concentration monitoring in vadose zone wells and the implication for detecting leakage events," *Journal of Environmental Management*, vol. 237, pp. 534–544, May., 2019. DOI: 10.1016/j.jenvman.2019.01.052.

46 K. Matheson, R. Manning, and S. Williams, "From brigade to service: An examination of the role of fire and rescue services in modern local government," *Local Government Studies*, vol. 37, no. 4, pp. 451–465, Aug., 2011. DOI: 10.1080/03003930.2011.588701.

47 R. J. Fischer, E. P. Halibozek, and D. C. Walters, "Contingency planning emergency response and safety," *Introduction to Security*, pp. 249–268, Oct., 2019. DOI: 10.1016/B978-0-12-805310-2.00011-1.

7

Carbon Capture and Sustainable Development Goals (SDGs)

7.1 Overview of the SDGs and Their Relevance to Carbon Capture

The Sustainable Development Goals (SDGs) consist of 17 strategic objectives aimed at addressing climate change, poverty, inequality, and environmental degradation, ensuring sustainable development by 2030. While carbon capture is primarily linked to SDG 13: Climate Action, its significance extends beyond this single goal [1, 2]. Carbon capture impacts various areas, including energy systems, industrial development, designing cleaner and safer cities, and environmental protection, making it relevant to several SDGs. To evaluate carbon capture effectively, it is essential to consider it from a comprehensive and forward-looking perspective [3].

Carbon capture technologies play an important role in reducing harmful greenhouse gas emissions from large industrial sources such as power plants, steel factories, and cement production facilities [4]. By capturing CO_2 before it is released into the atmosphere, these technologies help industries move toward long-term decarbonization, making them a key solution for SDG 13: Climate Action [3].

At the same time, carbon capture also supports SDG 7: Affordable and Clean Energy [5]. It does this in two important ways. First, it enables the production of low-carbon fuels, such as blue hydrogen, where natural gas is converted into hydrogen while the CO_2 produced is captured instead of being released into the atmosphere. This allows industries to use cleaner fuels, in this case, hydrogen, without increasing emissions [6]. Second, carbon capture helps renewable energy systems by working together with bioenergy and renewable energy sources such as wind and solar [7].

For example, in bioenergy with carbon capture (BECCS), biomass is used to generate energy, and the CO_2 released is captured and stored, resulting in net-negative emissions [8]. Additionally, CCS can also work together with renewable energy sources like wind and solar. In this case, clean electricity from wind or solar can be used to power parts of the CCS process, reducing the need to burn fossil fuels. This makes the entire system cleaner and helps ensure that carbon capture does not create additional emissions while operating [9].

As shown in Figure 7.1, a modern energy plant combines carbon capture facilities with renewable energy sources, such as wind and solar power. The background of the image shows an industrial carbon capture facility equipped with equipment and pipes. In front of

Figure 7.1 An integrated low-carbon energy system comprising a carbon capture facility powered by renewable energy sources, including solar panels and wind turbines, demonstrates the synergy between CCS and clean energy technologies for completely environmentally friendly industrial setup.

the facility, solar panels have been installed to produce clean electricity from the sun. The plant is surrounded by a number of wind turbines, highlighting the importance of wind power as a renewable energy source. This figure illustrates how carbon capture and storage (CCS) can work in harmony with renewable energy systems to reduce greenhouse gas emissions and ensure a reliable and sustainable energy supply. Specifically, it is a hybrid low-carbon energy infrastructure that supports the transition to net-zero and SDG 7—Affordable and Clean Energy.

Additionally, carbon capture plays a significant role in achieving Sustainable Development Goal (SDG) 9: Industry, Innovation, and Infrastructure [10]. It promotes innovation and the advancement of modern industrial development. This includes the development of advanced technologies, such as artificial intelligence (AI) for real-time monitoring and additive manufacturing (3D printing) to create customized components that enhance efficiency and reduce costs [11]. It's important to emphasize that carbon capture encourages circular-economy models, where captured CO_2 is not viewed as waste but is reused to produce new products like fuels, building materials, and chemicals. Through Carbon Capture and Storage (CCS), we can establish strong, sustainable, and low-carbon industrial systems that are well-equipped for the future.

At a broader societal level, carbon capture can support sustainable cities and communities (SDG 11). It helps improve local air quality by reducing pollution from factories and power plants located near where people live. CCS can also support smart city strategies, where cities use modern technology to reduce emissions and plan cleaner, more sustainable infrastructure. However, CCS must be implemented responsibly [12]. If not managed well, it could lead to high energy consumption, competition for land, or unequal benefits, where some communities face the risks but do not receive the advantages. For this reason, governments and companies must design fair and careful policies to ensure CCS truly supports environmental and social sustainability.

7.2 SDG 13 (Climate Action)

SDG 13—Climate Action is about taking urgent steps to reduce greenhouse gas emissions and protect the planet from climate change [13]. Carbon capture technologies help achieve this goal by capturing CO_2 before it is released into the air from factories and power plants. In some systems, carbon capture can even remove CO_2 that is already in the atmosphere, helping to lower overall emissions [14]. This makes carbon capture an important tool for slowing global warming and supporting international climate targets.

Carbon capture is particularly important for industries that are challenging to decarbonize, such as cement, steel, chemical production, and fossil-fuel power plants. These sectors cannot easily rely solely on renewable energy to reduce emissions because their processes involve specific chemical reactions that still generate CO_2. Therefore, capturing CO_2 directly at the source before it is released into the atmosphere becomes a practical and effective solution [15].

Technologies such as post-combustion capture remove CO_2 from flue gases after energy production, while direct air capture (DAC) can extract CO_2 that is already present in the ambient air [14]. Another noteworthy approach is bioenergy with carbon capture and storage (BECCS), which generates energy from biomass and captures the resulting emissions, allowing for net-negative climate benefits [8]. Together, these carbon capture technologies play a vital role in reducing emissions and supporting global goals for net-zero or even negative carbon emissions and climate action targets.

The Intergovernmental Panel on Climate Change (IPCC) and the International Energy Agency (IEA) both agree that the world cannot reach its climate targets without carbon capture. These are two of the most trusted global organizations in climate and energy research. The IPCC provides scientific assessments on climate change for the United Nations, while the IEA advises governments on energy policy and clean energy transitions [16]. According to both organizations, carbon capture technologies are essential if we want to keep global temperature rise below 1.5–2°C, in line with the Paris Agreement goals [16, 17]. Carbon capture helps countries reduce emissions more quickly, especially during the time when renewable energy systems are still being expanded and installed [16]. It also acts as a backup solution for industries where full electrification is not yet possible, ensuring that emissions from heavy industries can still be significantly reduced.

Figure 7.2 shows the role of carbon capture in supporting SDG 13: Climate Action by illustrating the full process of capturing, transporting, and securely storing CO_2. On the

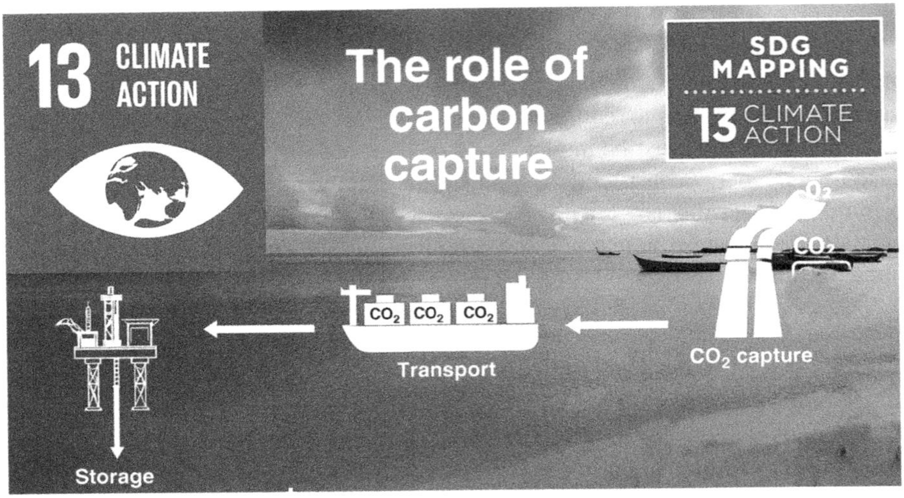

Figure 7.2 The role of carbon capture in achieving SDG 13 by capturing CO_2 emissions, transporting them safely, and storing them underground to prevent atmospheric release and support global climate targets.

right side, the process begins with CO_2 capture from industrial sources. It is followed by transportation via ship, and finally, storage deep underground, often in offshore geological formations. On the top left, the official SDG 13 Climate Action icon highlights the connection between carbon capture and the global effort to limit temperature rise to below 1.5–2°C, as outlined in the Paris Agreement [17]. The image illustrates how carbon capture is an essential solution to reducing emissions and preventing climate-related risks such as extreme weather events, rising sea levels, and environmental damage.

In many heavy industries, such as cement, steel, or chemical production, it is still not possible to completely switch to electricity from renewable energy sources like solar or wind. These industries produce carbon dioxide through processes or chemical reactions. In such cases, carbon capture works as a backup or support system. In fact, it allows these industries to keep operating while still reducing their emissions, even before full electrification or integration with renewable energy becomes available in the future [15].

In summary, carbon capture is an important technology that can contribute to the achievement of SDG 13: Climate Action [3]. The use of these technologies reduces CO_2 emissions from industries that are difficult to make environmentally friendly such as cement and steel even by integrating them with renewable energy sources [15]. However, carbon capture is not a replacement for renewable energy. Instead, it can combine solar, wind, hydrogen, and other clean technologies. By capturing CO_2 while renewable energy systems grow, carbon capture helps the world move faster toward net-zero emissions [7]. In this way, it serves as a bridge solution that makes deep decarbonization more realistic and achievable within the urgent timelines set by the global climate goals.

7.3 SDG 7 (Clean Energy) and Renewable Integration

SDG 7—Affordable and Clean Energy aims to make sure that everyone in the world can access energy that is reliable and does not harm the environment. The goal is to ensure that the energy is clean and sustainable for the future [18]. Renewable energy sources like solar, wind, and geothermal are growing quickly and are central to this goal. However, in many countries, it is still difficult to rely only on renewables at present. There are still some regions that require fossil fuels to maintain their energy supply, especially during times when renewable energy is not available, such as solar power at night or at low wind speeds [19].

This is where carbon capture (CCS) becomes very important. Carbon capture traps CO_2 emissions before they are released into the atmosphere. It allows existing power plants and heavy industries to continue operating but in a much cleaner way. By reducing emissions from fossil fuels, carbon capture helps countries move toward clean energy. Carbon capture acts as a bridge, supporting a smooth and realistic transition while renewable energy capacity continues to expand [20].

Although renewable energy sources such as solar and wind are expanding rapidly, they can be infrequent and variable and are influenced by weather conditions [21]. Carbon capture allows existing power plants, including natural gas and biomass plants, to produce heat and electricity with significantly reduced emissions. In response to the growing demand for renewable energy, carbon capture technology facilitates energy transitions that lead to a reduction of emissions and the prevention of climate change. Additionally, carbon capture contributes significantly to the production of low-carbon hydrogen, often called "blue hydrogen" [6]. This hydrogen can be used as a clean fuel in transportation, heating, and industrial energy, supporting SDG 7's objective to increase the availability of cleaner sources of energy.

When combined with bioenergy in BECCS systems, carbon capture can even be combined with renewable energy while removing CO_2 from the atmosphere, offering a strong complement to wind, solar, and hydropower [8]. In summary, carbon capture does not replace renewable energy. Instead, it supports producing clean hydrogen energy and promotes a clean energy transition. In fact, carbon capture enables countries to decarbonize faster while shifting to 100% renewable energy.

7.4 SDG 9 (Industry, Innovation, and Infrastructure)

SDG 9—Industry, Innovation, and Infrastructure emphasizes the development of robust industries, advanced technology, and sustainable infrastructure [22]. Carbon capture technologies strongly contribute to this goal by accelerating industrial transformation to low-carbon development in sectors that are traditionally difficult to decarbonize, such as cement, steel, chemicals, and power generation [15].

Besides reducing emissions, carbon capture and storage (CCS) is becoming a catalyst for industrial innovation, including the use of advanced technologies like digitalization that are transforming the design, construction, and operation of CCS plants [23]. As an example, additive manufacturing (AM, or 3D printing) is now used to manufacture components and facilities used in carbon capture plants, such as valves and heat exchangers. Through the use of additive manufacturing, these facilities can be produced with higher quality, tailored for specific applications , and manufactured at a faster rate than traditional manufacturing methods [24].

In addition, digitalization technologies such as AI, automation, sensors, and digital twins are being integrated into CCS facilities. These tools enable real-time monitoring, early fault detection, and predictive maintenance, helping to prevent failures before they occur. They also enhance energy efficiency, reduce operating costs, and significantly improve safety and reliability [23].

AI can be used to predict system performance , predict how well their CCS system will capture CO_2, allowing them to adjust the process before efficiency declines. AI can also control and adjust the amount of solvent used to absorb CO_2. This improves process efficiency while still capturing high levels of carbon dioxide. In addition, AI can analyze different geological storage locations for CO_2 storage. AI selects the most suitable option by analyzing its effectiveness in storing CO_2, the amount of CO_2 it can store, and the risk of leakage. AI analyzers are faster and more accurate than traditional methods [25].

Figure 7.3 shows a modern carbon capture or industrial plant that is being monitored digitally in real time. A plant engineer uses a tablet to access online data, while advanced digital systems, online sensors, and smart monitoring icons are displayed as holograms for operator use. As a result of digitalization, it is now possible to perform online monitoring and automated control either from a central control room or directly through a tablet operated in the field by the operator. This technology enables operators to detect leaks, continuously measure CO_2 process conditions such as pressure, temperature, and flow, automate valve operations with remote control from the control room, and optimize overall facility

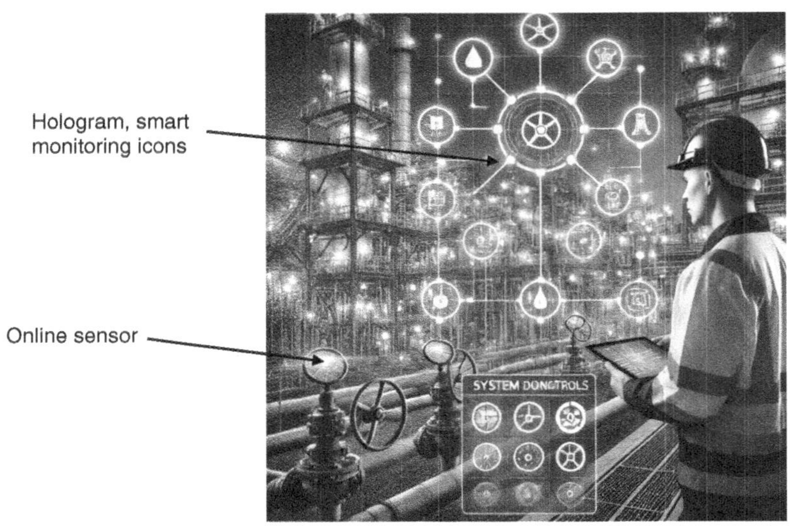

Figure 7.3 Digitalization in a carbon capture plant, including online sensors, smart monitoring, and real-time control using advanced digital technologies.

performance. As a result, the efficiency, safety, and reliability of plants, including carbon capture facilities, have significantly improved.

CCS is not only an emission-reduction technology but also a driver of new industrial value chains, especially through CO_2 utilization. In simple terms, a value chain refers to a series of steps from input to final product that add value at each stage of production. Regarding CCS, captured CO_2 is no longer treated as waste gas but becomes a valuable industrial feedstock, a raw material that is put into an industrial process to create a new product or energy. In addition to synthetic fuels, polymers, fertilizers, and construction materials such as aggregates and low-carbon concrete, it can be processed and converted [26].

Figure 7.4 illustrates the concept of carbon utilization. This is where CO_2 captured from an industrial facility is used as a valuable raw material rather than released into the atmosphere [26]. The diagram shows CO_2 captured from an industrial plant and directed toward different high-value product processes, including synthetic fuels, polymers, fertilizers, and construction materials. This visual highlights how CO_2 can become a useful input for manufacturing various materials. This supports the development of sustainable, circular industrial systems aligned with SDG 9 (Industry, Innovation, and Infrastructure).

Sustainable Development Goal 9 (SDG 9), Industry, Innovation, and Infrastructure, closely aligns with treating captured CO_2 as valuable industrial feedstock rather than as a waste product. The goal of this approach is to support the transition toward a circular and sustainable industrial economy, in which carbon is repurposed rather than discarded [22]. Converting CO_2 into useful products is sustainable because it prevents the gas from being released into the atmosphere and instead reuses it as a resource [26]. This reduces greenhouse gas emissions, lowers the need for fossil fuels, and supports a circular economy,

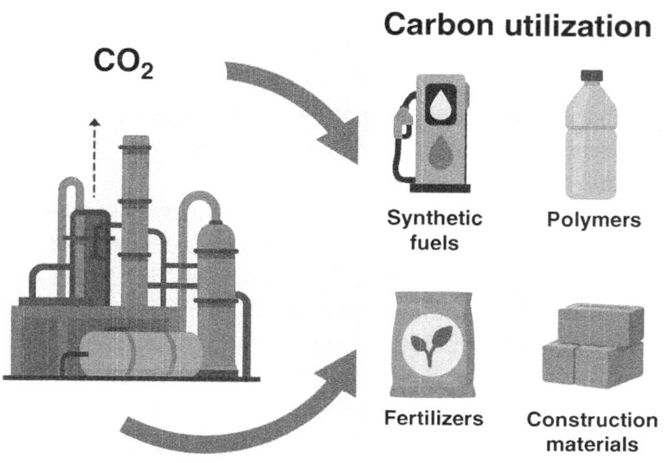

Figure 7.4 Conversion of captured CO_2 into value-added products such as synthetic fuels, polymers, fertilizers, and low-carbon construction materials. This process demonstrates the role of carbon utilization in enabling circular and sustainable industrial development.

where waste is turned into value instead of pollution. The advancement of carbon capture and utilization (CCU) technologies will enable industries to adopt innovative industries and manufacturing practices by converting CO_2 into synthetic fuels, polymers, and other high-value materials. Not only does this accelerate technological progress, but it also strengthens a resilient and future-proof industrial infrastructure, which is the foundation of Sustainable Development Goal 9.

7.5 SDG 11 (Sustainable Cities and Communities)

Sustainable Development Goal 11 (SDG 11) focuses on creating cities and communities that are clean, safe, and healthy for residents [27]. Carbon capture technology can contribute to this goal by lowering emissions from industries and power plants that pollute the air, especially those situated near or within urban areas. By capturing carbon dioxide (CO_2) before it is released into the atmosphere, cities can reduce pollution, enhance air quality, and protect public health [28].

Carbon capture plays a vital role in supporting sustainable city development in several ways. Firstly, it promotes greener industries by capturing CO_2 emissions from factories and power plants, allowing them to operate with significantly reduced environmental impact. Additionally, carbon capture enables cleaner transportation by converting captured CO_2 into synthetic low-carbon fuels for airplanes, ships, and heavy vehicles [29].

Furthermore, it contributes to sustainable infrastructure, as the captured CO_2 can be transformed into materials like low-carbon concrete and aggregates, which are essential for eco-friendly construction. Aggregates are raw materials used in construction, mainly in concrete, asphalt, and road building [30]. Through these initiatives, carbon capture helps cities become cleaner and healthier in terms of climate change.

Many large cities have industrial zones, waste-to-energy plants, district heating networks, and major transportation hubs, all of which produce significant amounts of CO_2. Instead of shutting these systems down or replacing them immediately with clean choices, carbon capture technology can be integrated directly into them. This approach allows cities to reduce emissions while maintaining essential services. Fully replacing existing infrastructure with new clean systems can be extremely costly, time-consuming, and may even lead to job losses or operational disruptions [20].

Waste-to-energy plants convert municipal waste into electricity and heat by burning it. When these facilities are equipped with carbon capture and storage (CCS) technology, they can effectively capture a significant portion of their CO_2 emissions. This allows them to manage urban waste safely while also providing local energy. In this way, carbon capture facilitates the decarbonization of existing infrastructure in cities without causing disruptions, making the transition to a sustainable, low-carbon future much smoother [31].

Figure 7.5 illustrates the integration of carbon capture technology with a waste-to-energy plant. On the left side, municipal waste is collected and directed to a recovery process, where it is converted into heat, fuel, and electricity [31, 32]. On the right, an industrial facility represents a waste-to-energy plant equipped with carbon capture and storage (CCS) technology, which captures carbon dioxide (CO_2) emissions rather than releasing them into the atmosphere. This figure clearly demonstrates that CCS significantly reduces CO_2

Figure 7.5 Integration of carbon capture technology with a waste-to-energy plant, where municipal waste is converted into heat, fuel, and power while CO_2 emissions from the waste are captured to reduce environmental impact.

emissions from the waste-heating process, making waste-to-energy a more sustainable solution for urban environments [31].

There are numerous advantages to sharing CO_2 transport and storage infrastructure, especially in urban areas like industrial ports. By connecting multiple industrial facilities, such as refineries, chemical plants, power stations, and waste-to-energy plants, to a single CO_2 pipeline network, this approach avoids the need to construct separate carbon transport and storage systems for each facility. This approach is referred to as the clustered or hub-based carbon capture and storage (CCS) method [33].

Figure 7.6 illustrates a hub-based carbon capture and storage (CCS) system. In this system, carbon dioxide (CO_2) emissions from multiple industrial facilities are collected and transported to a centralized CO_2 capture hub. The factories surrounding the hub represent various emission sources, such as refineries, power plants, and chemical facilities, each directing its CO_2 through pipelines toward the central capture location.

At the CO_2 capture hub, the gas is processed, purified, and conditioned to meet transport specifications. From this point, the captured CO_2 is transported through a main pipeline, which is clearly shown extending to the right of the figure. This pipeline leads to a storage site, depicted as an injection well, where CO_2 is safely injected deep underground into geological formations such as saline aquifers or depleted oil and gas reservoirs. This figure clearly illustrates the concept of collecting CO_2 from multiple sources, treating it centrally, and transporting it in one pipeline to permanent geological storage. This process forms the basis of a large-scale and cost-effective carbon capture infrastructure.

This shared model presents several significant advantages. First, it allows for cost-sharing and reduction; since the infrastructure is expensive to build, sharing it across multiple

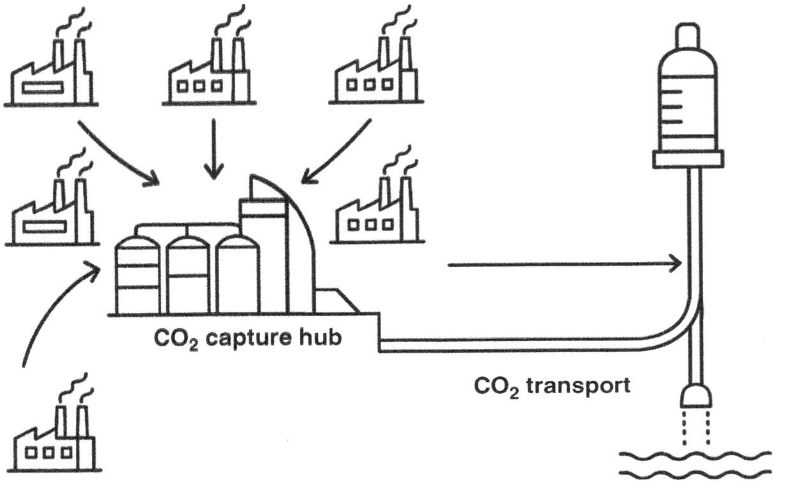

Figure 7.6 Illustration of a centralized CO_2 capture hub receiving emissions from multiple industrial facilities and transporting the captured CO_2 through a single pipeline for permanent underground storage.

industries makes it much more affordable. Additionally, it enhances space efficiency, which is crucial in densely populated urban and port areas where land is limited. Furthermore, this model supports centralized, coordinated safety and monitoring systems rather than spreading these efforts across a larger area. This makes the transport and storage of CO_2 more reliable and easier to regulate [33].

Overall, this shared infrastructure approach makes CCS much easier to apply in cities. Instead of each factory or plant managing carbon capture on its own, multiple industries can be decarbonized together using the same CO_2 pipeline and storage system. This makes the process more affordable, faster to implement, and easier to manage [33]. It helps cities cut emissions more efficiently, while also supporting smart urban planning and reducing costs for industries and governments.

Additionally, carbon capture helps cities become more climate-resilient and sustainable by supporting several key urban strategies. This means that carbon capture helps cities not only reduce emissions but also become stronger against climate change and more environmentally friendly in the long term. For example, CCS can be integrated into district heating systems, supplying clean heat to homes and buildings without adding CO_2 to the atmosphere. It also enables the production of low-carbon building materials, such as concrete and aggregates made using captured CO_2, which supports greener construction and infrastructure development [28].

Furthermore, captured CO_2 can be converted into clean fuels for city transport systems, including buses, ships, and even airplanes. This helps cities reduce emissions from the transportation sector that are traditionally very difficult to decarbonize. Through these applications, carbon capture contributes directly to cleaner energy fuel and low-emission transport, strengthening the long-term sustainability of urban communities [29].

Public safety, transparency, and social trust are essential factors to consider when carbon capture and storage (CCS) systems are installed in or near urban areas. Residents living close to these systems must feel assured that the technology is safe and that their concerns are addressed. Open and honest communication is necessary, not just regarding technical details, but also in addressing the community's worries [34].

It is essential for the carbon capture project team to actively involve the community and the public during the planning and decision-making process, rather than just informing them after decisions are made [34]. Additionally, project teams must implement strict safety regulations, continuous monitoring, and robust emergency response plans to manage risks effectively. If these elements are not carefully managed, public opposition may arise, which can delay or even stop the deployment of CCS. Therefore, earning and maintaining social acceptance is just as important as the technology itself.

Figure 7.7 depicts a community meeting focused on carbon capture and storage (CCS), highlighting the significance of public engagement, safety, and transparency when planning such projects near populated areas. At the front of the room, a female presenter is explaining the CCS project to a group of local residents. On the presentation screen, key points are listed: "Community Engagement," "Safety," and "Monitoring." These topics reflect the major concerns and priorities that communities usually have when CCS infrastructure is introduced in or near urban or semi-urban areas. The audience seems focused and engaged, indicating active public participation. This is essential for building trust, social acceptance, and ensuring residents feel informed and safe.

Figure 7.7 Community and public engagement meeting on carbon capture and storage (CCS), highlighting the importance of public safety, transparency, and active participation of local residents in the decision-making process.

7.6 Balancing Benefits and Challenges of CCS and SDGs

While carbon capture technologies offer significant support to the global climate goals and several SDGs, it is important to recognize that their deployment comes with practical challenges and compromises that must be carefully managed to ensure long-term sustainability and fairness [35].

One of the major concerns regarding carbon capture technologies is their high energy demand. Carbon capture systems, particularly DAC, require significant amounts of electricity and heat. If this energy is sourced from fossil fuels, the overall climate benefits of carbon capture may be diminished due to the emissions resulting from burning these fuels [36]. Therefore, it is essential that carbon capture and storage (CCS) is closely integrated with renewable energy sources to achieve meaningful reductions in emissions.

Another challenge is related to resource and land use, especially for Bioenergy with Carbon Capture and Storage (BECCS), which depends on large quantities of biomass sourced from agriculture and natural environments. Poor planning in this area could lead to conflicts with agriculture and food supply, water availability, and the health of natural ecosystems, particularly in developing regions [37].

Public acceptance is a crucial factor in the success of carbon capture and storage (CCS) projects. Even if the technology is effective, residents near CCS infrastructure, such as pipelines, storage sites, or industrial facilities, may have concerns about safety, the potential for leaks, environmental impacts, and how the CO_2 will be monitored over time. These concerns are understandable. To gain public trust, companies and governments must ensure clear and honest communication, involve local communities in decision-making, and follow strict safety regulations and inspections. People need to feel fully informed and protected, and they must be confident that the captured CO_2 will remain safely stored not just for a few years but for decades or even centuries [38].

For carbon capture to effectively contribute to all Sustainable Development Goals (SDGs), it must be implemented in a balanced and integrated manner. This requires collaboration across renewable energy, energy efficiency, digital monitoring, and circular-economy strategies—such as using captured carbon to produce fuels, building materials, polymers, and other valuable products [39, 40]. It is crucial that carbon capture not operate in isolation; it should be integrated with with complementary systems and technologies .

7.7 Key Highlights (Chapter Summary)

◇ **Chapter 7–Key Highlights:**

1) Carbon capture technologies directly support SDG 13 (Climate Action) by reducing or even removing CO_2 emissions from the atmosphere.
2) CCS also contributes to SDG 7 (Clean and Affordable Energy) by enabling low-carbon power generation such as hydrogen and supporting renewable energy integration.
3) SDG 9 (Industry, Innovation, and Infrastructure) benefits through technological advancements such as AI, digital twins, additive manufacturing, and CO_2 utilization industries.

4) CCS can help SDG 11 (Sustainable Cities and Communities) by reducing emissions from urban industries, waste-to-energy plants, and transportation hubs.
5) CCS is particularly crucial for eliminating pollutants from hard-to-decarbonize sectors such as cement, steel, and chemicals, where relying solely on renewable energy is insufficient.
6) One of the main challenges of CCS is that it requires a significant amount of energy, which must be sourced from renewable or low-carbon alternatives rather than fossil fuels to ensure that the climate benefits are not diminished.
7) Technologies like BECCS may create challenges concerning land, water, and resource overuse, necessitating robust sustainability guidelines.
8) Public acceptance and trust are crucial, as communities must feel confident about safety, transparency, and long-term CO_2 storage integrity.
9) The success of Carbon Capture and Storage (CCS) relies on its deployment in a balanced and coordinated manner. This means it should advance alongside renewable energy, be supported by digital technologies for monitoring and safety, be directed by strong and consistent government policies, and ensure equitable access and benefits for all regions worldwide.
10) CCS must be implemented within a reliable policy and regulatory framework to ensure safety, transparency, long-term successful monitoring during storage, and contributions to multiple SDGs simultaneously, rather than focusing solely on climate targets.

Keywords: Carbon Capture, Sustainable Development Goals (SDGs), Climate Action, Clean Energy Transition, Industrial Innovation, Carbon Neutrality, Circular Economy, Public Acceptance, Renewable Integration, Global Sustainability

References

1 W. L. Filho, T. Wall, A. L. Salvia, M. A. P. Dinis, and M. Mifsud, "The central role of climate action in achieving the United Nations' sustainable development goals," *Scientific Reports*, vol. 13, p. 20582, 2023. DOI: 10.1038/s41598-023-47746-w.

2 D. Chong, "The sustainable development goals and climate change," *Social Alternatives*, vol. 37, no. 1, pp. 43–48, Jan., 2018.

3 T. Mikunda, L. Brunner, E. Skylogianni, J. Monteiro, L. Rycroft, and J. Kemper, "Carbon capture and storage and the sustainable development goals," *International Journal of Greenhouse Gas Control*, vol. 108, 103318, Jun., 2021. DOI: 10.1016/j.ijggc.2021.103318.

4 M. Garcia and N. Berghout, "Toward a common method of cost-review for carbon capture technologies in the industrial sector: Cement and iron and steel plants," *International Journal of Greenhouse Gas Control*, vol. 87, pp. 142–158, Aug., 2019. DOI: 10.1016/j.ijggc.2019.05.005.

5 B. Ramasubramanian and S. Ramakrishna, "What's next for the sustainable development goals? Synergy and trade-offs in affordable and clean energy (SDG 7)," *Sustainable Earth Reviews*, vol. 6, no. 17, 2023. DOI: 10.1186/s42055-023-00069-0.

6 M. Riemer and V. Duscha, "Carbon capture in blue hydrogen production is not where it is supposed to be—evaluating the gap between practical experience and literature estimates," *Applied Energy*, vol. 349, p. 121622, Nov., 2023. DOI: 10.1016/j.apenergy.2023.121622.

7 A. N. Arnette, "Renewable energy and carbon capture and sequestration for a reduced carbon energy plan: An optimization model," *Renewable and Sustainable Energy Reviews*, vol. 70, pp. 254–265, Apr., 2017. DOI: 10.1016/j.rser.2016.11.218.

8 W. Burns and S. Nicholson, "Bioenergy and carbon capture with storage (BECCS): The prospects and challenges of an emerging climate policy response," *Journal of Environmental Studies and Sciences*, vol. 7, pp. 527–534, 2017. DOI: 10.1007/s13412-017-0445-6.

9 P. Viebahn, D. Vallentin, and S. Höller, "Integrated assessment of carbon capture and storage (CCS) in the German power sector and comparison with the deployment of renewable energies," *Applied Energy*, vol. 97, pp. 238–248, Sept., 2012. DOI: 10.1016/j.apenergy.2011.12.053.

10 J. Denoncourt, "Companies and UN 2030 sustainable development goal 9 industry, innovation and infrastructure," *Journal of Corporate Law Studies*, vol. 20, no. 1, pp. 199–235, 2020. DOI: 10.1080/14735970.2019.1652027.

11 U. M. Dilberoglu, B. Gharehpapagh, U. Yaman, and M. Dolen, "The role of additive manufacturing in the era of industry 4.0," *Procedia Manufacturing*, vol. 11, pp. 545–554, 2017. DOI: 10.1016/j.promfg.2017.07.148.

12 M. C. Ogwu and S. C. Izah, "Air pollution and the sustainable development goals," in *Sustainable Strategies for Air Pollution Mitigation*, M. C. Ogwu and S. C. Izah, Eds, The Handbook of Environmental Chemistry, Vol. 133. Cham, Switzerland: Springer, 2024, pp. 201–233, DOI: 10.1007/698_2024_1118.

13 S. Küfeoğlu, "SDG-13: Climate action," in *Emerging Technologies, Sustainable Development Goals Series*, S. Küfeoğlu, Ed. Cham, Switzerland: Springer, 2022, DOI: 10.1007/978-3-031-07127-0_15.

14 H. Xu, L. Yu, C. Chong, and F. Wang, "A comprehensive review on direct air carbon capture (DAC) technology by adsorption: From fundamentals to applications," *Energy Conversion and Management*, vol. 322, 119119, Dec., 2024. DOI: 10.1016/j.enconman.2024.119119.

15 S. Paltsev, J. Morris, H. Kheshgi, and H. Herzog, "Hard-to-abate sectors: The role of industrial carbon capture and storage (CCS) in emission mitigation," *Applied Energy*, vol. 300, 117322, Oct., 2021. DOI: 10.1016/j.apenergy.2021.117322.

16 H. Kheshgi, H. de Coninck, and J. Kessels, "Carbon dioxide capture and storage: Seven years after the IPCC special report," *Mitigation and Adaptation Strategies for Global Change*, vol. 17, pp. 563–567, 2012. DOI: 10.1007/s11027-012-9391-5.

17 J. Rogelj *et al.*, "Paris agreement climate proposals need a boost to keep warming well below 2 °C," *Nature*, vol. 534, pp. 631–639, 2016. DOI: 10.1038/nature18307.

18 S. Küfeoğlu, "SDG-7: Affordable and clean energy," in *Emerging Technologies, Sustainable Development Goals Series*, S. Küfeoğlu, Ed. Cham: Springer, 2022, DOI: 10.1007/978-3-031-07127-0_9.

19 F. Martins, C. Felgueiras, and M. Smitková, "Fossil fuel energy consumption in European countries," *Energy Procedia*, vol. 153, pp. 107–111, Oct., 2018. DOI: 10.1016/j.egypro.2018.10.050.

20 H. C. Lau, S. Ramakrishna, K. Zhang, and A. V. Radhamani, "The role of carbon capture and storage in the energy transition," *Energy & Fuels*, vol. 35, no. 9, pp. 7364–7386, Apr., 2021. DOI: 10.1021/acs.energyfuels.1c00032.

21 L. Liu *et al.*, "Optimizing wind/solar combinations at finer scales to mitigate renewable energy variability in China," *Renewable and Sustainable Energy Reviews*, vol. 132, 110151, Oct., 2020. DOI: 10.1016/j.rser.2020.110151.

22 S. Küfeoğlu, "SDG-9: Industry, innovation and infrastructure," in *Emerging Technologies, Sustainable Development Goals Series*, S. Küfeoğlu, Ed. Cham: Springer, 2022, DOI: 10.1007/978-3-031-07127-0_11.

23 S. Wang *et al.*, "Digitalization solution for carbon capture, transportation, utilization, and storage," *Transportation, Utilization, and Storage*, Dec., 2024.

24 A. Zhakeyev, P. Wang, L. Zhang, W. Shu, H. Wang, and J. Xuan, "Additive manufacturing: Unlocking the evolution of energy materials," *Advanced Science*, vol. 4, no. 10, 1700187, Oct., 2017. DOI: 10.1002/advs.201700187.

25 A. K. Priya, B. Devarajan, A. Alagumalai, and H. Song, "Artificial intelligence enabled carbon capture: A review," *Science of the Total Environment*, vol. 886, 163913, Aug., 2023. DOI: 10.1016/j.scitotenv.2023.163913.

26 M. Aresta and A. Dibenedetto, "Industrial utilization of carbon dioxide (CO_2)," in *Developments and Innovation in Carbon Dioxide (CO2) Capture and Storage Technology*, M. M. Maroto-Valer, Ed, Vol. 2. Woodhead Publishing Series in Energy, 2010, pp. 377–410, DOI: 10.1533/9781845699581.4.377.

27 T. Kellison, "An overview of sustainable development goal 11," in *The Routledge Handbook of Sport and Sustainable Development*, G. McCullough and B. Kellison, Eds. London: Routledge (Taylor & Francis Group), 2021, pp. 261–275.

28 Y. Alhorr, E. Eliskandarani, and E. Elsarrag, "Approaches to reducing carbon dioxide emissions in the built environment: Low carbon cities," *International Journal of Sustainable Built Environment*, vol. 3, no. 2, pp. 167–178, Dec., 2014. DOI: 10.1016/j.ijsbe.2014.11.003.

29 C. Kolster, E. Mechleri, S. Krevor, and N. Mac Dowell, "The role of CO_2 purification and transport networks in carbon capture and storage cost reduction," *International Journal of Greenhouse Gas Control*, vol. 58, pp. 127–141, Mar., 2017. DOI: 10.1016/j.ijggc.2017.01.014.

30 J.-L. Gálvez-Martos *et al.*, "Eco-efficiency of a novel construction material produced by carbon capture and utilization," *Journal of CO2 Utilization*, vol. 49, 101545, Jul., 2021. DOI: 10.1016/j.jcou.2021.101545.

31 P. Wienchol, A. Szlęk, and M. Ditaranto, "Waste-to-energy technology integrated with carbon capture—challenges and opportunities," *Energy*, vol. 198, 117352, May., 2020. DOI: 10.1016/j.energy.2020.117352.

32 W. A. Qazi, M. F. M. Abushammala, M.-H. Azam, and M. K. Younes, "Waste-to-energy technologies: A literature review," *The Journal of Solid Waste Technology and Management*, vol. 44, no. 4, pp. 387–409, Nov., 2018. DOI: 10.5276/JSWTM.2018.387.

33 T. Harkin, I. Filby, H. Sick, D. Manderson, and R. Ashton, "Development of a CO_2 specification for a CCS hub network," *Energy Procedia*, vol. 114, pp. 6708–6720, July., 2017. DOI: 10.1016/j.egypro.2017.03.1801.

34 F. Mulyasari, A. K. Harahap, A. O. Rio, R. Sule, and W. G. Kadir, "Potentials of the public engagement strategy for public acceptance and social license to operate: Case study of carbon capture, utilisation, and storage gundih pilot project in Indonesia," *International*

Journal of Greenhouse Gas Control, vol. 108, 103312, Jun., 2021. DOI: 10.1016/j.ijggc.2021.103312.

35 T. Wilberforce, A. Baroutaji, B. Soudan, A. H. Al-Alami, and A. G. Olabi, "Outlook of carbon capture technology and challenges," *Science of the Total Environment*, vol. 657, pp. 56–72, Mar., 2019. DOI: 10.1016/j.scitotenv.2018.11.424.

36 E. S. Rubin, H. Mantripragada, A. Marks, P. Versteeg, and J. Kitchin, "The outlook for improved carbon capture technology," *Progress in Energy and Combustion Science*, vol. 38, no. 5, pp. 630–671, Oct., 2012. DOI: 10.1016/j.pecs.2012.03.003.

37 P. C. Stoy *et al.*, "Opportunities and trade-offs among BECCS and the food, water, energy, biodiversity, and social systems nexus at regional scales," *Bioscience*, vol. 68, no. 2, pp. 100–111, Feb., 2018. DOI: 10.1093/biosci/bix145.

38 K. Arning, J. Offermann-van Heek, A. Sternberg, A. Bardow, and M. Ziefle, "Risk-benefit perceptions and public acceptance of carbon capture and utilization," *Environmental Innovation and Societal Transitions*, vol. 35, pp. 292–308, Jun., 2020. DOI: 10.1016/j.eist.2019.05.003.

39 X. Chen, X. Wu, and K. Y. Lee, "The mutual benefits of multi-energy carbon capture: Achieved by an artificial intelligent scheduling strategy," *Energy Conversion and Management*, vol. 233, 113856, Apr., 2021. DOI: 10.1016/j.enconman.2021.113856.

40 W. M. Budzianowski, "Implementing carbon capture, utilisation and storage in the circular economy," *International Journal of Global Warming*, vol. 12, no. 2, pp. 272–296, Jun., 2017. DOI: 10.1504/IJGW.2017.084510.

8

Carbon Capture in Developing vs. Developed Countries

8.1 Global Inequality in CCS Technology Access

Access to carbon capture technologies is not evenly distributed around the world, highlighting the long-standing differences between wealthier and poorer countries. Currently, most large carbon capture and storage (CCS) projects are located in North America, Western Europe, and parts of East Asia. These regions benefit from strong financial systems, well-developed industrial infrastructure, and advanced research institutions, enabling them to invest in high-cost, high-technology projects like CCS. Additionally, they have supportive government policies and clear regulations that encourage companies to implement carbon capture solutions [1].

Many developing and least-developed countries face significant challenges that hinder their ability to adopt CCS technology. They often struggle with a lack of access to skilled engineers and specialized technical training, along with unreliable funding mechanisms and limited international investments. Furthermore, their policy frameworks are frequently uncertain or underdeveloped. For instance, there may be no clear legal regulations for CO_2 storage, no established long-term carbon pricing system, and frequent changes in government climate policies. These factors create high risks for investors, making it difficult to plan long-term CCS projects [2]. Consequently, CCS deployment in these regions remains slow or primarily in the pilot stage, which contributes to a widening global divide in climate technology readiness.

This inequality is not only technical but also economic and political. High-income countries are often able to justify CCS investments as part of their long-term net-zero strategies. They have strong national policies and financial incentives, such as tax credits, carbon pricing schemes, or government funding programs that make CCS projects attractive to industries and investors. In contrast, many lower-income countries see CCS as financially unrealistic. Their immediate priorities are often centered on basic energy access, job creation, poverty reduction, and economic growth, rather than high-cost industrial decarbonization. Without affordable financing or clear policy support, CCS is viewed as a low priority compared to more urgent development needs [1, 2].

Moreover, geological constraints and data limitations exacerbate the challenges. Many developing countries do not have comprehensive geological studies to identify safe underground sites for CO_2 storage. Without this critical information, it becomes nearly impossible to design or approve CCS projects with confidence [3]. Additionally, the digital infrastructure in both non-developed and developing countries is often inadequate, hindering the deployment of modern technologies necessary for real-time monitoring, leak detection, and long-term safety verification of CO_2 storage sites [4].

In addition to challenges related to infrastructure and data, the institutional capacity of a country is another major factor that separates developed regions from developing ones. Institutional capacity refers to how strong, organized, and efficient a country's government systems and public institutions are, especially when it comes to managing complex projects like CCS. Developed countries usually have strong public institutions and well-established regulatory authorities. These institutions are able to enforce environmental standards, properly assess project risks, and coordinate different groups, such as industry, government agencies, and research institutions, to ensure smooth project implementation. In contrast, many developing countries struggle with weak coordination between key ministries, such as those responsible for energy, environment, finance, and industry. This often leads to delays, overlapping responsibilities, and weak enforcement of climate policies [5]. Overlapping responsibilities mean that two or more government departments, ministries, or agencies are responsible for the same task, but they are not clearly coordinated. As a result, decision-making becomes slow and uncertain, which discourages both domestic and international investors from committing to CCS projects [6].

Figure 8.1 illustrates the technological gap between developed and developing countries regarding the adoption of CCS. The top half of the figure shows a highly advanced CCS facility in a developed country, featuring modern industrial infrastructure, metallic pipelines, an organized layout, and a clearly marked sign. This facility represents a nation that possesses the financial resources, engineering expertise, and regulatory framework necessary to operate CCS at a full industrial scale. In contrast, the bottom half depicts a developing country where the sign mentions CCS, but no actual facility exists. Instead, the area appears rural, with basic surroundings and limited infrastructure, indicating that CCS is still in the planning or conceptual stage [2, 7]. Although the air in the bottom picture looks clean, reflecting a developing or rural area that has not yet been heavily industrialized, many of these regions are rapidly industrializing or planning large-scale steel, cement, and petrochemical projects. These industries will significantly increase CO_2 emissions in the near future, making the establishment of a carbon capture plant critically important. This visual comparison highlights how developed nations are already executing large-scale CCS projects, while developing countries remain far behind due to constraints in funding, infrastructure, governance, and technical readiness.

International organizations such as the International Energy Agency (IEA), the World Bank, and the United Nations Framework Convention on Climate Change (UNFCCC) have emphasized that CCS should not be restricted to a few advanced countries [8]. For CCS to effectively contribute to global decarbonization in a fair and inclusive manner, it must also be implemented in major emerging economies, including India, Indonesia, Brazil, and South Africa. Additionally, it should be established in regions like the Middle East and North Africa, where carbon dioxide emissions from industries such as steel, cement, and petrochemicals continue to rise [9].

However, expanding CCS to these regions requires more than just supplying the physical equipment or engineering expertise. It also demands long-term investments in strong

Figure 8.1 A contrast between CCS reality and readiness, while advanced economies already operate full-scale carbon capture facilities, many developing regions are still at the planning stage.

governance systems, reliable digital monitoring technology, and financial mechanisms. Without this kind of full support, CCS could end up increasing the gap between rich and developing countries, instead of helping to reduce it [10].

The unequal distribution of CCS readiness also introduces ethical concerns related to global climate justice. Developed countries have historically contributed the majority of global greenhouse gas emissions and now possess the wealth and technology to decarbonize rapidly using CCS and other advanced solutions. Meanwhile, developing nations, many of which have contributed far less to historical emissions, face increasing pressure to adopt expensive mitigation tools without having the same financial capacity or institutional readiness. This raises a key concern: CCS, if not equitably supported, may become another driver of environmental injustice, where wealthy regions advance toward net-zero while poorer nations are left behind or become dependent on foreign technologies and external funding [11].

8.2 Infrastructure Gaps in Developing Countries

A major barrier to the deployment of carbon capture technology in developing countries is the lack of essential infrastructure for safely transporting and storing CO_2. In many of these regions, the focus has traditionally been on building infrastructure for basic needs, such as electricity, water supply, and fuel distribution, rather than on advanced climate technologies. Consequently, the infrastructure necessary for CCS is often either inadequate or severely lacking [12].

Figure 8.2 provides a clear visual comparison of the difference in CCS readiness between developed and developing countries. On the left side, the photograph shows a modern, fully operational CCS facility in a developed country. The infrastructure is highly advanced, with organized metallic pipelines, pressure vessels, and industrial equipment. The facility is professionally secured with fencing and a clear identification sign, reflecting strong engineering capability, regulatory management, and established investment in carbon reduction technologies. In contrast, the right side of the image shows a typical industrial area in a developing country. There are factories and power lines, but no visible systems for capturing or storing CO_2. The area looks less advanced, with older buildings and unplanned open land. This shows that even though emissions are rising in many developing countries, the infrastructure needed for CCS is still missing due to financial, technical, and policy challenges. Overall, the picture effectively illustrates the global inequality in climate technology adoption. Developed countries are already implementing CCS at an industrial scale, while developing countries remain far behind, still lacking the essential infrastructure to begin large-scale deployment.

One of the most significant challenges is the lack of dedicated CO_2 pipeline networks in developing countries. In contrast, many developed countries, such as the United States and parts of Europe, already have CO_2 pipelines that have been in place for decades, initially built for enhanced oil recovery (EOR). These existing networks facilitate and reduce the cost for industries to connect carbon capture sites to storage locations. However, in developing countries, industrial facilities are often located far from suitable geological storage sites. This distance means that transporting compressed CO_2 over

Figure 8.2 A visual comparison of CCS readiness—a modern carbon capture facility in a developed country (left) vs. an industrial site in a developing region without CCS infrastructure (right). The picture illustrates the growing global gap in climate technology deployment between poor and rich countries.

long distances would require entirely new pipeline systems, representing a costly and complex investment [13].

Geological storage readiness is another challenge. Many developing countries have not yet completed detailed subsurface studies to identify safe, long-term CO_2 storage sites, such as deep Saline aquifers or depleted oil and gas reservoirs. Without accurate geological data for CO_2 storage, governments cannot issue permits for large-scale CCS projects, and investors cannot invest in them. In some countries, existing geological information for carbon storage is out-of-date, unclear, or controlled by companies that are not readily accessible for national climate planning [14].

Additionally, the renewable energy sources and modern technologies, such as the digital infrastructure required to operate CCS systems, are often insufficient in developing countries [15]. Carbon capture technology consumes significant energy, so it needs to be powered by reliable, low-carbon sources of heat and electricity, such as wind and solar. Unfortunately, these sources are not always available in developing countries that experience frequent power shortages or rely heavily on fossil fuels [16]. Using fossil fuels to power carbon capture plants releases substantial amounts of carbon dioxide into the atmosphere, diminishing the positive effects of these facilities [17]. Furthermore, digital systems for continuous monitoring, leak detection, and real-time data tracking remain underdeveloped in many low- and middle-income countries. As a result, the lack of these advanced technologies raises concerns about the safety and environmental monitoring of CCS plants [18].

Many developing countries are still in the early stages of preparing for CCS technology. They lack the essential infrastructure needed to support it. To become ready for CCS, these countries must take several important steps. First, they need to identify and investigate underground locations suitable for the safe storage of carbon dioxide (CO_2), known as geological storage. Next, they should plan and establish networks of pipelines to transport captured CO_2. Additionally, improvements are needed in their digital monitoring systems to safely track the CO_2. Finally, it is crucial to integrate CCS into their national energy strategies and long-term development plans [2].

8.3 Financial Barriers and the Role of Climate Finance

CCS is an expensive technology, and high upfront investment costs are among the most significant barriers to developing countries adopting it. A full-scale CCS project, including capture equipment, compression units, transport pipelines, and long-term storage facilities, can cost billions of dollars [19]. For many developing countries, this level of investment is difficult to justify when there are more urgent national priorities such as building hospitals, improving education, expanding energy access, and supporting economic growth.

In contrast, in many developed countries, governments provide strong financial incentives to encourage industries to invest in CCS technologies. One example is tax credits, which allow companies to reduce the amount of tax they must pay if they capture and store carbon dioxide [20]. For instance, the United States has the 45Q program, which offers a financial reward for every ton of CO_2 that is successfully captured and stored [21].

Another form of support in many developed countries is government subsidies. These are direct financial contributions from the government, such as grants or funding, that help reduce the high initial costs of setting up CCS infrastructure [22]. In addition to this, some countries use carbon pricing systems. These systems put a price on carbon emissions to encourage companies to reduce pollution. This could be through a carbon tax, where companies pay a fee for each ton of CO_2 they emit, or through an emissions trading system (ETS), where companies can buy or sell emission permits depending on how much they pollute.

An ETS is a method where the government sets a limit on how much CO_2 companies are allowed to emit and gives them a certain number of emission permits based on that. If a company emits less than its limit, it can sell its extra permits to other companies and earn money, but if it emits more than allowed, it must buy additional permits to avoid penalties. In this way, companies are financially encouraged to reduce their emissions, because polluting less can help them save or even make money [23].

Figure 8.3 illustrates how carbon trading works between two companies in a carbon market. EMITTER A produces more greenhouse gases than its allowed limit (shown by the horizontal dotted red line labeled "Allocated GHG emission units") and therefore exceeds its emission cap. GHG standards for greenhouse gas. To compensate for this excess pollution, it must purchase carbon credits. On the other hand, EMITTER B emits less than its allocated limit, meaning it has extra unused emission credits. EMITTER B can sell these extra credits to EMITTER A through the carbon market. In this way, the carbon market acts

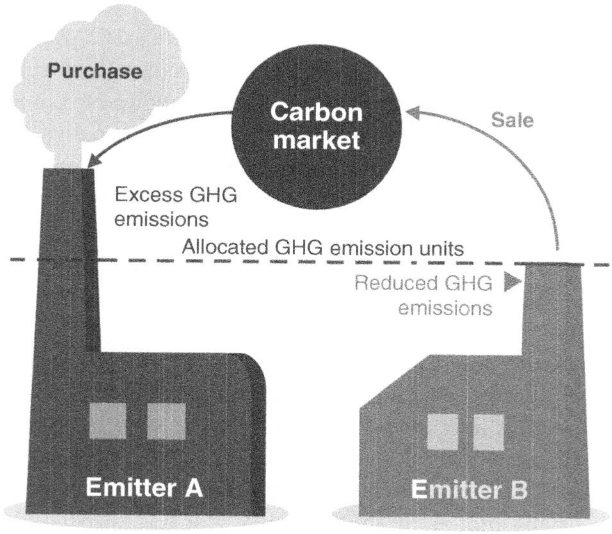

Figure 8.3 Schematic illustration of carbon trading between two emitters. EMITTER A exceeds its allocated GHG emission limit and purchases carbon credits from EMITTER B to compensate for its extra pollution, while EMITTER B emits below its limit and sells its surplus credits through the carbon market.

as a platform where companies with higher emissions pay, while cleaner companies earn money. The main goal of this system is to financially encourage industries to reduce emissions because polluters pay more, while those who reduce emissions are rewarded [23].

In contrast, many developing countries do not have strong carbon pricing policies, tax credit programs, or sufficient government funds to support CCS projects. Without these financial incentives, investing in CCS becomes risky and expensive for industries. As a result, companies in developing countries are often less motivated to adopt CCS technologies [2].

Another major challenge is that international private investors consider CCS projects in developing countries to be very risky. This is mainly because of political instability, uncertain or changing government regulations, and the fact that CCS projects usually take 20–30 years to complete. Investors and global banks prefer projects that give faster and safer returns, so without strong guarantees or government support, they avoid financing CCS in developing countries [24].

Because of this, climate finance from international organizations becomes very important. Institutions such as the Green Climate Fund (GCF), the World Bank, the Asian Development Bank (ADB), and the African Development Bank (AfDB) help by providing concessional loans (low-interest loans), grants (money that does not need to be repaid), or blended finance (a mix of public and private funding). These financial tools help reduce the risk for investors and make CCS projects more attractive and possible in developing countries [25].

Carbon credit markets are another important way to support the financing of CCS projects. Under Article 6 of the Paris Agreement, countries are allowed to cooperate and trade carbon emission reductions. This means that if one country helps another country reduce its emissions, it can legally claim part of those reductions as its own. For example, a developed country with high emissions can invest money or provide technology to build a CCS project in a developing country. The developing country benefits because it gets financial and technical support to reduce its emissions. In return, the developed country is allowed to count some of the captured CO_2 from that project toward its carbon reduction financial credit [26].

In simple terms, a country does not always need to reduce emissions inside its own borders. If it helps another country reduce emissions, it can "buy" or "earn" carbon credits and use them. This creates a global carbon market, where emission reductions can be bought and sold like a commodity. Countries or companies can get carbon credits in two main ways: earning them or buying them. They earn carbon credits when they directly reduce their own emissions or help reduce emissions in another country, for example by investing in a CCS project. On the other hand, they may buy carbon credits if they cannot reduce enough emissions themselves but still need to meet their climate targets. Buying credits means purchasing them from others who have already earned extra credits. In simple terms, earning credits comes from taking action, while buying credits is a way to support another companies' action to meet climate goals. In fact, buying credits means one company or country is supporting or paying money to another party that has successfully reduced CO_2 [27].

Figure 8.4 illustrates a formal bilateral carbon credit agreement under Article 6 of the Paris Agreement between a developed and a developing country [28]. Two professionals are seated at a negotiation desk, each representing their respective countries, indicated by the small national flags placed in front of them: one flag from a developed country (such as the United States) and the other from a developing country. In such agreements, the developed country provides financial support, technology, and technical expertise to help the developing country implement emission reduction projects, such as CCS. The developing

Figure 8.4 Bilateral carbon credit agreement under Article 6 of the Paris Agreement, where a developed country invests in a CCS or emission reduction project in a developing country in exchange for transferable carbon credits.

country hosts and operates the project, while the developed country receives carbon credits based on the amount of CO_2 it reduces or captures. These credits can then be used to meet the developed country's own climate targets in a cost-effective manner. This type of cooperation ensures mutual benefits—developed countries gain carbon credits to meet their climate targets more cost-effectively, while developing countries receive financial support, advanced technology, and technical training to reduce emissions and build low-carbon infrastructure [28].

This approach is especially useful for developing countries that do not have enough money, technology, or expertise to build CCS projects by themselves. Under Article 6 of the Paris Agreement, developed countries are allowed to invest in CCS projects located in developing countries. Based on this international agreement, the two countries signed a bilateral cooperation agreement or carbon credit transfer agreement. Through this agreement, the developed country provides the finance, technology, and training, while the developing country hosts the CCS project. In this system, the developed country signs a cooperation contract with the developing country and agrees to invest money, provide CCS equipment and technical expertise, and train local engineers and workers to operate the project. This investment directly creates local jobs, improves technical capacity, and supports long-term infrastructure development in the developing country. In exchange, the developed country earns the carbon credits from the captured CO_2 and uses them to meet its climate goals [28].

However, this system will only be successful if strong and transparent international rules are in place. These rules are important to prevent double-counting, meaning two countries should not claim the same CO_2 reduction [29]. The rules must also confirm that the captured CO_2 is real, permanent, and safely stored, and ensure that local people and the environment truly benefit from the project.

In summary, finance is not just about having money; it also involves trust, stable long-term policies, and creating a sense of safety for investors against risk. If these conditions are absent, investors are unlikely to support CCS projects [30]. Consequently, CCS is likely to remain in developed countries, where governments and financial systems are stronger and more reliable.

8.4 Policy Readiness and Institutional Capacity

The success of CCS depends a lot on strong government support and well-organized institutions. In this case, institutions mean government organizations, legal systems, and official agencies that make rules, guide national plans, and manage important projects like CCS [31]. Developed countries usually have clear long-term climate strategies, stable regulatory systems, and dedicated agencies responsible for climate and energy transitions. This allows them to create a predictable environment for industries and investors to invest in CCS with confidence.

In many developing countries, the policies required for CCS are still not fully established or clearly defined. There are no appropriate laws governing CO_2 storage, and carbon dioxide is not officially classified as either a "waste" or a "resource." Additionally, there is no single government agency responsible for approving and managing CCS projects [32].

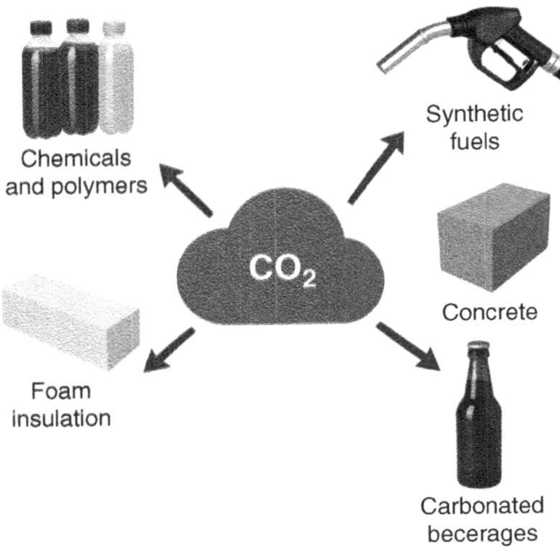

Figure 8.5 CO_2 is increasingly being used as a valuable raw material to produce useful products such as synthetic fuels, construction materials, insulation foams, chemicals, and carbonated beverages, demonstrating its role as a resource rather than a waste.

This lack of clarity results in confusion for industries and investors. In some cases, environmental regulations are outdated and do not even include CCS in their legislation, forcing every project to be reviewed on a case-by-case basis, which leads to long delays.

Figure 8.5 illustrates the growing shift in perspective toward treating CO_2 as a valuable resource rather than just a waste product. Instead of simply being released into the atmosphere, captured CO_2 can now be converted into useful products such as synthetic fuels, building materials like concrete, carbonated beverages, foam insulation, and industrial chemicals and polymers [33]. Foam insulation is a lightweight material used in buildings to reduce heat loss or heat gain, helping improve energy efficiency. It is commonly used inside walls, roofs, and floors to keep interiors warm in winter and cool in summer [34]. This visual emphasizes the concept of a circular carbon economy, where CO_2 is reused as a raw material in manufacturing plenty of useful materials and products. Carbon utilization supports both emission reduction and economic opportunity, especially for countries aiming to industrialize sustainably [35].

Another major challenge is institutional capacity. Institutional capacity refers to how well government ministries, such as energy, environment, finance, industry, and sometimes mining or oil and gas authorities, can coordinate and work together to plan, approve, and manage CCS projects. This makes it clear that the key issue is coordination and effective teamwork across multiple ministries, which is essential for CCS to succeed. In fact, CCS projects need cooperation between many government ministries, such as energy, environment, finance, industry, and sometimes mining or oil and gas authorities [36]. In many developing countries, these ministries do not work well together. This results in overlapping duties or missing responsibilities. Overlapping responsibilities occur when multiple government ministries or agencies are assigned the same task, but without proper coordination. This leads to slow and uncertain decision-making [37].

Without clear coordination or a national CCS strategy, industries and investors feel uncertain and are unwilling to make long-term investments. Many developing countries lack the necessary expertise to effectively manage CCS technologies. Proper management of CO_2 storage sites, addressing long-term risks associated with carbon capture facilities, such as potential leakage of carbon dioxide, and ensuring safety requires advanced technical and legal skills. While developed countries typically have specialized agencies with trained professionals, many developing nations do not have the skilled personnel or training programs needed to handle these complex technologies [38].

Encouragingly, some emerging economies (developing countries) have started taking important steps toward CCS development. Countries such as Indonesia, Malaysia, South Africa, and the United Arab Emirates are beginning to create national CCS strategies, review legal requirements, and build governance systems [7]. Building governance systems means setting up the official structures, rules, laws, and organizations needed to manage and control CCS projects properly [39]. However, overall progress remains slow and varies greatly from one region to another.

In summary, having clear policies, strong institutions, and appropriate regulatory expertise is just as essential as having the technology and finance for CCS projects. If government regulations are unclear, if various political and regulatory agencies fail to collaborate, or if there is a lack of qualified experts to handle regulations, CCS projects cannot advance, even when funding and advanced technology are available [39]. This illustrates that the successful deployment of CCS relies not only on engineering and investment but also on policy stability, good governance, and proficient regulatory management.

8.5 North–South Technology Transfer and Equity

In global discussions, the terms "Global North" and "Global South" are commonly used to describe the divide between developed and developing countries. The Global North generally refers to wealthier, industrialized nations such as those in North America, Western Europe, Japan, South Korea, and Australia. These countries typically have strong economies, advanced technology, and well-established infrastructure. In contrast, the Global South includes regions such as Africa, South Asia, Southeast Asia, Latin America, and parts of the Middle East, where many countries are still working to develop their industrial capacity, infrastructure, and governance systems. While the terms "North" and "South" are not strictly geographical, they highlight the economic and technological gap between the two groups, particularly in areas like climate technology, clean energy, and CCS deployment [40].

Technology transfer plays a vital role in the global expansion of CCS. Today, most advanced CCS technologies, including high-efficiency CO_2 capture units, digital monitoring and control systems, and sophisticated subsurface modeling software (used to simulate how CO_2 moves underground and ensure it remains safely stored without leaking), are developed and patented by companies and research institutions in North America, Europe, and East Asia. As a result, many developing countries do not have these technologies and must rely heavily on foreign expertise and on imported and licensed technologies and facilities. This technological imbalance creates a strong dependency of developing countries on developed nations and makes it difficult for them to independently implement large-scale CCS projects [41].

Technology transfer aims to bridge the gap between developed and developing countries by facilitating access to modern CCS technologies through international collaboration. This process can include training engineers, signing licensing agreements for the use of

Figure 8.6 Practical CCS technology transfer in action, showing experts from developed and developing countries collaborating in a control room during operation inside an industrial facility.

licensed technology, collaborating on joint research projects, and forming partnerships with global companies [41]. However, progress in technology transfer has been slow and limited in practice.

Figure 8.6 illustrates technology transfer and knowledge sharing between experts from developed and developing countries in the context of CCS. It shows an experienced engineer, from a developed country, explaining a CCS process flow diagram (PFD) displayed on a computer screen to two engineers who appear to be from developing regions [42]. The setting is inside a control room of an industrial plant, suggesting real operational training rather than a theoretical classroom setting. This image effectively represents the practical side of North–South cooperation, where advanced CCS knowledge and operational skills are being transferred through direct training, guidance, and mentorship.

One major barrier is intellectual property protection. Many companies from developed countries own key CCS technologies. They are not willing to share them easily unless they receive strong legal guarantees that their technology will not be copied or misused, or unless they gain clear financial benefits [43]. Another challenge is that local industries in developing countries may not yet have sufficient technical skills or experience to use and operate these advanced technologies immediately. As a result, even when technology is available, it may take time to be fully adopted and implemented [2].

For technology transfer to be effective, developing countries need more than just imported equipment. They must also receive knowledge transfer, which involves training local engineers, establishing research and innovation centers, and building national expertise to manage and regulate CCS projects independently [44]. Without this focus on capacity-building of developing countries, these countries risk becoming permanently dependent on foreign technology.

There are growing concerns about fairness and climate justice. Developed countries are responsible for the majority of the world's CO_2 emissions over the past century due to their early industrialization. Because of this historical responsibility, they have a greater obligation to support developing countries in their efforts to decarbonize [45]. This support should go beyond mere promises; it must include fair financial aid, access to modern CCS technologies, and robust capacity-building programs such as training and skill development in local communities. It would be unjust for developed countries to demand that developing nations implement costly CCS solutions without providing substantial and practical assistance to make it feasible [41].

International enterprises such as the UNFCCC Technology Mechanism (which supports developing countries in accessing climate technologies), Mission Innovation (a global collaboration among governments to accelerate clean energy innovation), and Japan's Joint Crediting Mechanism (JCM) (which allows Japan to finance emission reduction projects abroad in exchange for carbon credits) are positive steps toward supporting CCS deployment in developing countries [46].

However, current efforts are still not sufficient to close the global CCS readiness gap. For technology transfer to be truly fair and effective, it must be designed to empower developing countries by building local skills, expertise, and ownership rather than creating new forms of dependency on foreign technology and financing. In summary, fair technology transfer is essential to ensure that CCS does not become a solution limited only to wealthy nations. It must enable developing countries to build long-term technical capacity and independence, while also supporting principles of climate justice and the shared global responsibility to reduce emissions [41].

8.6 Case Studies: Contrasting Regional Approaches Toward CCS

Different regions around the world are approaching CCS technologies in various ways. These methods are influenced by their priorities between economic efficiency and environmental cleanup, the readiness of their infrastructure, and their policy frameworks. While some areas are advancing CCS projects and technologies with rapid implementation, others remain in the initial planning or pilot phases [14].

8.6.1 North America (United States and Canada)

The United States is currently leading the world in building and using CCS projects. One big reason is money, in a way that the U.S. government gives companies financial rewards for capturing carbon dioxide through something called the 45Q tax credit. This basically means companies get paid for every ton of CO_2 they capture and safely store underground, so it becomes profitable to do CCS [21].

The United States has a considerable advantage in CCS because it already has thousands of kilometers of CO_2 pipelines and extensive experience using CO_2 for EOR [47]. EOR is a process in which CO_2 is injected underground to help extract more oil from the reservoir by pressurizing it. As a result, the necessary transportation and injection technologies and infrastructures were in place in the United States before CCS became a key climate strategy.

Canada is also very active in CCS, especially in the province of Alberta, which is similar to Texas in the United States. It has a strong oil and gas industry, skilled engineers, and clear laws and regulations that make it easier for companies to shift and start CCS projects. The government also supports CCS because it reduces emissions from heavy industries like oil, gas, and steel [38]. Overall, both the United States and Canada see CCS as essential for reaching net-zero emissions, especially because they have many industries that are hard to decarbonize, like oil, gas, cement, and chemicals.

8.6.2 Europe (Norway, United Kingdom, and the Netherlands)

The European Union has very strong climate laws through the EU Green Deal. The EU Green Deal is the European Union's main plan to fight climate change and make Europe the first climate-neutral continent by 2050. This means to shift to net-zero greenhouse gas emissions strategy. One of the main considerations is the EU ETS, where companies must pay money for every ton of CO_2 they emit. This makes CO_2 emissions expensive, so capturing and storing CO_2 with CCS becomes financially attractive [48].

Norway is leading with the Northern Lights project, which is one of the world's first open-access CO_2 storage systems. This means any company from any European country can ship its captured CO_2 to Norway for safe storage under the sea, and they don't need to build their own storage site [49].

Figure 8.7 shows the onshore CO_2 receiving and temporary storage terminal of the Northern Lights CCS project located at Øygarden on the west coast of Norway. The facility is seen positioned along the coastline, where incoming CO_2 will arrive by ship, be offloaded in the foreground, and transferred to the intermediate white storage tanks shown to the left. The CO_2 is briefly processed and conditioned in the temporary

Figure 8.7 View of the Northern Lights CCS onshore terminal at Øygarden, Norway, showing the marine offloading sites and intermediate CO_2 storage tanks.

handling area located just behind the jetty. From this point, the CO_2 will be conditioned and transported via pipeline to an offshore injection site in the North Sea for permanent geological storage.

The United Kingdom and the Netherlands are building industrial CCS clusters that are areas where many factories share the same pipelines and offshore storage infrastructure. Instead of each company doing CCS alone, they share the costs and speed up deployment [50].

A key difference is that Europe's primary goal is to reduce industrial emissions, not to increase oil production. The strategy for building CCS differs in the United States, where some projects use CO_2 for EOR [47, 49, 50]. Europe focuses first on climate impacts, especially from hard-to-abate industries such as steel, cement, chemicals, and energy-intensive industries.

8.6.3 Middle East (UAE, Saudi Arabia, and Qatar)

Countries like the United Arab Emirates (UAE) and Saudi Arabia are investing major resources into CCS to reduce emissions from their oil and gas industries, while still remaining important energy suppliers to the world. These countries already have advanced technical knowledge, strong industry infrastructure, and large financial capacity, which allows them to scale CCS very quickly. However, most of their existing CCS projects are related to EOR. This means the captured CO_2 is injected into oil fields' reservoirs to produce more oil, not mainly for long-term carbon dioxide storage for climate benefits. This approach supports their economy, but it does not always guarantee permanent CO_2 storage for climate benefits [51].

Recently, countries like the UAE have started to change their strategy. They now aim to become global hubs for CO_2 storage by capturing and storing carbon dioxide not only from their own industries but also from other countries. This approach is similar to the role that Norway is playing in Europe. This strategy reflects their efforts to position themselves as leaders in climate solutions, moving beyond just being oil producers, as part of their commitments to achieve net-zero emissions [51, 52].

8.6.4 Asia-Pacific (China, Japan, and Indonesia)

China has launched multiple CCS pilot projects, mainly focused on coal power plants and heavy industries like steel and cement. However, wider deployment is still slow due to high costs and uncertainty about long-term CO_2 storage rules. China is technically capable but still building the right economic and policy framework for large-scale CCS [53].

Japan is taking a different approach. Instead of focusing only on domestic CCS, it is forming international partnerships, especially with Southeast Asian countries, through the JCM. Under this scheme, Japan invests in CCS projects abroad and earns carbon credits in return, helping both Japan and developing countries reach their climate goals [54].

Indonesia and Malaysia are seen as future CCS hubs, mainly because they have huge offshore geological storage potential. International energy companies are already exploring storage sites there. However, these countries are still in the early stages of creating clear laws, regulations, and business models for CCS, so large-scale projects are still in preparation [55].

8.6.5 Africa and Latin America

Most countries in Africa and Latin America are still in the feasibility study phase of CCS development, with no major large-scale operational projects currently in place. The primary challenge is limited funding and infrastructure, as CCS requires significant investment in pipelines, storage sites, and regulatory frameworks, which many developing countries in Africa and Latin America do not have yet [56].

South Africa and Brazil are currently the most advanced countries in Africa and Latin America in terms of CCS development. However, both face significant barriers to large-scale deployment. In South Africa, CCS efforts are mainly focused on the coal-based power sector, where emissions are high. Research and feasibility studies are ongoing, but progress remains slow due to economic limitations and continued uncertainty in government policy and long-term regulatory commitment [56, 57].

Brazil, on the other hand, has strong technical potential, particularly for offshore CO_2 storage in depleted oil and gas fields. However, despite this advantage, the country is held back by regulatory complexity, insufficient financial incentives, and governance challenges that prevent CCS from moving beyond the early project stages [57].

Overall, many nations in Africa and Latin America prioritize renewable energy expansion over CCS, as technologies like solar, wind, and hydropower are more affordable, easier to implement, and better aligned with development and energy access goals. As a result, CCS is generally seen as a longer-term option, rather than an immediate climate strategy in these regions.

8.7 Conclusion

CCS is moving forward quickly in developed, wealthy, industrialized countries, mainly because they already have advanced infrastructure, strong government support, and clear climate policies. These countries can offer financial structure, tax credits, and investment security, which makes companies confident to build CCS projects.

On the other hand, many developing countries are moving much more slowly. They often lack the pipelines, storage sites, funding, and technical expertise needed to develop CCS. In addition, regulations are not yet fully developed, and governments may have more urgent priorities such as energy access, poverty reduction, and basic infrastructure. As a result, CCS is not yet a realistic near-term goal for many of them.

This situation could lead to a serious global inequality problem. If CCS becomes a tool used mainly by rich countries, it may allow them to meet climate goals faster, while poorer countries fall behind. That is why international cooperation between developed and developing countries, and technology sharing are critical. Without support, developing nations remain left out of the global net-zero transition goal. To make climate action fair and effective, CCS must be deployed not only in wealthy regions but also in developing economies.

8.8 Key Highlights (Chapter Summary)

◇ Chapter 8 – Key Highlights:

1) CCS deployment is advancing fastest in developed countries with strong infrastructure, government support, and financial structure.
2) Developing countries remain in early stages of CCS technology, mainly due to limited funding, weak infrastructure, and policy gaps.
3) The United States is the global leader in CCS deployment, supported by strong incentives such as the 45Q tax credit and an existing CO_2 pipeline network.
4) Europe leads through strong climate policy, driven by the EU Green Deal and the Emissions Trading System (ETS), focusing mainly on industrial decarbonization rather than using carbon dioxide for oil recovery.
5) The United States and Europe are currently leading global CCS investments and large-scale project implementation.
6) Africa and Latin America are currently behind in CCS deployment, with most projects still in the early planning, feasibility, or pilot stages. These regions do not yet have large-scale operational CCS facilities like those seen in the U.S. or Europe.
7) A major reason for this slow progress in CCS development in developing countries is the lack of financial support and unclear government regulations. CCS is an expensive technology, and many developing countries do not have the budget or policy systems needed to attract private investment from industry.
8) If only rich countries move ahead with CCS technology development while poorer nations are left behind, it could increase global inequality and climate injustice.
9) To reduce global inequality and address climate injustice, technology transfer and climate finance from wealthier nations to developing countries are crucial.
10) International cooperation and fair climate partnerships between nations are necessary to develop CCS technology and plants globally.

Keywords: Carbon Management, Industrial Decarbonization, CO_2 Storage Hubs, Cross-Border CO_2 Transport, Climate Policy Frameworks, Environmental Sustainability, International Climate Agreements, Clean Technology Deployment, Hard-To-Abate Sectors, Carbon Pricing Mechanisms, Geologic Storage, Low-Carbon Economy, Climate Justice, Green Investment, CCS Infrastructure Development, Multinational Collaboration

References

1 M. T. Heil and Q. T. Wodon, "Inequality in CO_2 emissions between poor and rich countries," *The Journal of Environment & Development*, vol. 6, no. 4, pp. 426–452, Dec., 1997. DOI: 10.1177/107049659700600404.

2 R. Shirmohammadi, A. Aslani, and R. Ghasempour, "Challenges of carbon capture technologies deployment in developing countries," *Sustainable Energy Technologies and Assessments*, vol. 42, 100837, Dec., 2020. DOI: 10.1016/j.seta.2020.100837.

3 N. Kulichenko and E. Ereira, *Carbon Capture and Storage in Developing Countries: A Perspective on Barriers to Deployment*. Washington, DC, USA: World Bank Publications, 2012.

4 D. G. Alemneh and S. K. Hastings, "Developing the ICT infrastructure for Africa: Overview of barriers to harnessing the full power of the internet," *Journal of Education for Library and Information Science*, vol. 47, no. 1, pp. 4–16, 2006.

5 A. Watkins, T. Papaioannou, J. Mugwagwa, and D. Kale, "National innovation systems and the intermediary role of industry associations in building institutional capacities for innovation in developing countries: A critical review of the literature," *Research Policy*, vol. 44, no. 8, pp. 1407–1418, Oct., 2015. DOI: 10.1016/j.respol.2015.05.004.

6 D. Allain-Dupré, "Assigning responsibilities across levels of government: Trends, challenges and guidelines for policy-makers," OECD Working Paper, 2018.

7 M. Farooq *et al.*, "Carbon capture for sustainable environment in developing countries," in *Energy and Environmental Security in Developing Countries*, M. Asif, Ed. Cham: Springer, 2021, pp. 525–544, DOI: 10.1007/978-3-030-63654-8_21.

8 M. M. Naser and P. Pearce, "Evolution of the international climate change policy and processes: UNFCCC to Paris Agreement," Oxford Research Encyclopedia of Environmental Science, Nov. 22, 2022. DOI: 10.1093/acrefore/9780199389414.013.422.

9 K. Itaoka, T. Hahimoto, M. Akai, S. Aoyama, C. Shinohara, and H. Hori koshi, "Towards an implementation of carbon dioxide capture and storage: aligning climate change mitigation with principles of justice and fairness," in Proc. *17th Greenhouse Gas Control Technologies Conf. (GHGT-17)*, Oct. 20–24, 2024. SSRN, Jan. 2025.

10 B. K. Sovacool, "Expanding carbon removal to the Global South: Thematic concerns on systems, justice, and climate governance," *Energy and Climate Change*, vol. 4, Art. no. 100103, Dec. 2023. 10.1016/j.egycc.2023.100103.

11 C. Arlota and H. K. de Medeiros Costa, "Climate change, carbon capture and storage (CCS), energy transition, and justice: Where we are now, and where are (should be) we headed?," in *Carbon Capture and Storage in International Energy Policy and Law*, H. K. de Medeiros Costa and C. Arlota, Eds. Elsevier, 2021, pp. 385–393, DOI: 10.1016/B978-0-323-85250-0.00019-0.

12 M. Y. Ibrahim, Z. Ghazali, and H. U. Rahman, "The feasibility of carbon capturing, storage and utilization projects in developing countries: A case of Malaysia," *International Journal of Economics and Financial Issues*, vol. 6, no. 3, pp. 6–11, May., 2016.

13 V. E. Onyebuchi, A. Kolios, D. P. Hanak, C. Biliyok, and V. Manovic, "A systematic review of key challenges of CO_2 transport via pipelines," *Renewable and Sustainable Energy Reviews*, vol. 81, pp. 2563–2583, Jan., 2018. DOI: 10.1016/j.rser.2017.06.064.

14 C. P. Consoli, I. Havercroft, and L. Irlam, "Carbon capture and storage readiness index: Comparative review of global progress towards wide-scale deployment," *Energy Procedia*, vol. 114, pp. 7348–7355, Jul., 2017. DOI: 10.1016/j.egypro.2017.03.1585.

15 C. J. Nosike, "Digitalization in developing countries: opportunities and challenges," *Nigerian Journal of Arts and Humanities*, vol. 4, no. 1, pp. 2955–0343, 2024.

16 S. Tabrizian, "Technological innovation to achieve sustainable development—renewable energy technologies diffusion in developing countries," *Sustainable Development*, vol. 27, no. 3, pp. 537–544, May/Jun., 2019. DOI: 10.1002/sd.1918.

17 A. R. G. Heesterman, "Renewable energy supply and carbon capture: Capturing all the carbon dioxide at zero cost," *Clean Technologies and Environmental Policy*, vol. 21, pp. 1177–1191, 2019. DOI: 10.1007/s10098-019-01716-x.

18 A. Chawla, Y. Arellano, M. V. Johansson, H. Darvishi, K. Shuneen, and M. Vitali, "IoT-based monitoring in carbon capture and storage systems," *IEEE Internet of Things Magazine*, vol. 5, no. 4, pp. 106–111, 2023. DOI: 10.1109/IOTM.2023.10012404.

19 S. P. Filippov, "The economics of carbon dioxide capture and storage technologies (review)," *Thermal Engineering*, vol. 69, pp. 738–750, 2022. DOI: 10.1134/S0040601522100020.

20 M. Mohsin, F. Taghizadeh-Hesary, N. Panthamit, S. Anwar, Q. Abbas, and X. V. Vo, "Developing low carbon finance index: Evidence from developed and developing economies," *Finance Research Letters*, vol. 43, 101520, Nov., 2021. DOI: 10.1016/j.frl.2020.101520.

21 R. A. Esposito, V. A. Kuuskraa, C. G. Rossman, and M. M. Corser, "Reconsidering CCS in the US fossil-fuel fired electricity industry under section 45Q tax credits," *Greenhouse Gases: Science and Technology*, vol. 9, no. 6, pp. 1288–1301, Dec., 2019. DOI: 10.1002/ghg.1925.

22 J. Sekera and A. Lichtenberger, "Assessing carbon capture: Public policy, science, and societal need," *Biophysical Economics and Sustainability*, vol. 5, no. 3, p. 14, 2020. DOI: 10.1007/s41247-020-00080-5.

23 J. K. Boyce, "Carbon pricing: Effectiveness and equity," *Ecological Economics*, vol. 150, pp. 52–61, Aug., 2018. DOI: 10.1016/j.ecolecon.2018.03.030.

24 N. Ameli, O. Dessens, M. Winning, *et al.*, "Higher cost of finance exacerbates a climate investment trap in developing economies," *Nature Communications*, vol. 12, no. 1, p. 4046, 2021. DOI: 10.1038/s41467-021-24305-3.

25 B. Reinsberg, I. Shishlov, K. Michaelowa, and A. Michaelowa, "Climate change-related trust funds at the multilateral development banks. Zurich open repository and archive. 2020.

26 A. Michaelowa, M. Honegger, M. Poralla, M. Winkler, S. Dalfiume, and A. Nayak, "International carbon markets for carbon dioxide removal," *PLOS Climate*, vol. 2, no. 5, Art. no. e0000118, May. 2023. 10.1371/journal.pclm.0000118.

27 M. Verma, "Navigating the world of carbon credits: Strategies for emissions reduction and market participation," *International Journal of Trend in Scientific Research and Development (IJTSRD)*, vol. 7, no. 5, pp. 259–264, Sep.–Oct., 2023.

28 G. Cook, P. Zakkour, S. Neades, and T. Dixon, "CCS under Article 6 of the Paris Agreement," *International Journal of Greenhouse Gas Control*, vol. 134, 104110, May., 2024. DOI: 10.1016/j.ijggc.2024.104110.

29 L. Schneider *et al.*, "Double counting and the Paris Agreement rulebook: Poor emissions accounting could undermine carbon markets," *Science*, vol. 366, no. 6462, pp. 180–183, Oct., 2019. DOI: 10.1126/science.aay8750.

30 J. Byrd and E. S. Cooperman, "Investors and stranded asset risk: Evidence from shareholder responses to carbon capture and sequestration (CCS) events," *Journal of Sustainable Finance & Investment*, vol. 8, no. 2, pp. 185–202, 2018. DOI: 10.1080/20430795.2017.1418063.

31 S. van Egmond and M. P. Hekkert, "Analysis of a prominent carbon storage project failure—the role of the national government as initiator and decision maker in the barendrecht case," *International Journal of Greenhouse Gas Control*, vol. 34, pp. 1–11, Mar., 2015. DOI: 10.1016/j.ijggc.2014.12.014.

32 P. Fragkos, "Assessing the role of carbon capture and storage in mitigation pathways of developing economies," *Energies*, vol. 14, Mar., 2021. DOI: 10.3390/en14071879.

33 S. M. Alhamadi, K. Al-Garadi and A. A. Alghamdi, "CO_2 utilization for environmental innovation: From waste stream to sustainable solutions," Presented at the SPE Energy Transition Symposium, Houston, Texas, USA, Aug. 2024, Paper No. SPE-221423-MS, DOI: 10.2118/221423-MS

34 C. Forest, P. Chaumont, P. Cassagnau, B. Swoboda, and P. Sonntag, "Polymer nano-foams for insulating applications prepared from CO_2 foaming," *Progress in Polymer Science*, vol. 41, pp. 122–145, Feb., 2015. DOI: 10.1016/j.progpolymsci.2014.07.001.

35 P. Tcvettkov, A. Cherepovitsyn, and S. Fedoseev, "The changing role of CO_2 in the transition to a circular economy: Review of carbon sequestration projects," *Sustainability*, vol. 11, no. 20, p. 5834, Oct., 2019. DOI: 10.3390/su11205834.

36 M. J. Gidden *et al.*, "Fairness and feasibility in deep mitigation pathways with novel carbon dioxide removal considering institutional capacity to mitigate," *Environmental Research Letters*, vol. 18, no. 7, 074006, Jun., 2023. DOI: 10.1088/1748-9326/acd8d5.

37 B. Reinsberg and O. Westerwinter, "Institutional overlap in global governance and the design of intergovernmental organizations," *Review of International Organizations*, vol. 18, pp. 693–724, 2023. DOI: 10.1007/s11558-023-09488-2.

38 S. A. Usas and L. Ricardez-Sandoval, "An optimal sustainable planning strategy for national carbon capture deployment: A review on the state of CO_2 capture in Canada," *The Canadian Journal of Chemical Engineering*, vol. 102, no. 7, pp. 2332–2351, Jul., 2024. DOI: 10.1002/cjce.25249.

39 J. Van der Heijden, "The new governance for low-carbon buildings: Mapping, exploring, interrogating," *Building Research and Information*, vol. 44, no. 5–6, pp. 575–584, 2016. DOI: 10.1080/09613218.2016.1159394.

40 H. Confaria, M. M. Godinho, and L. Wang, "Determinants of citation impact: A comparative analysis of the Global South versus the Global North," *Research Policy*, vol. 46, no. 1, pp. 265–279, Feb., 2017. DOI: 10.1016/j.respol.2016.11.004.

41 J. Kirchherr and F. Urban, "Technology transfer and cooperation for low carbon energy technology: Analysing 30 years of scholarship and proposing a research agenda," *Energy Policy*, vol. 119, pp. 600–609, Aug., 2018. DOI: 10.1016/j.enpol.2018.05.001.

42 M. Wang, A. S. Joel, C. Ramshaw, D. Eimer, and N. M. Musa, "Process intensification for post-combustion CO_2 capture with chemical absorption: A critical review," *Applied Energy*, vol. 158, pp. 275–291, Nov., 2015. DOI: 10.1016/j.apenergy.2015.08.083.

43 M. Rimmer, "The Paris agreement: Intellectual property, technology transfer, and climate change," in *Intellectual Property and Clean Energy*, M. Rimmer, Ed. Singapore: Springer, 2018, p. 22, DOI: 10.1007/978-981-13-2155-9_2.

44 A. M. Ragossnig and G. Vujić, "Challenges in technology transfer from developed to developing countries," *Waste Management & Research*, vol. 33, pp. 93–95, Feb., 2015. DOI: 10.1177/0734242X15569403.

45 F. Sultana, “Critical climate justice,” *The Geographical Journal*, vol. 188, no. 1, pp. 118–124, Mar., 2022. DOI: 10.1111/geoj.12417.

46 J. Lipponen, S. McCulloch, S. Keeling, T. Stanley, N. Berghout, and T. Berly, “The politics of large-scale CCS deployment,” *Energy Procedia*, vol. 114, pp. 7581–7595, Jul., 2017. DOI: 10.1016/j.egypro.2017.03.1890.

47 R. W. J. Edwards and M. A. Celia, “Infrastructure to enable deployment of carbon capture, utilization, and storage in the United States,” *Proceedings of the National Academy of Sciences of the United States of America (PNAS)*, vol. 115, no. 38, pp. E8815–E8824, Sep., 2018. DOI: 10.1073/pnas.1806504115.

48 R. C. Pietzcker, S. Osorio, and R. Rodrigues, “Tightening EU ETS targets in line with the European green deal: Impacts on the decarbonization of the EU power sector,” *Applied Energy*, vol. 293, 116914, Jul., 2021. DOI: 10.1016/j.apenergy.2021.116914.

49 A.-K. Furre, R. Meneguolo, P. Ringrose, and S. Kassold, “Building confidence in CCS: From sleipner to the Northern lights project,” *First Break*, vol. 37, no. 7, pp. 81–87, Jul., 2019. DOI: 10.3997/1365-2397.n0038.

50 M. Covarrubias, I. Kustova, M. Elkerbout, L. Hatton and C. Henderson, "Advanced carbon capture for steel industries integrated in CCUS clusters: Report on comparative analysis of industrial CCUS policy instrumentation (Task 5.3.2)," D5.3, pp. 1–79, 2023

51 J. Dargin, “The pathway to a green Gulf: A review and analysis of the evolution of Saudi Arabia, Qatar, and the United Arab Emirates’ climate change positions,” *Carbon and Climate Law Review*, vol. 15, pp. 313–341, 2021.

52 N. Darraj, M. Mahmoud Abdelrahman, and M. Y. Alkih, “Regional review for large-scale deployment of carbon capture and storage (CCS) in the Middle East,” Presented at the Mediterranean Offshore Conference, Alexandria, Egypt, Oct. 2024, Paper No. SPE-223358-MS, DOI: 10.2118/223358-MS

53 J. Yao, H. Han, Y. Yang, Y. Song, and G. Li, “A review of recent progress of carbon capture, utilization, and storage (CCUS) in China,” *Applied Sciences*, vol. 13, no. 2, 1169, 2023. DOI: 10.3390/app13021169.

54 N. A. Pambudi, A. Chapman, A. Sarifudin, and I. W. Kuncoro, “The current status of carbon capture and storage development in Japan: Potency, policy, demonstration projects, implication, and scenario model in emission reduction,” *Energy Sources, Part B:, Planning, and Policy*, vol. 19, no. 1, 2308839, 2024. DOI: 10.1080/15567249.2024.2308839.

55 N. H. Muhd Nor et al., “Carbon sequestration and carbon capture and storage (CCS) in Southeast Asia,” in *Journal of Physics: Conference Series*, vol. 725, 2016 International Congress on Theoretical and Applied Mathematics, Physics and Chemistry, Bandung, Indonesia, Apr. 2016. 012010, DOI: 10.1088/1742-6596/725/1/012010.

56 Y.-C. Ko, K. Zigan, and Y.-L. Liu, “Carbon capture and storage in South Africa: A technological innovation system with a political economy focus,” *Technological Forecasting and Social Change*, vol. 166, 120633, May., 2021. DOI: 10.1016/j.techfore.2021.120633.

57 M. R. Pelissari, S. G. Relva, and D. Peyerl, “Possibilities for carbon capture, utilization, and storage in Brazil,” in *Energy Transition in Brazil*, D. Peyerl, S. Relva, and V. Da Silva, Eds. Cham: Springer, 2023, pp. 141–157, DOI: 10.1007/978-3-031-21033-4_9.

9

Policy Mechanisms and Regulatory Frameworks

9.1 Overview of Carbon Policy Mechanisms

Carbon policy mechanisms are government tools designed to guide and influence how industries reduce their greenhouse gas emissions [1]. These tools are especially important because technologies like carbon capture and storage (CCS) are still very expensive and not immediately profitable without government support [2]. The main purpose of carbon policy is to make polluting more costly and make clean solutions more economically attractive. In other words, governments use these policies to push industries away from emitting CO_2 and pull them toward cleaner options through financial pressure or rewards [3].

There are several main types of policy approaches that governments use to support CCS and reduce emissions.

1) **Market-based Mechanisms:** These include tools such as carbon taxes and emissions trading systems (ETS). They put a price on every ton of CO_2 emitted, making pollution financially costly. In the ETS system, companies that emit less than their allocated carbon allowance can sell their unused emissions permits to others. This creates a financial encouragement to cut emissions or invest in cleaner technologies like CCS, rather than paying the cost of emitting more CO_2 [4].
2) **Regulatory Mechanisms:** These are strict rules made by the government that industries must legally follow. Examples include emission performance standards, which set a maximum limit on how much CO_2 a factory or power plant is allowed to release. In many cases, industries are given flexibility, meaning they can choose any low-carbon technology, such as CCS, energy efficiency improvements, or switching to hydrogen fuel, to achieve that target. However, in some situations, the government uses technology requirements, where a specific solution is mandatory. These rules do not give companies a choice to ignore them; they must meet the minimum environmental standard. The main idea is that pollution is no longer optional, and industries must comply or face penalties [5].
3) **Financial Support Instruments:** These include subsidies (which help lower the cost of CCS equipment or operation), tax credits (which reduce the amount of tax a company must pay when it captures and stores CO_2), grants (nonrepayable financial support from the government), and direct public funding (where the government pays part

of the project cost directly). The main purpose of these financial tools is to lower the cost and reduce the financial risk for companies. Since CCS is very expensive and not always profitable in the beginning, many companies would not invest without government support. By offering financial help, governments make CCS more affordable and attractive, which encourages industries to start building and deploying the technology sooner [6, 7].

In addition to national policies, international agreements such as the Paris Agreement play an important role by guiding global cooperation on CCS development [8]. Under the Paris Agreement, every country must submit a national climate plan called a Nationally Determined Contribution (NDC). In these plans, countries explain how they will reduce their CO_2 emissions [9]. Many governments do not rely on just one policy tool; instead, they combine several mechanisms, such as carbon pricing, regulations, and financial support [4, 6, 7]. This approach, known as a policy mix, is often more effective because it helps address multiple barriers to CCS development simultaneously [10]. Overall, carbon policy mechanisms are essential for scaling up CCS. They help create a carbon market, attract industry investment, and reduce the long-term financial risks of adopting new technologies.

Figure 9.1 shows the main categories of carbon policy mechanisms used by governments to accelerate industrial decarbonization. These include market-based instruments that put a price on CO_2 emissions, regulatory measures that set mandatory emission limits, financial incentives that lower the cost of clean technologies such as CCS, and international agreements that coordinate global climate action.

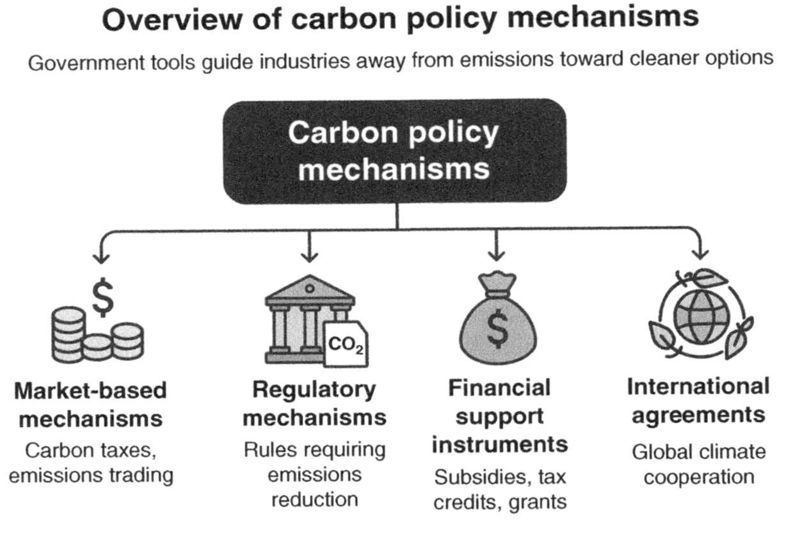

Figure 9.1 Key categories of carbon policy mechanisms used to accelerate industrial decarbonization and support technologies such as CCS.

9.2 Carbon Pricing: Taxes and Emissions Trading Systems

Carbon pricing is a policy tool that puts a financial cost on CO_2 emissions. The goal is to make pollution more expensive so that companies have a strong reason to reduce emissions or use cleaner technologies, such as CCS. There are two main types of carbon pricing systems used around the world. Today, more than 70 countries use some form of carbon pricing [11]. There are two main types of carbon pricing used globally: carbon taxes and ETS.

A carbon tax is a policy that establishes a fixed price per ton of carbon dioxide (CO_2) emitted into the atmosphere. This means companies must pay directly based on the level of their pollution. One major advantage of a carbon tax is its simplicity: it is predictable and easy for businesses to manage, as they can calculate their emissions and the corresponding costs. Countries like Sweden, Canada, and South Africa have implemented this approach. However, while a carbon tax increases the cost of pollution, it does not obligate companies to reduce their emissions by a specific amount. As a result, if the tax rate is too low, companies may choose to pay the tax rather than invest in reducing their emissions [12].

An ETS, also known as cap-and-trade, operates differently from a carbon tax. In this system, the government establishes a cap on total emissions and issues emissions permits to companies, which can be bought or traded in a carbon market. Companies that emit less than their allocated limit can sell their surplus allowances for profit, while those that exceed their allowable emissions must purchase additional permits [4]. The EU Emissions Trading System (EU ETS) is the largest and most well-known example of this approach. One of the main advantages of an ETS is that it ensures a specific emissions target is met [13].

Figure 9.2 shows how carbon pricing works as a financial policy to make companies reduce their CO_2 emissions or adopt cleaner technologies such as CCS. The figure illustrates the two main carbon pricing instruments used worldwide: carbon tax, where companies pay a fixed cost per ton of CO_2 emitted, and ETS, where the government sets a cap on

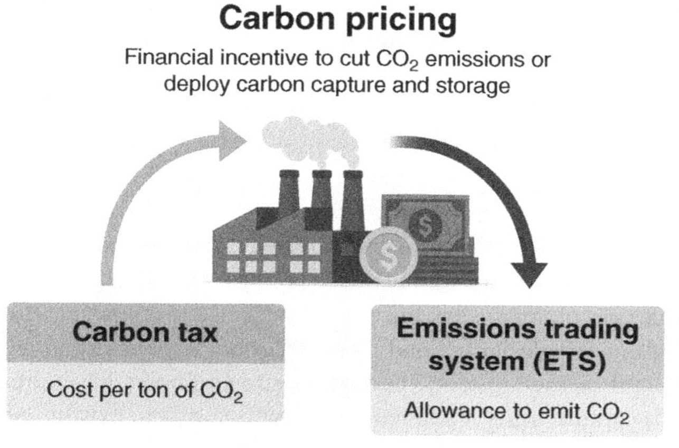

Figure 9.2 Carbon pricing mechanisms—comparison of carbon tax and emissions trading system (ETS) as financial tools to encourage CO_2 reduction.

total emissions and companies must buy or trade allowances to emit CO_2. Both mechanisms create a financial incentive to cut emissions, either by avoiding tax payments or by selling unused emission allowances. Thus, carbon pricing rewards companies that actively reduce their carbon footprint.

Both carbon taxes and ETS reward companies that actively reduce their emissions, including those utilizing CCS technology. In a carbon tax system, deploying CCS reduces companies' taxable emissions, helping them lower their tax burden. In an ETS, captured CO_2 can generate additional carbon credits that can be sold to other companies, providing financial incentives to implement CCS [12, 13].

9.3 Regulatory Mandates and Technology Standards

Regulatory mandates and technology standards are strict government rules that companies are legally required to follow. These rules do not offer financial encouragement. Instead, they force companies to reduce emissions or use cleaner technologies within a specified timeframe [14]. Regulatory mandates differ from carbon pricing in that carbon pricing aims to promote cleaner environmental practices by making pollution costly. In contrast, regulatory policies provide no options for noncompliance with climate goals. It means companies that fail to adhere to these environmental regulations may incur penalties, face legal action, or risk being shut down [11, 14].

These policies are crucial when rapid emission reductions are necessary, particularly in industries that generate the highest levels of pollution, such as steel manufacturing, cement production, and oil refining. These sectors emit significant amounts of CO_2 and often rely on expensive technologies, like CCS, to lower their emissions [15]. In these cases, carbon taxes or ETS may not be sufficient, as companies might choose to pay the tax and continue their polluting practices.

Regulatory mandates, however, leave no option to delay or avoid action. They force industries to comply by law, ensuring that emission cuts happen quickly and within a fixed timeline. This makes regulatory policies more effective and reliable for hard-to-abate industries where voluntary action or market-based incentives might be too slow or insufficient [14]. For example, a government may require that all new coal or gas power plants must include CCS from the start, or that all industries must reach net-zero emissions by 2050.

In some countries, new fossil fuel projects are not even approved unless they can prove they are "CCS-ready," meaning the facility is already designed to easily add CCS in the future [16]. For example, in the European Union, new large power plants must be built so that carbon capture equipment can be added later [17]. This is called being "CCS-ready." It means the plant's layout, space, and piping system are already designed to connect to future carbon capture units without major redesigning and reconstruction. A good example is a modern natural gas power plant in the United Kingdom, where engineers have already reserved space for future CO_2 capture units, included connection points to existing pipelines for easy CCS integration, and planned a dedicated route to transport the captured CO_2 to an offshore storage site via pipeline. By designing the facility this way, the company can easily install CCS in the future when carbon prices or regulations make it necessary.

This approach saves money, avoids redesigning the plant later, and helps make a faster transition to lower emissions [18].

Technology standards are a bit different from regulatory mandates and emission limits. Instead of focusing on how much CO_2 a company is allowed to emit, they focus on what kind of technology must be used. For example, a government may require factories to install modern pollution control equipment or to use technology that captures a minimum percentage of emissions. These rules are about using specific clean technology, not just reaching a certain emission number [19]. Some countries require new industrial facilities to be built with future CCS adaptation capability, even if CCS is not installed immediately.

Regulatory policies or mandates are often used and prioritized because they guarantee real emission reductions. Unlike carbon taxes or trading systems, companies cannot simply pay to continue polluting. They are legally required to upgrade their technology or reduce emissions directly. This makes regulatory mandates highly effective, especially when governments need fast and certain results.

Figure 9.3 illustrates how regulatory mandates and technology standards serve as legally enforced tools that compel industries to reduce their emissions. Both approaches fall under legal requirements, meaning companies cannot ignore them; they are obligated to comply with the law. These regulations may either establish strict emission limits or require industries to utilize specific clean technologies, such as CCS or pollution control equipment. In both scenarios, these rules exert direct pressure on industrial facilities to take action, resulting in mandatory compliance that ultimately leads to reduced CO_2 emissions. The figure emphasizes that this policy pathway is not voluntary and does not rely on financial incentives; rather, it is driven by strict legal and environmental obligations that must be adhered to.

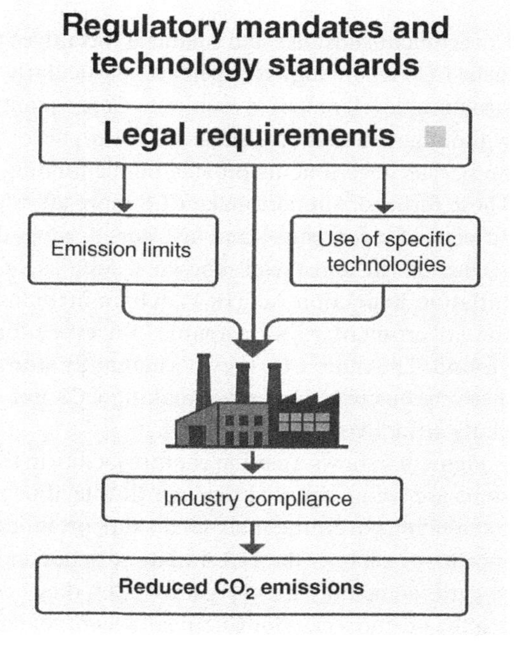

Figure 9.3 Regulatory mandates and technology standards as legally binding climate policies that enforce industrial compliance to achieve CO_2 emission reductions.

However, there are also challenges with regulatory mandates. These rules can be very expensive and difficult to follow for industries, especially in developing countries that have limited financial resources. Companies in developing countries often face limited financial resources due to low profit margins, restricted access to capital and bank loans, and high borrowing costs [14]. For example, if a government forces all cement factories to install CCS technology by a certain year, this can be extremely expensive. A single CCS system for a cement plant can cost hundreds of millions of dollars, excluding energy use and operational costs [20]. Thus, if governments do not provide proper financial support, strict regulations may become a burden. In some cases, industries might even shift their business and operations to countries with weaker environmental rules, which is known as carbon leakage. So, carbon leakage is a situation where climate policy in one country unintentionally pushes pollution to another country, instead of reducing it worldwide [21]. For this reason, many experts argue that regulatory mandates should always be combined with subsidies and financial support [22]. This approach ensures that industries are not only forced to act but also supported in making the transition to cleaner technologies like CCS.

In summary, regulatory mandates and technology standards are powerful policy tools because they legally require industries to cut emissions rather than relying on voluntary action. They are essential for high-emission sectors that are slow or reluctant to act on climate change, helping accelerate the deployment of CCS. When combined with financial incentives and government support, these regulations can drive fast and reliable emission reductions, ensuring that climate action happens on time and with greater certainty.

9.4 Government Subsidies and Incentives

Government subsidies and financial incentives play a crucial role in facilitating CCS projects. CCS technology is expensive, particularly in its early stages when significant infrastructure development is required . Consequently, many industries are reluctant to invest without government assistance to cover part of the costs. To relieve this financial burden, numerous governments provide public funding, tax credits, grants, or low-interest loans. These forms of support make CCS more affordable and less risky, making it much more attractive for companies and investors to adopt this technology [23].

One of the most well-known examples of financial support is the United States Inflation Reduction Act (IRA), which includes the 45Q tax credit. Under this program, the government pays companies for every ton of CO_2 they capture and store underground. The more CO_2 they permanently store, the more money they earn. This creates a strong business incentive, making CCS not only a climate solution but also a financially attractive investment [24].

Figure 9.4 shows a carbon capture facility in the background, where industrial CO_2 emissions are being captured through CCS facilities. In the foreground, a person holds a sign referencing government tax-credit support for carbon reduction, symbolizing the financial incentives such as the U.S. Inflation Reduction Act (IRA) 45Q tax credit. This program rewards companies for every ton of CO_2 they permanently store underground, strengthening the business case for CCS deployment for contributing to climate-mitigation goals.

Figure 9.4 Illustration of the U.S. 45Q incentive mechanism. It shows a carbon capture plant supported by government tax-credit payments for each ton of CO_2 permanently stored underground. The image highlights the link between carbon-emission reduction and financial incentives that make carbon capture and storage (CCS) economically attractive.

In Europe, support for CCS comes mainly through direct public funding and grants. One of the most important programs is the EU Innovation Fund, which provides billions of euros to finance large-scale CCS and other clean energy projects. The money for this fund comes from revenues generated by the EU ETS, meaning it is funded by companies that pay for their CO_2 emissions [25]. Unlike the United States, which uses tax credits, Europe prefers upfront funding, which helps reduce the financial risk for industries at the early stage of project development.

Subsidies are not limited to giving direct money. In some cases, governments also offer guaranteed CO_2 storage contracts, where the government takes responsibility for the long-term safety of the stored CO_2. This means that if CO_2 leaks in the future or if something goes wrong underground, the government, and not the company, will handle the problem. This is important because companies worry about what might happen 20–50 years after the CO_2 is stored. They do not want to be legally or financially responsible forever [26]. By taking over this long-term liability and storage risk, the government reduces fear and uncertainty, making companies much more willing to invest in CCS projects.

In addition, governments may support CCS by offering low-interest loans, research and development funding, or creating public–private partnerships in which both the government and private companies share the costs [27]. These policy tools are essential for building expensive infrastructure, such as CO_2 transport pipelines and underground storage sites, which are too costly for any one company to make alone.

Figure 9.5 shows how carbon trading works under an ETS. There are two factories: the one on the left releases more CO_2 than allowed, so it must buy emission permits to cover its extra pollution. The factory on the right produces less CO_2 than the emission cap, so it has spare permits and can sell them. Money flows from the high-emission factory on the left to the low-emission factory on the right, and in return, the low-emission factory transfers its unused permit. This system encourages companies to reduce pollution in a way that those who pollute less can earn money, while high polluters must pay more.

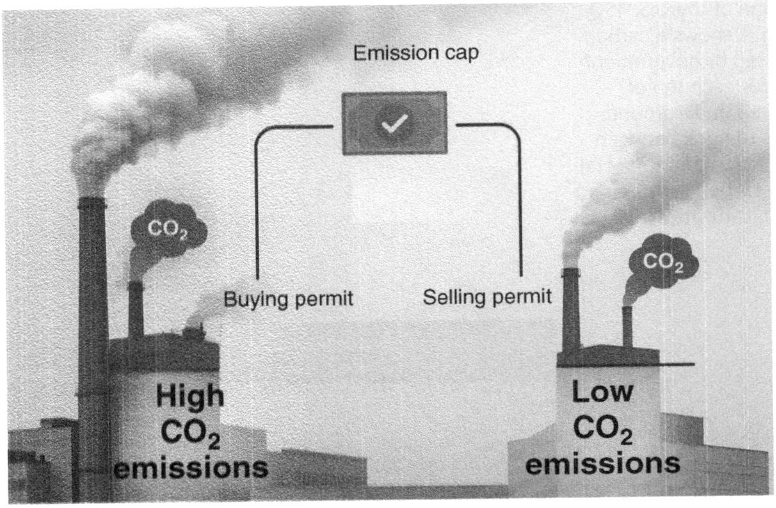

Figure 9.5 Carbon trading: high-emission industry buys permits from low-emission industry under an ETS cap.

These partnerships are essential because no single company wants to take on all the financial risk for such a large project [28]. For example, in Norway's Northern Lights project, the Norwegian government pays for part of the pipeline and storage system, while companies like Equinor, Shell, and TotalEnergies handle the technical development [29]. By sharing the cost and risk, the project becomes much more realistic and attractive for industry.

In developing countries, international financial support is essential. Organizations such as the World Bank, the Asian Development Bank, and the Green Climate Fund (GCF) provide subsidies, grants, and low-interest loans to assist these nations in investing in clean technologies like CCS [30]. Many developing countries do not have enough national funding to finance expensive CCS projects. Therefore, foreign financing and global climate funds are crucial. Without this external support, many developing nations would struggle to participate in international efforts to reduce emissions, even if they are willing to take action [31].

Overall, government subsidies and financial incentives play a crucial role in making CCS projects possible. They reduce financial risk, increase investor confidence, and encourage early adoption of the technology. This support is especially important because CCS is expensive and not always profitable in the early stages. Without government assistance, most CCS projects would struggle to move forward, particularly in regions where carbon prices are low or environmental regulations are not very strict [32].

9.5 International Frameworks and Agreements

International frameworks and agreements are essential for the global implementation of CCS. Climate change is a worldwide issue. Therefore, no single country can effectively address this challenge on its own. CCS projects necessitate high levels of coordination, and mutual support among nations, especially between developed and developing countries. These agreements provide the foundation for global collaboration by establishing clear climate goals, such as those outlined in the Paris Agreement, which encourage countries to include CCS in their long-term decarbonization strategies [33].

The most important global climate agreement is the Paris Agreement (2015), created under the United Nations Framework Convention on Climate Change (UNFCCC). It does not specifically require countries to use CCS, but it strongly encourages all nations to work toward net-zero emissions by around 2050 [34]. Each country must submit its own plan for reducing emissions, called NDCs, and CCS is one of the technologies that countries are allowed to include in these plans.

A key part of the Paris Agreement is Article 6, which creates a system for carbon trading between countries. This means that if one country develops CCS projects and removes or stores carbon, it can earn carbon credits. These credits can then be sold to another country that needs help reaching its climate targets. In this way, countries with strong CCS capabilities can support others, while also reducing global emissions more efficiently [35].

Another major international mechanism is the European Union Emissions Trading System (EU ETS), the largest carbon market in the world. It works by placing a price on CO_2 emissions, which means companies must pay for every tonne of carbon they release. However, if a company captures and stores CO_2 using CCS technology, it can avoid paying or even earn financial rewards. This makes CCS much more economically attractive for industries operating in Europe [36].

The EU ETS is not just about pricing emissions, and it also supports the creation of shared CCS infrastructure across borders. This means CO_2 captured in one European country can legally and safely be transported to another country for permanent storage. A good example is Norway's Northern Lights project, which is designed to receive CO_2 from several European nations and store it beneath the North Sea. This cross-border approach helps countries work together to meet climate goals more efficiently [36].

The United States supports CCS not only through domestic laws, such as the IRA, which offers large tax credits for each tonne of CO_2 captured, but also through international collaboration. The United States is actively involved in global technology partnerships that promote CCS development around the world [37]. For example, the United States participates in Asia-Pacific cooperation programs to help emerging economies adopt CCS technologies [38]. Emerging economies are countries that are not yet fully developed but are growing quickly and becoming more industrialized and modern. They are between poor (developing) countries and rich (developed) countries. Through these partnerships, the United States shares technical expertise, and funding support, to make it easier for developing countries to deploy CCS as part of their climate strategies.

Figure 9.6 shows a conceptual illustration of global cooperation for climate action and carbon emission reduction. The Earth is depicted as being supported by two hands, symbolizing shared international responsibility and partnership in addressing climate change. Keywords such as Paris Agreement, Climate, International Frameworks, Developed, Support, and Storage highlight critical themes related to global climate governance and the role of collaboration in advancing CCS initiatives. The image emphasizes that achieving global emissions reduction targets requires joint effort, mutual support between developed and developing nations, and adherence to international climate agreements.

Figure 9.6 Global cooperation and international climate frameworks supporting carbon capture and climate action under the Paris Agreement.

For many developing countries, building CCS projects is very difficult because of a lack of enough money, technology, or trained experts [39]. That is why international financial institutions like the World Bank, Green Climate Fund (GCF), and Asian Development Bank (ADB) are very important. These organizations give financial support in the form of grants and low-interest loans. They help reduce the high initial cost of CCS projects, which makes it easier for developing governments and industries in developing countries to take action on climate change [40].

In addition to money, these institutions also provide technical support and risk protection. These organizations reduce the financial risk of CCS projects by offering guarantees, low-interest loans, and cost-sharing. If the project fails or loses money, they help cover part of the loss. This gives governments and investors confidence to move forward. They also send experts to help with technical, safety, and regulatory matters. They also support training programs so local engineers and workers can learn to operate CCS systems. By sharing risks and giving confidence to investors, these organizations help developing countries use CCS to reduce their emissions. Without this help, many developing countries would struggle to adopt CCS technology [41].

In summary, international frameworks and agreements play a vital role in making CCS a real global solution. They establish common rules and standards among nations, facilitate cross-border carbon trading, and enable countries to work together to achieve clean air rather than work in isolation. Additionally, these frameworks support shared infrastructure, allowing CO_2

to be transported from one country to another for safe storage [42]. Most importantly, they provide financial and technical support so that CCS is not only available to rich nations.

9.6 Evaluating Policy Effectiveness and Gaps

Many countries have adopted carbon pricing, regulations, and financial incentives to promote CCS. However, the effectiveness of these policies varies by region [43]. In advanced economies such as the United States, Canada, Norway, the United Kingdom, and the European Union, strong and well-designed policy frameworks have facilitated the development of large-scale CCS projects. These regions provide a combination of clear regulations, high carbon prices, generous tax credits, and direct public funding [44]. This mix of support makes CCS financially attractive and less risky for companies. As a result, industries in these regions are more confident to invest in CCS, leading to real project deployment rather than just pilot studies or research plans.

However, in many other countries, especially in the developing world, CCS deployment is still very slow. The main reason is that policy frameworks are weak or inconsistent. In many cases, carbon prices, whether as a tax or a reward, are too low to make CCS financially worthwhile, and regulations are unclear or not yet established. Financial assistances are also not strong enough to cover the high upfront costs of building CCS projects. In some countries, governments have announced long-term climate goals, but they have not provided detailed laws, and strategies, to support CCS in practice. This creates a high level of uncertainty for industries and investors, making them unwilling to commit to expensive projects. As a result, progress remains slow despite growing awareness of the importance of CCS [45].

Another major challenge is the lack of clear rules for long-term liability and monitoring. In many countries, it is unclear who will be responsible for the stored CO_2 after 20 or 30 years. Is this responsibility with the company that operated the project or the government? This legal uncertainty makes companies worried about future risks, so they hesitate to invest.

Figure 9.7 shows how carbon dioxide (CO_2) is stored deep underground in a saline aquifer and monitored from the surface. At the surface, an industrial facility is connected to an injection well that carries captured CO_2 into a deep geological formation. A technician with monitoring equipment makes continuous observation needed to ensure the CO_2 remains securely trapped beneath rock layers. However, in many real-world CCS projects, there is ongoing uncertainty about who is responsible for this long-term monitoring once the CO_2 has been stored. In some cases, private companies manage monitoring for a limited period after injection stops, while in others, the responsibility may transfer to the government. This legal ambiguity, regarding who must perform and finance long-term CO_2 monitoring, creates hesitation among companies to invest in CCS projects. That is because these companies fear potential future costs if leakage or storage issues arise decades later.

Additionally, bureaucratic delays, such as slow permit approvals, complex paperwork, and lengthy approval processes, make it challenging to start CCS projects quickly [46]. Many developing countries also suffer from weak institutional capacity, meaning government agencies may lack sufficient staff, expertise, or coordination to manage large, complex projects like CCS [47]. On top of that, there is often a shortage of skilled workers and engineers who are trained in CCS technologies [48]. These practical challenges significantly slow down CCS development, even when governments are interested in pursuing CCS projects.

Figure 9.7 Cross-sectional view of CO_2 storage in a deep saline aquifer with monitoring from the surface. The picture shows CO_2 injected from an industrial facility into a porous rock layer, while monitoring equipment at the surface tracks storage performance. In many real-world projects, uncertainty remains over which party, the operating company or the government, is responsible for long-term monitoring of stored carbon dioxide after injection ends.

Despite these challenges, global interest in CCS is steadily growing. Increasing numbers of countries are now developing national CCS strategies, updating their regulations, and even exploring regional or cross-border CO_2 storage networks, similar to the projects already seen in Europe. However, for CCS to really rise worldwide, especially beyond advanced economies, stronger policies are needed [49]. Governments must provide clear regulations, stable and strong financial support, and long-term guarantees, particularly for emerging economies that lack sufficient funding and technical expertise. Without this support, global CCS deployment will remain uneven between developed and developing countries and will be too slow to meet climate goals.

In conclusion, while current policies have enabled significant CCS progress in certain advanced countries, major gaps exist worldwide. Many countries, especially in the developing world, lack clear regulations, government financial support, and long-term planning needed to deploy CCS at scale. Closing these gaps will require stronger, more precise policy frameworks, including clearer regulations, financial commitments, and defined rules for long-term monitoring of carbon dioxide storage [50]. In addition, greater international cooperation and support will be essential to ensure that CCS is not limited to wealthy nations but becomes a real global climate solution.

9.7 Key Highlights (Chapter Summary)

◇ **Chapter 9—Key Highlights:**

1) Carbon capture and storage (CCS) requires firm policy and financial support because the technology is expensive and unlikely to become profitable without government support.
2) Governments use different policy tools such as carbon taxes, emissions trading systems (ETS), regulations, and financial incentives to encourage industries to start CCS projects.
3) Carbon pricing, through taxes or trading systems, makes CO_2 emissions more costly financially and rewards companies that capture or reduce emissions.
4) Combining regulatory mandates and technology standards means the government legally requires industries and companies to reduce their emissions to a specific limit, and to use a particular tool or technology, such as CCS or other low-carbon technologies, within a fixed timeline.
5) Government subsidies, tax credits, and funding programs help reduce the financial risks and high initial costs of CCS projects. As a result, carbon capture projects become more attractive to investors and industries.
6) Policies like the U.S. Inflation Reduction Act (IRA) and the EU Innovation Fund are strong examples of how financial incentives can accelerate CCS deployment. The IRA supports CCS mainly through generous tax credits. However, the EU Innovation Fund works alongside the EU Emissions Trading System (ETS).
7) International frameworks, including the Paris Agreement, provide long-term direction for reducing emissions, encourage cooperation between countries, and support carbon trading mechanisms that can help advance CCS deployment.
8) Developing countries rely heavily on international finance from institutions such as the World Bank and Green Climate Fund to implement CCS, as they lack domestic funding.
9) Carbon capture policy effectiveness varies across regions, with advanced countries moving faster while many developing countries still face unclear regulations, lack of skilled workforce, and funding difficulties.
10) Future CCS success will depend on stronger global cooperation, clear and stable policies, technology transfer between developing and developed countries, and inclusive support for developing economies to close the global deployment gap.

Keywords: Carbon Policy, Carbon Pricing, Emissions Trading, Carbon Tax, Regulatory Mandates, Technology Standards, Government Incentives, Tax Credits, International Agreements, Climate Finance, CCS Deployment, Policy Frameworks, Net-Zero Strategies, Public-Private Partnerships, Environmental Regulation

References

1 A. Holland-Cunz and C. Baatz, "How to govern carbon dioxide removal: An assessment framework for policy instruments," *Climate Policy*, pp. 1–16, 2025. DOI: 10.1080/14693062.2025.2459315.

2 Y. M. Yetişkin, F. G. Uçtuğ, and N. Mac Dowell, "The role of policy instruments on deployment of carbon capture, storage and utilization technologies: A case study in Türkiye," *International Journal of Greenhouse Gas Control*, vol. 130 104004, Dec. 2023, 10.1016/j.ijggc.2023.104004, 104004.

3 A. Baranzini, J. C. J. M. van den Bergh, S. Carattini, R. B. Howarth, E. Padilla, and J. Roca, "Carbon pricing in climate policy: Seven reasons, complementary instruments, and political economy considerations," *Wiley Interdisciplinary Reviews: Climate Change*, vol. 8, no. 4, e462, Jul./Aug., 2017. DOI: 10.1002/wcc.462.

4 E. Narassimhan, K. S. Gallagher, S. Koester, and J. R. Alejo, "Carbon pricing in practice: A review of existing emissions trading systems," *Climate Policy*, vol. 18, no. 8, pp. 967–991, 2018. DOI: 10.1080/14693062.2018.1467827.

5 T. Dixon, S. T. McCoy, and I. Havercroft, "Legal and regulatory developments on CCS," *International Journal of Greenhouse Gas Control*, vol. 40, pp. 431–448, Sep., 2015. DOI: 10.1016/j.ijggc.2015.05.024.

6 P. Perreault, N. B. Kummamuru, A. Gonzalez Quiroga, and S. Lenaerts, "CO_2 capture initiatives: Are governments, society, industry and the financial sector ready?" *Current Opinion in Chemical Engineering*, vol. 38 100874, Dec. 2022, 10.1016/j.coche.2022.100874, 100874.

7 A. Ogihara, "Mapping the necessary policy instruments to unlock the potentials of private finance for carbon capture and storage technologies," in *Financing for Low-carbon Energy Transition*, V. Anbumozhi, K. Kalirajan, and F. Kimura, Eds. Singapore: Springer, 2018, pp. 115–132, DOI: 10.1007/978-981-10-8582-6_5.

8 R. S. Haszeldine, S. Flude, G. Johnson, and V. Scott, "Negative emissions technologies and carbon capture and storage to achieve the Paris Agreement commitments," *Philosophical Transactions of the Royal Society A*, vol. 376, no. 2119, Apr., 2018. DOI: 10.1098/rsta.2016.0447.

9 M. Winning *et al.*, "Nationally determined contributions under the Paris Agreement and the costs of delayed action," *Climate Policy*, vol. 19, no. 8, pp. 947–958, 2019. DOI: 10.1080/14693062.2019.1615858.

10 J. Oda and K. Akimoto, "Preliminary study on policy mix effects on economic viability of carbon capture and storage project in Japanese steel industry," *Energy Procedia*, vol. 114, pp. 7615–7624, Jul., 2017. DOI: 10.1016/j.egypro.2017.03.1894.

11 J. K. Boyce, "Carbon pricing: Effectiveness and equity," *Ecological Economics*, vol. 150, pp. 52–61 , Aug., 2018. DOI: 10.1016/j.ecolecon.2018.03.030.

12 E. Haites, "Carbon taxes and greenhouse gas emissions trading systems: What have we learned?," *Climate Policy*, vol. 18, no. 8, pp. 955–966, 2018. DOI: 10.1080/14693062.2018.1492897.

13 J. Teixidó, S. F. Verde, and F. Nicolli, "The impact of the EU emissions trading system on low-carbon technological change: The empirical evidence," *Ecological Economics*, vol. 164 106347, Oct. 2019, 10.1016/j.ecolecon.2019.06.002, 106347.

14 H. G. Al-Thani and R. J. Isaifan, "Policies and regulations for sustainable clean air: An overview," in *Sustainable Strategies for Air Pollution Mitigation*, M. C. Ogwu and S. C. Izah, Eds, The Handbook of Environmental Chemistry, Vol. 133. Cham, Switzerland: Springer, 2024, pp. 1–22, DOI: 10.1007/698_2024_1093.

15 K. Buznitsky, S. Verma, and M. P. Nitzsche, "Decarbonizing industry: Policy approaches to eliminate hard-to-abate emissions," *MIT Science Policy Review*, vol. 5, pp. 58–70, 2024.

16 R. T. Dahowski *et al.*, "The impact of CCS readiness on the evolution of China's electric power sector," *Energy Procedia*, vol. 114, pp. 6631–6637, Jul., 2017. DOI: 10.1016/j.egypro.2017.03.1817.

17 D. Bellotti, A. Sorce, M. Rivarolo, and L. Magistri, "Techno-economic analysis for the integration of a power to fuel system with a CCS coal power plant," *Journal of CO2 Utilization*, vol. 33, pp. 262–272, Oct., 2019. DOI: 10.1016/j.jcou.2019.05.019.

18 S. Babaee and D. H. Loughlin, "Exploring the role of natural gas power plants with carbon capture and storage as a bridge to a low-carbon future," *Clean Technologies and Environmental Policy*, vol. 20, pp. 379–391, 2018. DOI: 10.1007/s10098-017-1479-x.

19 Z. Khanam, F. M. Sultana, and F. Mushtaq, "Environmental pollution control measures and strategies: An overview of recent developments," in *Geospatial Analytics for Environmental Pollution Modeling*, F. Mushtaq, M. Farooq, A. B. Mukherjee, and M. Ghosh, Eds (nee Lala). Cham, Switzerland: Springer, 2023, pp. 327–348, DOI: 10.1007/978-3-031-45300-7_15.

20 F. Kazemifar, "A review of technologies for carbon capture, sequestration, and utilization: Cost, capacity, and technology readiness," *Greenhouse Gases: Science and Technology*, vol. 12, no. 1, pp. 200–230, Feb., 2022. DOI: 10.1002/ghg.2131.

21 A. Gianoli and F. Bravo, "Carbon tax, carbon leakage and the theory of induced innovation in the decarbonisation of industrial processes: The case of the Port of Rotterdam," *Sustainability*, vol. 12, no. 18, 7667, Sep., 2020. DOI: 10.3390/su12187667.

22 R. Golombek, M. Greaker, and S. Kverndokk, and others, "Policies to promote carbon capture and storage technologies," *Environmental and Resource Economics*, vol. 85, pp. 267–302, 2023. DOI: 10.1007/s10640-023-00767-5.

23 A. M. van Valkengoed and E. van der Werff, "Are subsidies for climate action effective? Two case studies in the Netherlands," *Environmental Science & Policy*, vol. 127, pp. 137–145, Jan., 2022. DOI: 10.1016/j.envsci.2021.09.018.

24 E. Grubert and F. Sawyer, "US power sector carbon capture and storage under the inflation reduction act could be costly with limited or negative abatement potential," *Environmental Research: Infrastructure and Sustainability*, vol. 31, Art. no. 015008, Mar. 2023, 10.1088/2634-4505/acbed9.

25 T. Mikunda and L. Rycroft, "Briefing on the EU Innovation Fund and the implications for CCUS projects–first report on the thematic working group on policy, regulation and public perception," CCUS Project Network, policy paper, Jan. 2020.

26 R. J. Heffron *et al.*, "Ownership, risk and the law for a CO_2 transport network for carbon capture and storage in the European Union," *Journal of Energy & Natural Resources Law*, vol. 36, no. 4, pp. 433–462, 2018. DOI: 10.1080/02646811.2018.1442215.

27 A. Zapantis, A. Townsend, and D. Rassool, "Policy priorities to incentivise large scale deployment of CCS," Thought Leadership Report, Global CCS Institute, 2019.

28 P. Larkin *et al.*, "An integrated risk assessment and management framework for carbon capture and storage: A Canadian perspective," *International Journal of Risk Assessment* *and Management*, vol. 22, no. 3–4, pp. 464–508, Oct., 2019. DOI: 10.1504/IJRAM.2019.103336.

29 A. Tsvetkova and A. Middleton, "Carbon capture, transport, and storage projects in Norwegian seabed: Sustainable implications and challenges of new green technologies rooted in the past," in *Carbon Capture, Transport, and Storage Projects in Norwegian Seabed*. London: Routledge, 2023, pp. 223–247.

30 L. Cui, Y. Sun, M. Song, and L. Zhu, "Co-financing in the green climate fund: Lessons from the global environment facility," *Climate Policy*, vol. 20, no. 1, pp. 95–108, 2019. DOI: 10.1080/14693062.2019.1690968.

31 J. B. Smith, T. Dickinson, and J. D. Donahue, "Development and climate change adaptation funding: Coordination and integration," in *International Climate Finance*, J. E. Stiglitz and N. Stern, Eds. Routledge, 2016, pp. 72–89.

32 W. Zhang, C. Dai, X. Luo, and X. Ou, "Policy incentives in carbon capture utilization and storage (CCUS) investment based on real options analysis," *Clean Technologies and Environmental Policy*, vol. 23, pp. 1311–1326, 2021. DOI: 10.1007/s10098-021-02025-y.

33 P. Larkin, W. Leiss, and D. Krewski, "Risk assessment and management frameworks for carbon capture and geological storage: A global perspective," *International Journal of Risk Assessment and Management*, vol. 22, no. 3–4, pp. 254–285, 2019. DOI: 10.1504/IJRAM.2019.103332.

34 F. Bakhtiari, "International cooperative initiatives and the United Nations Framework Convention on Climate Change," *Climate Policy*, vol. 18, no. 5, pp. 655–663, 2018. DOI: 10.1080/14693062.2017.1321522.

35 G. Cook, P. Zakkour, S. Neades, and T. Dixon, "CCS under article 6 of the Paris agreement," *International Journal of Greenhouse Gas Control*, vol. 134, May 2024, Art no. 104110, 10.1016/j.ijggc.2024.104110.

36 A. D. Ellerman, C. Marcantonini, and A. Zaklan, "The European Union Emissions Trading System: Ten years and counting," *Review of Environmental Economics and Policy*, vol. 10, no. 1, pp. 89–107, 2016. DOI: 10.1093/reep/rev014.

37 L. Beck, "Carbon capture and storage in the USA: The role of US innovation leadership in climate-technology commercialization," *Clean Energy*, vol. 4, no. 1, pp. 2–11 , Mar., 2020. DOI: 10.1093/ce/zkz031.

38 Z. Wang, W. He, and K. Chen, "The integrated efficiency of economic development and CO_2 emissions among Asia Pacific Economic Cooperation members," *Journal of Cleaner Production*, vol. 131, pp. 765–772, Sept., 2016. DOI: 10.1016/j.jclepro.2016.04.097.

39 R. Shirmohammadi, A. Aslani, and R. Ghasempour, "Challenges of carbon capture technologies deployment in developing countries," *Sustainable Energy Technologies and Assessments*, vol. 42 , Dec. 2020, Art no. 100837, 10.1016/j.seta.2020.100837.

40 A. Gilbert, "International financial institutions ask to contribute to climate protection," *Finance: Theory and Practice*, vol. 25, no. 4, pp. 6–23, 2021.

41 P. Warren, "The role of climate finance beyond renewables: Demand-side management and carbon capture, usage and storage," *Climate Policy*, vol. 19, no. 7, pp. 861–877, 2019. DOI: 10.1080/14693062.2019.1605330.

42 A. Yadav, "A review: Climate policy and international agreements," in *A Textbook of Climatology*. New Delhi, India: Wisdom Press, 2022, ch. 17, pp. 116–123.

43 J. Hassler, P. Krusell, C. Olovsson, and M. Reiter, "On the effectiveness of climate policies," IIES Working Paper, no. 53, pp. 1–54, 2020.

44 C. Bataille, H. Waisman, M. Colombier, L. Segafredo, J. Williams, and F. Jotzo, "The need for national deep decarbonization pathways for effective climate policy," *Climate Policy*, vol. 16, no. Suppl. 1, pp. S7–S26, 2016. DOI: 10.1080/14693062.2016.1173005.

45 S. Dissanayake, R. Mahadevan, and J. Asafu-Adjaye, "Evaluating the efficiency of carbon emissions policies in a large emitting developing country," *Energy Policy*, vol. 136, Jan. 2020, Art no. 111080, 10.1016/j.enpol.2019.111080.

46 S. T. Anderson, "Risk, liability, and economic issues with long-term CO_2 storage—a review," *Natural Resources Research*, vol. 26, pp. 89–112, 2017. DOI: 10.1007/s11053-016-9303-6.

47 A. A. Adenle, D. T. Manning, and J. Arbiol, "Mitigating climate change in Africa: Barriers to financing low-carbon development," *World Development*, vol. 100, pp. 123–132, Dec., 2017. DOI: 10.1016/j.worlddev.2017.07.033.

48 H. Corbett, C. Calvillo, A. Katris, P. Gonzalez-Martinez, and E. Lucas, Policy brief: Providing certainty and acting with urgency—addressing skills and worker shortages for industrial decarbonization, University of Strathclyde, Glasgow, U.K., Centre for Energy Policy, 2023.

49 B. Page *et al.*, The global status of CCS 2020: Vital to achieve net zero. Global CCS Institute, Dec. 2020.

50 L. B. Baptista *et al.*, "Good practice policies to bridge the emissions gap in key countries," *Global Environmental Change*, vol. 73, p. 102472, Mar., 2022. DOI: 10.1016/j.gloenvcha.2022.102472.

10

Community Engagement and Ethical Dimensions of Carbon Projects

10.1 Ethics of Large-Scale Environmental Technologies

Large environmental technologies like carbon capture and storage (CCS) affect not only the planet but also people and communities. These technologies reduce emissions and help fight climate change, which benefits everyone around the world by protecting the environment and future generations [1]. These technologies help reduce emissions and combat climate change, benefiting everyone worldwide by protecting the environment for current and future generations. However, some negative impacts are often felt in the locations where these projects are implemented [2].

Communities may face various risks, including changes to land and water use, the construction of new industrial facilities near their homes, or restrictions on land use due to CCS operations [3]. Additionally, residents may experience uncertainty regarding safety concerns, the long-term storage of CO_2, and how the project may affect their health, livelihoods, and overall environment [4].

Because of this, it is important to make sure CCS projects are fair and transparent. Communities should receive clear and honest information about the project, its risks, and its benefits. They should have real opportunities to participate in discussions and decision-making, not just be informed after decisions are made. Local people must also be protected through strong safety standards, long-term CO_2 monitoring after storage, and guaranteed rights so they are not left to carry unfair risks. Guaranteed rights in this context mean basic protections that communities must receive by law or policy when a CCS project is planned near them [5]. In this way, the benefits of climate action are shared fairly, not only by the world but also by the people who live near these projects.

Ethical thinking in CCS starts with a basic rule: **do no harm**. CCS projects are meant to help the planet, so they should not create new environmental or social problems while trying to fight climate change [6]. This means, for example, carbon dioxide transport pipelines and storage sites must be safely designed and carefully monitored to prevent leaks or long-term damage [7, 8]. Communities should also receive clear information about safety plans and emergency protections so they feel secure and included.

Another essential principle is **fairness and justice**. The benefits and risks of CCS should be shared fairly. It is unethical if some communities bear the risks, such as land-use changes, noise, or safety concerns, while other regions, companies, or countries reap economic or climate benefits. Fairness means that local people should have a voice, receive benefits, and not be left with problems while others profit [9].

CCS projects must also consider **future generations**, not only people today. Carbon stored underground can remain there for hundreds of years, so long-term monitoring and engineering are necessary [10]. This means someone must check the site, maintain the equipment, and ensure it remains safe long after the project is finished. This raises important questions: Who will be responsible for the site in the future? Who will make sure the storage stays safe after the company or project team leaves?

These responsibilities are **ethical**, not only technical, meaning that they are about doing what is right, fair, and responsible, not just following engineering rules. A really responsible CCS project plans for long-term safety, establishes clear rules for who will look after the site in the future, and ensures the community is protected even after the project ends. In this context, community means the people and groups who live in, use, or are affected by the area around a CCS project. In addition, building a CCS plant must guarantee that future generations are kept safe just as much as the people living near the site today [6, 10].

Finally, large environmental technologies must be **open and inclusive**. This means the public should clearly understand what the project is, how it works, and what risks and benefits it brings. People who live near CCS sites should have a real voice, not just be informed after decisions are made. Good communication, honest information, and opportunities to participate help build trust and avoid conflict in CCS projects [11].

Figure 10.1 illustrates the social dimension of CCS projects, where local communities, stakeholders, and industry representatives engage in dialogue regarding potential benefits,

Figure 10.1 Community members gathering near a carbon capture and storage (CCS) plant to discuss the social impacts and local concerns associated with the project.

concerns, and long-term impacts of a CCS project. Such interactions are essential for transparency, trust-building, and obtaining social acceptance to operate the facilities.

In short, CCS must do more than work technically. It must also be socially responsible. Projects should respect local people's values, treat people fairly, and show that protecting communities is just as important as reducing emissions. When communities are informed, engaged, and respected, CCS projects are more likely to succeed and support a fair and sustainable future.

10.2 Procedural Justice and Inclusive Governance

Procedural justice involves making decisions in a fair, transparent, and respectful manner. In the context of carbon capture projects, this means that local communities must have a real opportunity to participate in decisions that impact their land, health, and future. Rather than simply being informed about what will happen, community members should be invited to meetings, actively engaged in discussions, and given plenty of time to ask questions and express their concerns [12].

Inclusive governance ensures that everyone, especially local residents and vulnerable groups, can engage in the planning and approval processes of the project, not just government officials or company representatives. When people are included, and their voices are heard, the resulting decisions are more equitable, trustworthy, and successful [13].

In the past, many large infrastructure projects were planned and built without listening to the people who lived nearby. This often created mistrust, strong public opposition, and even social conflict. To avoid repeating these mistakes, CCS projects must involve communities from the beginning, and not after decisions are already made [14].

Early engagement of people in CCS projects, providing clear information, answering their questions, and allowing them to share their opinions before plans are finalized, is essential. Public meetings, information sessions, and ongoing conversations help communities express concerns and understand the risks and benefits of the projects [15]. When people feel heard and respected, projects are more likely to be accepted and successful.

Inclusive governance involves ensuring that everyone has the opportunity to express their views and be heard, not just powerful or well-connected groups. This includes Indigenous communities, rural residents, women, young people, and low-income households. Often, these groups are excluded or lack sufficient information to participate meaningfully [13].

To ensure fair participation, information must be communicated in a way that is easy to understand. This involves using clear language, offering assistance to explain technical issues, and providing materials in multiple languages when needed, so that everyone is included [16]. When communities understand the project and feel comfortable asking questions, they are more capable of sharing their opinions and interests.

True procedural justice goes beyond simply checking off boxes or holding a single meeting with the public. It involves establishing a continuous partnership between project leaders and the community. This means actively listening to people's concerns, valuing their feedback, addressing their questions, and making adjustments to plans when necessary. When communities feel respected and truly involved, they are more likely to trust the process. This trust helps reduce conflict, improve cooperation, and increase the chances that carbon projects will be accepted and successful over the long term [13, 16].

Figure 10.2 illustrates the core principles required to achieve just and inclusive governance in CCS projects. The central circle represents the goal of "Just & Inclusive CCS Governance," while the surrounding circles highlight key elements that support this outcome. These include transparency, access to information, accountability, and recognition of local voices, which together ensure communities are fully informed and able to contribute meaningfully. Additional principles such as participation, early engagement, community trust, and equity and fairness emphasize the need for continuous dialogue, respect for diverse perspectives, and fair treatment of all stakeholders [17]. By integrating these interconnected components, CCS projects can build public confidence and trust, reduce conflict, and promote long-term social acceptance.

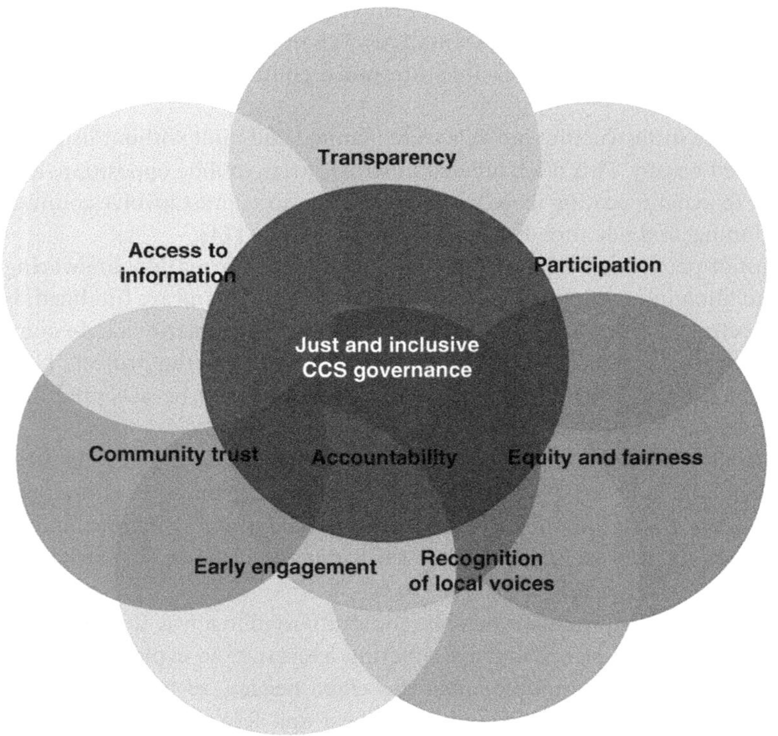

Figure 10.2 Key elements of just and inclusive governance in CCS projects demonstrate that transparency, participation, accountability, equity, trust, and early public and community engagement support public and community involvement and fair decision-making.

10.3 Informed Consent and Community Rights

Informed consent means that communities should be given full and honest information about a carbon capture project before it starts, and they should have the right to decide whether they want the project to happen on their land or in their area [18]. This includes understanding what the project involves, why it is being done, and how it may affect their lives, environment, and future.

It is not enough to simply tell people that a project is planned. People must have the chance to learn about it in a clear and easy-to-understand way, ask questions, and receive truthful answers. They should be informed about both the possible benefits, such as jobs or investment in the area, and the potential risks, such as safety concerns or long-term environmental impacts, such as carbon dioxide leakage from underground storage sites [19].

Informed consent also means communities should have the time and support they need to understand what the project means for them fully. Communications may include meetings, translated materials, advice, and transparent information sharing. Only when people actually understand the project and have sufficient knowledge can they make a fair, informed decision [19]. Above all, informed consent respects the community's right to say "yes," "no," or request changes before a project moves forward. It ensures that local voices are valued and that decisions are made with fairness, transparency, and respect for the people who will be most affected [4].

For carbon capture projects, informed consent means giving communities a full and understandable explanation of how the project will work from start to finish. People need to know how the CO_2 will be transported, where it will be stored, and how the storage site will be monitored over time to make sure it stays safe. They should also clearly understand who is responsible for safety, who they can contact with questions or concerns, and what actions will be taken if something goes wrong [4, 19].

Communities must also be informed about the long-term plan for the site, including how long it will be supervised and what protections will stay in place. This process also requires discussing the environmental and social impacts of the project, such as how the land will be used, whether there will be noise or construction activity, and whether the project could limit future land use [4, 19].

Respecting community rights means recognizing and honoring local people's land ownership, cultural traditions, and ways of life, especially for Indigenous and rural communities. For many of these groups, land is not just a place to live, but it represents identity, history, and often spiritual meaning. It may include sacred areas, family-farming land, or locations used for generations. When these deep connections are ignored, communities can feel disrespected and unsafe. This can lead to frustration, mistrust, conflict, and long-term damage to relationships between local people and project developers [20].

To avoid these issues mentioned in previous paragraph, carbon projects must approach communities with respect. This includes listening to their concerns, recognizing their rights, and involving them in decisions that affect their land and future. A key principle that supports this is Free, Prior, and Informed Consent (FPIC), which is recognized in international law. Free means communities must make decisions without pressure, force, or manipulation. They should not be forced or influenced by power, money, or political pressure. Prior means consultations must happen before project approvals, contracts, or construction begin. Communities must have enough time to consider the information and discuss it. Informed means people must receive complete, accurate, and understandable information about the project, including risks, benefits, long-term impacts, and responsibilities. Information should be provided in clear language and local languages when needed. Consent means communities have the right to approve, reject, or request changes to a project. Consent is a decision-making process, not just a one-time signature, and it should be respected throughout the project's lifetime. By respecting community rights and following FPIC principles, project developers can build trust, avoid conflict, and create solutions that benefit both the community and the project [21].

Figure 10.3 provides a simplified illustration of the CCS process, along with key social and ethical principles that support its responsible implementation. On the left side, the process shows how CO_2 is captured from an industrial facility, transported via pipelines, and stored deep underground in geological formations. On the right, essential community-focused principles are highlighted, including FPIC, as well as values such as respect and transparency. This figure emphasizes that successful CCS projects require not only effective technical systems for capture, transport, and storage but also robust social guidelines that encourage public trust in both CCS initiatives and their developers. It is crucial to include the public in decision-making processes and ensure meaningful community participation throughout the project.

In simple terms, informed consent is about respect, honesty, and transparency. Communities should never be rushed, pressured, or persuaded with unrealistic promises. Instead, they should have the time, information, and support they need to understand the

Figure 10.3 Overview of the carbon capture and storage (CCS) process and key social principles for ethical deployment. The left side illustrates CO_2 capture, transport, and deep geological storage. The right side highlights essential community-focused values, including Free, Prior, and Informed Consent (FPIC), respect, and transparency. The picture emphasizes that successful CCS projects require both robust technical engineering systems and strong social engagement and trust-building.

project and make their own decision. In other words, informed consent is not only a legal or procedural requirement, but it is a way of building strong, fair, and lasting relationships between project developers and the people who are most affected by their decisions [22].

10.4 Trust-Building and Long-Term Commitments

Trust is a crucial foundation for the success of carbon capture projects. When people believe that project developers, companies, and government authorities are honest and responsible, they are more willing to accept and support the project. However, trust is not built quickly; it develops over time through open communication, honesty, and consistent positive actions [23]. To build trust, communities need to see that project leaders really listen to their concerns, provide clear and accurate information, and follow through on their commitments. It is important to answer questions honestly, even when the responses may not favor the companies involved [24]. Trust increases when people feel respected, included, and confident that their safety and environment are being protected [25]. When communities trust the process and the people behind it, they are far more likely to become supportive partners and active participants in long-term carbon management efforts.

For carbon projects, building trust should start long before construction begins or before receiving project permits and approvals. It should commence at the very first discussion about the project. Project leaders and developers need to be open and honest about their plans, the reasons for proposing the project, and the potential risks and benefits involved. This includes clearly explaining any uncertainties, challenges, or limitations, rather than concealing information or only sharing positive aspects [1].

Communities quickly lose trust when they feel they're not receiving the full story. This often happens when information is presented in complex technical language that is hard for them to understand, or when decisions seem to be made without their input or involvement. When people feel excluded or confused, they are more likely to become skeptical or oppose the project [26]. To build confidence, communication must be clear, regular, and respectful. Information should be provided in simple language. Equally important is genuine listening, giving people opportunities to ask questions, express concerns, and provide feedback, and then showing that their input matters.

Long-term commitment is extremely important because CCS projects operate for many years, sometimes even decades. Communities need to know that companies and government agencies will not disappear once the construction is finished. Instead, they must continue monitoring the storage site, handling leaks or required changes, and responding quickly if any problems arise [27].

People feel safer and more confident when there are clear plans showing who is responsible in the long run, how monitoring will be performed, and what steps will be taken if something goes wrong [4]. This includes establishing emergency response procedures, ongoing inspections, transparent reporting, and clear communication channels for the public [28]. Long-term commitment shows that project leaders are not only interested in building the project, but also in protecting local people and the environment for many years to come. When communities see this responsibility and reliability, they are far more likely to trust the project and support it over its lifetime [17].

Another important part of building trust is keeping promises. When a project team promises jobs, environmental protection, or community benefits, those commitments must be delivered. Communities pay close attention to whether actions match words. If people see that promises are not kept, or benefits do not reach local residents as expected, trust can quickly disappear. Broken promises do more than harm one project; they can also affect how people view the entire CCS sector. If one project fails to deliver, future projects may face greater public and community resistance, even if they are well-planned and well-executed. To avoid this, project developers must set honest, realistic expectations from the beginning and provide regular updates on progress. Honoring commitments, reporting results openly, and fixing problems when they arise show that a project cares about long-term community well-being, not just short-term success [29].

Figure 10.4 shows a simplified overview of the CCS process alongside key principles for building public trust. On the left, CO_2 is captured at an industrial facility, transported through a pipeline, and injected into a secure underground storage formation, protected by a caprock layer and monitored over time. On the right, the figure highlights two essential elements for ethical CCS deployment: public trust and long-term commitment. These emphasize that successful CCS projects require not only technical reliability and safe geological storage, but also transparency, consistent communication to build public trust, and long-term responsibility to the communities that live near storage sites.

In summary, building trust is not a one-time task, and it requires ongoing effort throughout the entire lifetime of a CCS project. Trust grows when project developers act

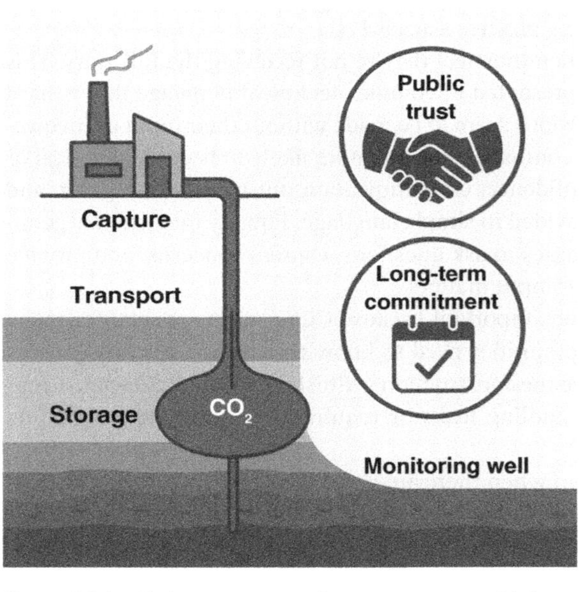

Figure 10.4 Carbon capture and storage system with key social principles. The figure illustrates the technical steps of CCS: capture, transport, and secure underground storage. Alongside are the social foundations required for success of CCS project: public trust and long-term community commitment. Together, the picture highlights the need for both strong engineering design and responsible community engagement.

consistently, communicate honestly, and demonstrate real commitment to the community over many years. When communities feel genuinely respected, supported, and included from the early planning stages through long-term monitoring, they are more likely to have a positive view of the project [23]. In this supportive environment, CCS projects are more likely to gain social acceptance and achieve long-term success.

10.5 Benefit-Sharing Mechanisms

Benefit-sharing means that communities affected by carbon capture projects should receive real and fair advantages, not just carry the risks or inconveniences. This means that local people should benefit from the project in meaningful ways, such as job opportunities, improved infrastructure, environmental protection efforts, training programs, or community development support. When communities see that a project brings positive change to their area, they are more likely to support it and trust the project teams involved. Benefit-sharing is not distributing gifts to public or charity; it is a principle of fairness that recognizes communities as partners. This principle ensures that those who live closest to the project, and who are most affected by it, also share in the advantages. By treating communities as partners and ensuring they benefit alongside industry and government, carbon projects can build stronger public relationships and earn long-term social acceptance [1, 4].

Benefits can take many forms based on the community's needs and priorities. One of the most significant benefits is local employment in carbon capture projects. When community members are trained and hired to work on these projects, during construction, operation, and long-term storage monitoring, they acquire new skills, achieve stable income, and gain more career opportunities. Providing skills training, internships, and educational programs helps young people prepare for future clean-energy jobs and builds local knowledge and skill capacity. This means that even after the project ends, the community continues to benefit from the knowledge, experience, and professional opportunities acquired [30, 31].

Benefit-sharing can also include improving local infrastructure and services. For example, projects may help build or upgrade roads, water systems, schools, or health services. These improvements can be especially valuable in rural or underserved areas, where infrastructure investment is often limited. By supporting local employment, education, and essential services, CCS projects demonstrate that they are not only using land and resources but also giving back to the community, long-lasting positive benefits [1].

Financial benefits are also an essential part of benefit-sharing. Some carbon projects provide direct economic support to communities, such as funds for community development, carbon-capture project revenue-sharing agreements with the community, or lease payments for project land or underground storage. These financial arrangements help ensure that local people receive a fair share of the economic value created by the project. In many cases, these funds can be used to support local priorities, such as improving public services or investing in local education and health facilities. When financial benefits are shared openly and fairly, communities can experience long-term economic improvements that last beyond the project duration [32]. For Indigenous communities and traditional landowners, benefit-sharing must respect cultural practices and land-use traditions. For example, some communities rely on traditional hunting and fishing areas, sacred sites, or land used for ceremonies and gathering medicinal plants. Fair agreements should protect these places, ensure continued access, and support cultural activities [33]. In conclusion, when financial benefits are transparent, fair, and respectful of culture and tradition, they help build trust and strengthen relationships between project developers and the public, and ensure communities benefit from hosting carbon storage projects [34].

Most importantly, benefit-sharing must be fair and long-term. Communities should not receive benefits only at the beginning of a project, or be given short-term incentives just to secure project approval. Promises made during planning must continue throughout the entire lifetime of the CCS project, which may last for decades. Precise and reliable agreements are essential. Benefits should be written down by the project developers and agreed upon by both the project developers and the community [35]. There should also be regular reporting by project developers to the public, so communities can see how benefits are being delivered and whether commitments are being met. Figure 10.5 shows how benefit-sharing ensures that local communities receive real and lasting advantages from CCS projects. When communities gain jobs, training, shared revenues, better infrastructure, and environmental improvements, they are more likely to support and participate in climate projects, such as CCS.

In summary, fair benefit-sharing plays a key role in building positive and lasting relationships with communities. When local people feel that a project brings real value to their lives, not only risks or disruption, they are more likely to support it and feel proud to be part of it. Fair benefit-sharing strengthens public trust in the projects, supports local economic

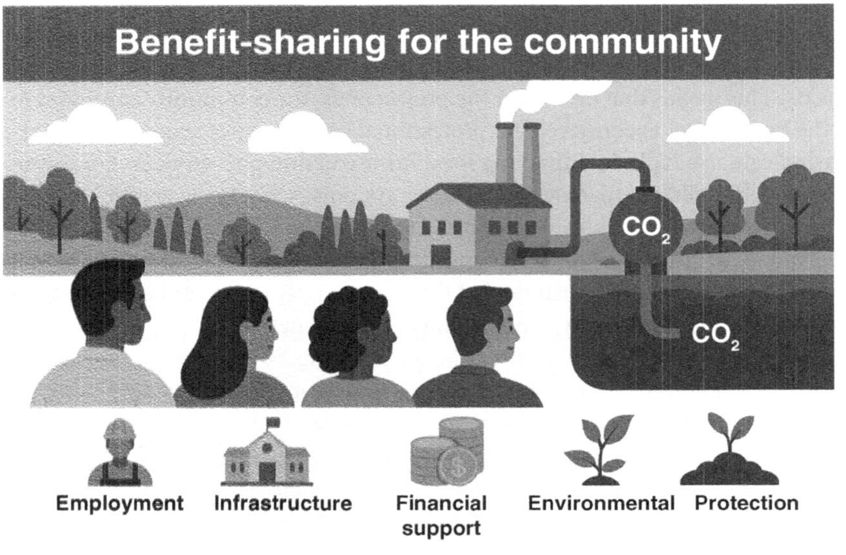

Figure 10.5 Community benefit-sharing in carbon capture and storage (CCS) projects. The diagram illustrates how a CCS project can deliver social benefits through local employment, infrastructure investment, financial support to locals, and environmental protection and improvements. These benefits help build trust, strengthen community partnerships with CCS project, and support social acceptance of CCS projects.

and social development, and helps build a strong partnership between the project and the community [36]. When communities see steady improvements in areas such as jobs, skills, infrastructure, and financial opportunities, they begin to view the project as a shared success rather than something problematic imposed from the outside. This creates a foundation for long-term cooperation and social acceptance.

10.6 Avoiding Green Colonialism and Ethical Missteps

As the world moves toward low-carbon technologies, there is growing concern about green colonialism. This refers to situations where climate-related projects primarily benefit wealthy countries or corporations rather than the local communities in which they are implemented [37]. This issue is particularly significant in developing regions, where communities may already face economic or social disadvantages.

In carbon capture projects, green colonialism can occur when project plans are created without involving local people, when land is used without fair and transparent agreements with locals and communities, or when communities are expected to accept risks while most of the financial and environmental benefits go elsewhere [38].

To prevent issues related to green colonialism, carbon projects should be designed with the active participation of the public and local communities, ensuring fair benefit-sharing and respectful engagement. It is essential to prioritize respect, fairness, and equality from the beginning. Developers and governments must include local voices right from the beginning, rather than consulting them after decisions have already been made [39].

Communities should have the authority to influence how the project is designed, where it is located, and how benefits and risks are distributed. Projects should not be planned under the assumption that local approval or compliance can be secured without proper public and community consultation. Instead, they must be based on mutual agreement, fairness, and true partnership [40].

Successful carbon projects are built on collaboration rather than pressure. When communities are treated as equal partners with the government and project developers, their knowledge and rights are respected, and they are given significant roles in the projects, these initiatives become more just, gain trust, and are more likely to succeed in the long term [41].

An ethical approach also requires full transparency about who benefits from a project and who carries the burdens. It is not fair or responsible to place CO_2 storage facilities in poorer regions simply because land is cheaper or regulations are weaker [42]. Doing so shifts risks onto communities that may already face disadvantages.

In the same way, it is unethical to promise jobs, development, or compensation that never arrive, or to walk away from a site once construction is finished, leaving local people to face possible environmental or safety risks in the future. If a carbon project claims to support sustainability, it must also ensure social justice and long-term protection for the communities involved [4]. Responsible carbon projects must respect and protect community rights, land, culture, and the natural environment. This means honest communication, fair agreements, and ongoing responsibility, and not just during construction, but for the entire lifespan of the project [5].

Figure 10.6 Community engagement in clean-energy planning, illustrating transparent dialogue and local participation in renewable and carbon project decisions.

True ethical practice also means investing in local people and their future, not just delivering short-term or symbolic benefits [43]. Communities should gain real skills, jobs, and long-lasting opportunities from carbon projects. Capacity building, such as training programs, technical education, and support for local businesses, helps ensure that communities can participate in the clean-energy projects, both now and in the future [44].

Figure 10.6 shows a community meeting taking place in a rural village, where a group of local residents sits together outdoors while listening to a speaker. The presenter stands beside a simple whiteboard illustration of wind turbines and solar panels, explaining clean-energy plans in a way that is accessible and participatory. Behind the group, real wind turbines and solar panels are visible, symbolizing a future in which renewable technologies are integrated into local landscapes. The setting under a large tree, with traditional village homes and natural scenery in the background, reflects a community-centered approach to environmental projects. The scene highlights respectful engagement, transparency, and shared decision-making, illustrating how sustainable energy projects can be introduced ethically, with full involvement and empowerment of local people.

In simple terms, carbon capture projects should support communities, not take advantage of them. Local people should benefit from cleaner air, jobs, education, and long-term opportunities, and not be left with risks, disruption, or broken promises. When CCS projects are carried out with fairness, transparency, and respect, they can strengthen communities while also helping the world meet climate goals. In this way, climate action and social justice move forward together.

10.7 Key Highlights (Chapter Summary)

◇ **Chapter 10—Key Highlights:**

1) Carbon capture projects must be evaluated not only on technical and economic grounds but also on ethical and community impacts.
2) Large-scale environmental technologies like CCS require fairness, transparency, and responsibility to avoid harming local populations.
3) Procedural justice ensures communities are included in CCS projects' decision-making and have real opportunities to participate in planning processes.
4) Community engagement means involving people early in CCS projects, listening to their concerns, and addressing them honestly.
5) Informed consent requires that communities receive clear, understandable information and have the right to approve or reject projects that affect their land and lives.
6) Respecting community rights includes understanding their cultural ties to land, respecting Indigenous values, and protecting traditional practices.
7) Trust-building relies on transparency, continuous communication, and fulfilling long-term commitments throughout the life of the project.
8) Benefit-sharing ensures local people receive fair economic and social advantages such as jobs, training, or shared revenue from CCS projects.
9) Ethical CCS development must avoid "green colonialism," where projects benefit only government and project developers while leaving local communities with risks or minimal gains.
10) A socially responsible CCS strategy requires fairness, long-term commitment of project developers to public and communities, capacity-building, and partnership with communities to ensure mutual benefit and lasting support.

Keywords: Community Engagement, Environmental Ethics, Procedural Justice, Informed Consent, Community Rights, Trust-Building, Long-Term Commitment, Benefit-Sharing, Social License, Green Colonialism, Public Participation, Ethical Governance, Local Empowerment, Climate Justice, Community and Public Inclusion

References

1 J. A. E. Nielson, K. Stavrianakis, and Z. Morrison, "Community acceptance and social impacts of carbon capture, utilization and storage projects: A systematic meta-narrative literature review," *PLoS One*, vol. 17, no. 8, Aug., 2022. DOI: 10.1371/journal.pone.0272409.

2 R. Wennersten, Q. Sun, and H. Li, "The future potential for carbon capture and storage in climate change mitigation—an overview from perspectives of technology, economy and risk," *Journal of Cleaner Production*, vol. 103, pp. 724–736, Sept., 2015. DOI: 10.1016/j.jclepro.2014.09.023.

3 L. Rosa, D. L. Sanchez, G. Realmonte, D. Baldocchi, and P. D'Odorico, "The water footprint of carbon capture and storage technologies," *Renewable and Sustainable Energy Reviews*, vol. 138, p. 110511, Mar., 2021. DOI: 10.1016/j.rser.2020.110511.

4 P. Tchetverikov, A. Cherepovitsyn, and S. Fedoseev, "Public perception of carbon capture and storage: A state-of-the-art overview," *Heliyon*, vol. 5, no. 12, p. e02845, Dec., 2019. DOI: 10.1016/j.heliyon.2019.e02845.

5 S. Nawaz and T. Satterfield, "Towards just, responsible, and socially viable carbon removal: Lessons from offshore DACCS research for early-stage carbon removal projects," *Environmental Science & Policy*, vol. 151, p. 103633, Jan., 2024. DOI: 10.1016/j.envsci.2023.103633.

6 D. Scott, "Chapter 14: Ethics of climate engineering: Chemical capture of carbon dioxide from air," in *Ethics of Chemistry*, M. A. B. Whitaker and P. Mahaffy, Eds. World Scientific Publishing, 2021, pp. 375–401, DOI: 10.1142/9789811233548_0014.

7 A. K. Sleiti and W. A. Al-Ammari, "Chapter 10—CO_2 transportation: Safety regulations and energy requirement," in *Emerging Carbon Capture Technologies: Towards a Sustainable Future*, A. K. Haghi and J. M. Smith, Eds. Elsevier, 2022, pp. 279–319, DOI: 10.1016/B978-0-323-89782-2.00010-2.

8 C. Cao, H. Liu, Z. Hou, F. Mehmood, J. Liao, and W. Feng, "A review of CO_2 storage in view of safety and cost-effectiveness," *Energies*, vol. 13, no. 3, p. 600, 2020. DOI: 10.3390/en13030600.

9 K. Itaoka, T. Hahimoto, M. Akai, S. Aoyama, C. Shinohara, and H. Hori Koshi, "Towards an implementation of carbon dioxide capture and storage: Aligning climate change mitigation with principles of justice and fairness," *SSRN Electronic Journal*, Dec. 24,, 2025. DOI: 10.2139/ssrn.5071026.

10 W. Sajjad, "Carbon capturing technology: A step for sustainable future," *NUST Journal of Natural Sciences*, vol. 6, no. 1, Oct., 2021. DOI: 10.53992/njns.v6i1.69.

11 O. A. Shobande, L. Ogbefun, and A. K. Tiwari, "Re-evaluating the impacts of green innovations and renewable energy on carbon neutrality: Does social inclusiveness really matters?," *Journal of Environmental Management*, vol. 336, 117670, Jun., 2023. DOI: 10.1016/j.jenvman.2023.117670.

12 T. Arko, "Violations of procedural justice in carbon offset projects," *SSRN Electronic Journal*, p. 18 pages, Oct. 8, 2025. DOI: 10.2139/ssrn.5570941.

13 L. Fritz, C. M. Baum, S. Low, and B. K. Sovacool, "Public engagement for inclusive and sustainable governance of climate interventions," *Nature Communications*, vol. 15, p. 4168, 2024. DOI: 10.1038/s41467-024-48510-y.

14 L. S. Greenberger, "Effective communications for project success," *Remediation: The Journal of Environmental Cleanup Costs, Technologies & Techniques*, vol. 26, no. 2, pp. 121–128, Spring, 2016. DOI: 10.1002/rem.21463.

15 P. Parmiter and R. Bell, Public perception of CCS: A review of public engagement for CCS projects. 2nd Report of the Thematic Working Group on Policy, Regulation and Public Perception, EU CCUS Projects Network, May 2020, Project ID: SCCS0208.

16 F. J. Coyle, "'Best practice' community dialogue: The promise of a small-scale deliberative engagement around the siting of a carbon dioxide capture and storage (CCS) facility," *Int. J. Greenh. Gas Control*, vol. 45, pp. 233–244, Feb., 2016. DOI: 10.1016/j.ijggc.2015.12.006.

17 W. Leiss and P. Larkin, "Risk communication and public engagement in CCS projects: The foundations of public acceptability," *International Journal of Risk Assessment and Management*, vol. 22, no. 3–4, pp. 384–403, Oct., 2019. DOI: 10.1504/IJRAM.2019.103339.

18 R. Kent and R. Hannay, "Explaining 'carbon' in community sequestration projects: A key element in the creation of local carbon knowledges," *Environmental Communication*, vol. 14, no. 3, pp. 364–377, 2020. DOI: 10.1080/17524032.2019.1673459.

19 Y. Vasilev, A. Cherepovitsyn, A. Tsvetkova, and N. Komendantova, "Promoting public awareness of carbon capture and storage technologies in the Russian federation: A system of educational activities," *Energies*, vol. 14, no. 5, 1408, Mar., 2021. DOI: 10.3390/en14051408.

20 L. Godden and M. Tehan, "REDD+: climate justice and indigenous and local community rights in an era of climate disruption," *Journal of Energy & Natural Resources Law*, vol. 34, no. 1, pp. 95–108, 2016. DOI: 10.1080/02646811.2016.1121620.

21 M. Raftopoulos and D. Short, "Implementing free prior and informed consent: The United Nations declaration on the rights of indigenous peoples (2007), the challenges of REDD+ and the case for the precautionary principle," *The International Journal of Human Rights*, vol. 23, no. 1–2, pp. 87–103, 2019. DOI: 10.1080/13642987.2019.1579990.

22 R. Pahuja, "Ethical dimensions of informed consent: Principles, processes, and case studies," in *Ethical Frameworks in Special Education: A Guide for Researchers*, S. K. Singh and R. K. Sharma, Eds, Vol. 2. Springer, 2024, p. 116.

23 L. Yang, X. Zhang, and K. J. McAlinden, "The effect of trust on people's acceptance of CCS (carbon capture and storage) technologies: Evidence from a survey in the People's Republic of China," *Energy*, vol. 96, pp. 69–79, Feb., 2016. DOI: 10.1016/j.energy.2015.12.044.

24 S. V. Fedoseev and P. S. Tcvetkov, "Key factors of public perception of carbon dioxide capture and storage projects," *Journal of Mining Institute (Записки Горного института)*, vol. 237, pp. 361–368, 2019.

25 Y. S. Melhem and M. F. Al Qudah, "Work engagement: Trust and respect to engage your people," *Indian Journal of Science and Technology*, vol. 12, no. 17, pp. 1–13, 2019.

26 J. Offermann-van Heek, K. Arning, A. Linzenich, and M. Ziefle, "Trust and distrust in carbon capture and utilization industry as relevant factors for the acceptance of carbon-based products," *Frontiers in Energy Research*, vol. 6, p. 73, Aug., 2018. DOI: 10.3389/fenrg.2018.00073.

27 B. Chu *et al.*, "A long-term impact assessment of carbon capture (storage) investment conducted by conventional power company on sustainable development," *Applied Energy*, vol. 358, p. 122567, Mar., 2024. DOI: 10.1016/j.apenergy.2023.122567.

28 R. D. Dela Cruz, L. T. Enopia, and M. Saura, "Review of major chemical safety incidents in the global carbon capture industry," Saltegra Technical Report, pp. 1–12, 2023. [Online]. Available: https://saltegra.com/wp-content/uploads/Review-of-Major-Chemical-Safety-Incidents-in-the-Global-Carbon-Capture-Industry.pdf

29 D. Hillson and P. Simon, *Practical Project Risk Management: The ATOM Methodology*, 2nd ed.. San Francisco, CA, USA: Berrett-Koehler Publishers, 2020.

30 Z. R. Yelebe and R. J. Samuel, "Benefits and challenges of implementing carbon capture and sequestration technology in Nigeria," *The International Journal of Engineering and Science (IJES)*, vol. 4, no. 5, pp. 42–49, 2015.

31 K. Turner, J. Race, O. Alabi, A. Katris, and K. Swales, "Policy options for funding carbon capture in regional industrial clusters: What are the impacts and trade-offs involved in compensating industry competitiveness loss?," *Ecological Economics*, vol. 184, p. 106978, Jun., 2021. DOI: 10.1016/j.ecolecon.2021.106978.

32 E. Hirsch and T. Foust, "Policies and programs available in the United States in support of carbon capture and utilization," *Energy Law Journal*, vol. 41, p. 91, 2020.

33 S. W. Ririhena, H. Tambajong, H. Oja, and P. A. Moento, "Establishing land rights: The vital role of indigenous community institutions in controlling land ownership certificates," *International Journal of Social Science and Business*, vol. 8, no. 1, pp. 182–189, Feb., 2024. DOI: 10.23887/ijssb.v8i1.67942.

34 M. Meitern, "Unlocking carbon finance: Empowering energy communities for mutual benefit," *Renewable and Sustainable Energy Reviews*, vol. 199, p. 114499, Jul., 2024. DOI: 10.1016/j.rser.2024.114499.

35 S. S. Dhillion, "Benefit sharing for project risk–conflict reduction and fostering sustainable development: Current understanding and mechanisms," in *Natural Resource Conflicts and Sustainable Development*, Series: Earthscan Studies in Natural Resource Management. Abingdon, Oxon ; New York, NY: Routledge, 2019, pp. 147–165.

36 F. Vanclay, F. Vanclay, A. M. Esteves, "Benefit-sharing and enhancing outcomes for project-affected communities," in M. S. Reed and L. Stringer *Handbook of Social Impact Assessment and Management*, F. Vanclay and A. M. Esteves, Eds. Cheltenham, UK: Edward Elgar Publishing, 2024, pp. 426–443, 10.4337/9781802208870.00037.

37 C. Zografos, "The contradictions of green new deals: green sacrifice and colonialism," *Soundings*, vol. 80, pp. 37–50, Summer, 2022. DOI: 10.3898/SOUN.80.03.2022.

38 N. Redvers *et al.*, "Carbon markets: A new form of colonialism for indigenous peoples?," *The Lancet Planetary Health*, vol. 9, no. 5, pp. E421–E430, May, 2025. DOI: 10.1016/S2542-5196(25)00086-5.

39 N. Tabassum, H. Zafarullah, and A. S. Huque, "Climate capitalism or carbon colonialism? The critical features of climate change adaptation and mitigation policies," in *Handbook of Development Policy*, H. Zafarullah and A. S. Huque, Eds. Cheltenham, UK: Edward Elgar Publishing, 2021, pp. 313–323, DOI: 10.4337/9781839100871.00035.

40 F. Hunter, *Community Power Structure: A Study of Decision Makers*. Chapel Hill, NC, USA: University of North Carolina Press, 2017.

41 F. Mulyasari, A. K. Harahap, A. O. Rio, R. Sule, and W. G. A. Kadir, "Potentials of the public engagement strategy for public acceptance and social license to operate: Case study of carbon capture, utilisation, and storage gundih pilot project in Indonesia," *Int. J. Greenh. Gas Control*, vol. 108, 103312, June, 2021. DOI: 10.1016/j.ijggc.2021.103312.

42 S. T. Anderson, "Risk, liability, and economic issues with long-term CO_2 storage—a review," *Natural Resources Research*, vol. 26, pp. 89–112, 2017. DOI: 10.1007/s11053-016-9303-6.

43 H. Schübel, The ethics of carbon dioxide removal: Differentiating agents and responsibilities to remove carbon dioxide, Doctoral thesis, Univ. of Fribourg, Fribourg, Switzerland, 2024.

44 L. Mabon, S. Shackley, and K. Boot, "Deliberative decarbonisation? Assessing the potential of an ethical governance framework for low-carbon energy through the case of carbon dioxide capture and storage," *Environment and Planning C: Politics and Spaces*, vol. 33, no. 2, pp. 256–271, 2015. DOI: 10.1068/c12133.

11

Carbon Capture and the Water–Energy Nexus

11.1 Understanding the Water–Energy Nexus

The water–energy nexus refers to the strong link between water systems and energy systems. In simple terms, water is needed to produce energy, and energy is needed to supply and manage water [1]. For example, power plants need water for cooling and processing, while water treatment plants and pumps need electricity to clean and deliver water to homes and industries [2]. Because of this connection, pressure on one resource often creates pressure on the other.

This link between water and energy becomes especially important when discussing carbon capture technologies [1]. Many carbon capture and storage (CCS) processes require large amounts of water for cooling, steam production, and chemical reactions. In addition, they also use additional energy to capture, compress, and store carbon. If water is scarce or energy supplies are limited, CCS systems may become harder to operate efficiently [3].

When carbon capture equipment is added to power plants or factories, it requires extra energy to run. More electricity and heat are needed to capture, move, and store CO_2 safely. This means the facility has to use more energy than before, which can put pressure on local power systems [4, 5]. At the same time, some processes, like cooling and steam generation, can increase overall water needs. If water and energy resources are not carefully managed, there is a risk of putting stress on both the energy system and local water supplies [6].

Figure 11.1 illustrates the water–energy nexus in carbon capture, showing how CCS depends on both water and energy in a circular relationship. Carbon capture systems require large amounts of energy to run capture and compression processes, and this energy use increases water demand for cooling and processing [4]. At the same time, water availability affects energy production, efficiency, and the reliability of CCS operations [7]. In simple terms, more CCS activity needs more energy, and more energy use requires more water. This means CCS planning must consider local water resources and energy supply to avoid operational and environmental challenges [8].

When planning carbon capture projects, engineers and decision-makers must think not only about reducing emissions but also about how the technology affects energy demand, water availability, and long-term resource sustainability. Considering the water–energy nexus is essential to ensure that CCS deployment does not create new environmental pressures while solving climate challenges [9].

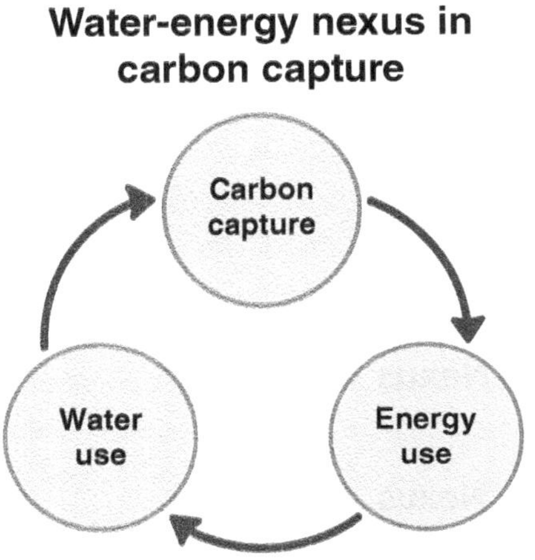

Figure 11.1 Water–energy nexus in carbon capture systems, illustrating the interdependent cycle where carbon capture processes require energy. Increased energy demand raises water use, and water availability influences energy efficiency and CCS performance.

In regions with relatively abundant water supplies and robust energy systems, such as parts of Northern Europe, Canada, and the U.S. Gulf Coast, water–energy demands may be manageable due to strong infrastructure and reliable resource availability. Conversely, in water-stressed or energy-constrained areas, including the Middle East, North Africa, and arid regions, insufficient planning could negatively affect communities, agriculture, and ecosystems [10].

Therefore, the water–energy nexus reminds us that carbon capture should be planned carefully and responsibly [11]. CCS systems need to help reduce CO_2 emissions without creating new problems for local water or energy resources. In practice, this means understanding how much water and energy a project will need, and making sure those needs can be met sustainably. If CCS projects are placed in areas already facing water shortages or weak energy supply, they could increase pressure on local communities, agriculture, and the environment.

11.2 Water Demands of Carbon Capture Processes

Carbon capture technologies can use a considerable amount of water, depending on how they are designed, and where they are installed. In many cases, water is needed to support key processes such as cooling equipment, producing steam, preparing or handling chemicals for capture, and keeping the system clean and functioning safely. Because of this, water often plays a crucial role in carbon capture operations [3].

In places that already have dry climates or limited water, this issue becomes even more important. If a carbon capture project needs more water than the local area can supply, it can put pressure on existing water resources [10]. This may reduce the water available for farming, drinking, and natural ecosystems, creating competition between the CCS project and the needs of people, agriculture, and the environment [12, 13].

Therefore, careful planning is necessary to make sure carbon capture systems do not put extra pressure on communities or natural resources [9]. Evaluating water demand early, exploring technologies that use less water, and considering alternative cooling or recycling systems can help ensure that CCS projects support climate goals without creating new environmental challenges, especially in regions with water shortages [14].

In post-combustion carbon capture systems, which are widely used in power plants, large cooling units are often needed to remove heat from the process. These carbon capture systems may rely on cooling towers and cooling heat exchangers, all of which depend on a steady and reliable water supply [15]. Many post-combustion systems use liquid solvents (e.g. Amines) to absorb CO_2 from the flue or exhaust gas. A common example is amine-based capture. These solvents require water to prepare and dilute the chemical mixture [16]. In addition, the solvent must be heated to release the captured CO_2 and regenerate the solution for reuse, and this heat typically comes from steam, which is produced using water [17].

Figure 11.2 shows an industrial post-combustion carbon capture facility equipped with absorber and regeneration towers, process heat exchangers, and extensive piping systems. These systems require both significant energy, primarily for solvent regeneration, pumping, and compression, and substantial water use for cooling, steam generation, and solvent preparation [18]. As discussed in this section, carbon capture units can place additional demand on local water resources, particularly in areas with limited supply or competing needs, such as agriculture and community use.

While some of the water in the carbon capture system can be cleaned and reused, not all of it can be recovered. Losses due to evaporation and chemical reactions mean that fresh water needs to be continuously added. Therefore, post-combustion capture must be carefully planned in areas with limited water resources to ensure that the technology operates without putting stress on local communities or natural ecosystems [3].

Figure 11.2 Industrial post-combustion carbon capture unit showing several facilities and piping. These carbon capture systems require significant energy and cooling water, highlighting the water–energy nexus in carbon capture operations.

Water demand varies significantly depending on the carbon capture technology. Some systems are inherently less water-intensive than others. For example, solid sorbent-based capture typically uses less water than solvent-based amine systems, as it relies on solid materials or membranes rather than water-rich chemical solvents [19].

Direct air capture (DAC) systems also vary widely in their water requirements. Some DAC technologies use liquid chemical solutions that need water to absorb CO_2, resulting in relatively high water demand. Others are designed to operate with minimal water use by using solid sorbents or by recycling water within the system [20, 21]. In certain cases, especially in humid environments, DAC units can even produce water as a by-product through condensation.

In regions already experiencing drought, limited freshwater supplies, or competing water needs from agriculture and communities, high water consumption can pose major challenges. Therefore, water availability should be assessed early in project planning [22]. Approaches such as using recycled or seawater, adopting dry-cooling technology (which uses air instead of water for cooling), and improving process efficiency can help reduce the burden on local water resources [23].

Because of these differences, it is important to select and design carbon capture technologies based on local water availability and environmental conditions. Matching the technology to regional water resources helps ensure that carbon capture supports climate goals without putting additional pressure on local supplies [6]. In summary, understanding and managing water demand is essential for responsible CCS deployment. Projects that plan water use early and prioritize water-efficient technologies are more likely to be successful, particularly in regions with limited natural resources.

11.3 Energy Requirements and Carbon Intensity

Carbon capture systems require additional energy to function, meaning that a power plant or industrial site must produce more energy than it did before. This extra energy is necessary to operate the capture equipment, compress the CO_2 for transport or storage, and move it through pipelines. In systems that use solvents or sorbents, more energy is also needed to heat and regenerate the materials for reuse. As a result of these processes, carbon capture can significantly increase a facility's energy demand. In many cases, energy usage may rise by 15–50%, depending on the specific carbon capture technology, the type of plant, and how efficiently the carbon capture system is designed and operated. This additional energy requirement is often referred to as the "energy penalty" of carbon capture. Understanding this energy increase is important because it affects energy use, operating costs, and the overall environmental impact of the system [24].

Figure 11.3 illustrates a typical industrial compression unit used in CCS systems to increase the pressure of captured CO_2. After CO_2 is separated from flue gas, it needs to be compressed by the compressor to a high pressure for efficient transportation and injection into geological formations [25].

The equipment consists of an electric motor (on the right) connected to a compressor unit (on the left) via a coupling, both mounted on a steel skid [26]. In CCS applications, these units consume a significant amount of energy, contributing substantially to the

Figure 11.3 Motor-driven CO_2 compression unit used in CCS systems. Compressors pressurize CO_2 for transport and storage and contribute to the energy penalty of CCS.

overall energy penalty associated with CO_2 capture, transport, and storage. The compressor is typically powered by electricity, which is the most common and efficient energy source for CO_2 compression in CCS facilities [27]. Compressors of this type are essential components in large-scale CCS operations, especially when CO_2 needs to be delivered to pipeline pressures or injected underground in supercritical form.

An energy penalty refers to the additional power required for a plant to operate its carbon capture system [28]. If this extra energy comes from fossil fuels, it can diminish the overall environmental benefits of capturing CO_2, as burning more fuel generates more emissions. Essentially, some of the CO_2 that is captured can be counterbalanced by the emissions produced to power the system. To ensure that carbon capture remains effective, it is crucial to consider how the additional energy is generated. Utilizing cleaner energy sources, recovering waste heat, or improving system efficiency can help maintain significant climate benefits when implementing carbon capture technology [29].

To maximize the climate benefits of carbon capture, it should be powered by clean, low-carbon energy whenever possible. Using renewable or low-emission energy sources minimizes additional emissions from operating the capture system, thereby enhancing the overall environmental effectiveness [29]. Examples of such energy sources include solar, wind, hydropower, and geothermal energy, as well as other low-carbon electricity systems. In many industrial settings, a practical solution is to utilize waste heat; this is the heat already produced during normal operations that would otherwise be released unused. By capturing and reusing this heat, facilities can avoid generating extra energy solely for CO_2 capture, which reduces fuel consumption and lowers operating costs. In simple terms, combining carbon capture with clean energy sources or recycled heat makes the system more efficient and environmentally friendly [29, 30].

Figure 11.4 illustrates a conceptual integrated energy landscape that combines carbon capture technology with renewable energy generation. The image features two carbon capture units, which represent industrial facilities designed to capture CO_2 emissions. Surrounding these facilities are wind turbines and solar photovoltaic panels, demonstrating how low-carbon electricity and heat can power the capture, compression, and

Figure 11.4 Renewable-powered carbon capture system. A conceptual industrial CCS facility supplied by wind and solar energy. Integrating renewable electricity with CO_2 capture and compression systems reduces lifecycle emissions.

processing operations in CCS plants. This setup presents a future path in which renewable energy supports carbon capture efforts, ultimately helping reduce the overall carbon footprint of industrial activities and enabling near-zero-emission energy by replacing fossil fuels with renewable energy [29].

One important method to reduce energy consumption is by improving the efficiency of the CO_2 capture process itself. Researchers and engineers are working on developing new materials and technologies that can capture CO_2 with less heat and electricity. Examples of these innovations include solid sorbents, advanced solvents, and improved heat-exchanger designs that enhance heat efficiency. The goal of these technologies is to lower energy usage while still effectively capturing significant amounts of CO_2. Improving the efficiency of the carbon capture process can help facilities lower operating costs, reduce fuel consumption, and decrease emissions. Over time, ongoing innovation will make carbon capture systems cleaner, more affordable, and more sustainable. This progress will facilitate industries in achieving their net-zero goals [31].

In summary, carbon capture systems often need a considerable amount of additional energy [4]. However, this challenge can be managed with effective strategies. By combining CCS with renewable energy sources, using waste heat generated from industrial processes, and consistently enhancing technology efficiency, it is possible to minimize the extra energy requirements. To achieve real climate benefits, the energy used for carbon capture must come from clean and efficient sources, not high-carbon fuels [29, 30]. When CCS is supported by renewable energy and smart and efficient system design and operation, such as digitalization, it can effectively reduce emissions without increasing pollution or putting pressure on energy supplies.

11.4 Geographic and Regional Constraints

The success of carbon capture projects is greatly influenced by local geography and regional conditions. Different areas have varying access to essential resources, including water, energy, land, and suitable underground storage sites. Consequently, some regions may be more naturally suited for carbon capture than others [32].

For example, regions with reliable energy supplies, access to water resources, and safe geological formations for CO_2 storage are more likely to support CCS projects that are both economically and environmentally effective. In contrast, areas with limited water availability, weak energy infrastructure, or unsuitable geological conditions may pose higher costs, greater technical challenges, or increased environmental risks for the CCS project [33].

Figure 11.5 illustrates how local geography plays a critical role in determining the feasibility and long-term sustainability of CCS projects. The image contrasts two distinct environments: a green, water-rich region on the left and an arid desert landscape on the right. Although the same CCS facility and pipeline network appear in both settings, the surrounding conditions differ substantially. In the green region, abundant water resources, stable soil conditions, and moderate temperatures create a more favorable

Figure 11.5 Comparison of a carbon capture and storage (CCS) facility located in two contrasting environments: a green, water-rich region and an arid desert landscape.

environment for CCS operation and management, supporting effective cooling, reliable water supply, and more predictable long-term pipeline performance with reduced thermal stresses. In contrast, the desert setting presents significant challenges, including limited water availability, high evaporation rates, and extreme climatic conditions that can increase operational difficulties and strain local ecosystems. By introducing the same CCS project across two contrasting landscapes, the figure underscores the importance of adapting CCS design and planning to regional water resources, climate conditions, and environmental constraints.

Identifying local conditions early in the planning process is crucial. By analyzing the environmental, social, and geological characteristics of each location, developers can create carbon capture projects that are safe, efficient, and sustainable [34]. This approach helps prevent unnecessary pressure on local communities and natural resources.

In regions with stable freshwater supplies and strong energy infrastructure, carbon capture projects can usually be developed with fewer obstacles. However, in arid or drought-prone regions, water is already limited. In these places, high water demand from carbon capture systems may compete with essential needs such as farming, drinking water, and protecting natural ecosystems [3]. Regions like the Middle East, parts of Africa, and the southwestern United States are already experiencing severe water stress, which makes water-intensive CCS technologies much harder to deploy responsibly.

In such regions, CCS deployment must be carefully planned to avoid intensifying local water stress [3]. Options include using alternative water sources, such as treated wastewater, industrial recycled water, or seawater where coastal access is available, rather than relying on fresh groundwater or municipal supplies. Facilities may also adopt water-saving process designs, including dry-cooling systems that use air instead of water for cooling. In addition, closed-loop water recycling systems can be implemented to capture, treat, and reuse process water within the plant, significantly reducing overall consumption and discharge [35]. Another strategy is to prioritize low-water or no-water capture technologies, such as advanced solid sorbents or membrane-based processes, which require minimal water input and therefore place far less pressure on local resources [36].

Energy availability, such as electricity, also varies significantly across regions. Countries that face limited or unreliable electricity supply may struggle to support the additional power demand that carbon capture systems require [37]. In many developing areas, electricity access is still being expanded to households, schools, hospitals, and local industries. Introducing energy-intensive carbon capture systems without adequate planning could place extra pressure on power grids, increase electricity prices, or slow the progress of energy access for communities and essential services [38]. To avoid such challenges, CCS deployment in regions with limited energy capacity must be aligned with grid capacity or grid upgrades, and investments in renewable power generation.

Geological conditions are a critical factor in selecting suitable carbon storage sites. CCS is only effective in regions that possess secure underground formations capable of storing CO_2, such as deep saline aquifers or depleted oil and gas reservoirs. In some regions, these storage resources are abundant and located close to major industrial facilities, making CCS deployment more practical and cost-efficient [39]. However, many areas lack appropriate geological formations or have storage sites situated far from emission sources. In such cases, captured CO_2 must be transported over long distances by pipelines, ships, or other

Figure 11.6 Carbon capture facility located close to the storage site, enabling short-distance CO_2 transport via pipeline, while ships in the background illustrate long-distance transport options for offshore or remote storage locations.

means. Greater transport distances increase both project costs and technical complexity, which can become a significant barrier to carbon capture implementation [40].

Figure 11.6 shows two different paths for transporting captured carbon dioxide from an industrial carbon capture facility to a permanent storage site. On the left side, the capture plant is located very close to the geological storage formation, allowing the CO_2 to be transported through a short, direct pipeline that enters the subsurface only a few meters away from the facility. This represents projects where capture and storage occur on the same site or within a very short distance. On the right side, the coastline and large CO_2 transport ships illustrate situations where storage sites are located far offshore or at long distances from the point of capture. In such cases, the captured CO_2 must be transported by ship to remote storage reservoirs. The picture, therefore, contrasts short-distance pipeline transport with long-distance ship-based transport, highlighting how the location of the storage site influences the design and complexity of the CO_2 transportation system.

Geography plays a crucial role in the design of carbon capture systems. These systems cannot be standardized for every region; instead, they must be tailored to local conditions, including water resources, energy availability, climate, and geological storage options. Some areas may already have the ideal conditions for CCS, while others might need more careful planning or different technological approaches. By aligning carbon capture solutions with the specific realities of each region, decision-makers can prevent unnecessary stress on water resources, energy systems, and local communities [41]. Thoughtful planning and appropriate technology selection are essential to ensuring that CCS effectively supports climate goals in a fair, sustainable, and practical manner, without creating additional burdens for regions that are already facing resource constraints.

11.5 Design Strategies for Resource Optimization

To ensure that carbon capture systems are really sustainable, it is essential to design them for maximum efficiency in water and energy use. Achieving sustainability requires careful planning by engineers and decision-makers to optimize every aspect of the process. Resource optimization involves maximizing the benefits of a system while minimizing the consumption of water and energy, particularly in areas where these resources are already scarce [27].

Efficient design helps reduce operating costs, lowers environmental impact, and ensures that carbon capture does not compete with essential needs such as drinking water, agriculture, or electricity for homes and schools. By emphasizing more effective and innovative engineering and technological improvements, CCS can support climate goals while minimizing additional stress on local communities and natural systems [42].

A major strategy for making carbon capture more sustainable is improving energy efficiency. This means designing systems that need less power to remove CO_2 from emissions, transport, and inject to the storage sites. New technologies, such as advanced solvents, solid sorbents, membrane systems, and improved heat-recovery designs, are being developed to reduce energy use while still capturing large amounts of CO_2 effectively [27, 36].

A key strategy for making carbon capture more sustainable is to improve energy efficiency. This involves designing systems that use less power to remove CO_2 from emissions, transport it, and inject it into storage sites [43]. New technologies, such as advanced solvents, solid sorbents, membrane systems, and enhanced heat recovery designs, are being developed to reduce energy consumption while still capturing significant amounts of CO_2 effectively [30, 36].

Another important approach is to capture and reuse waste heat that is already produced in industrial facilities. Instead of letting this heat escape into the air, it can be used to help power the capture process, lowering the need for additional fuel and reducing operating costs [44]. In addition, integrating renewable energy sources, such as solar, wind, or geothermal power, can further reduce emissions from the capture system. By replacing fossil fuels with clean energy, carbon capture facilities can operate with a much lower carbon footprint [29].

Water efficiency is just as critical as energy efficiency in sustainable carbon capture system design. Modern CCS facilities increasingly adopt dry-cooling systems (using air instead of water), hybrid cooling technologies (reducing but not eliminating water use), and water-recycling systems (treating and reusing process water) to minimize freshwater consumption while maintaining reliable performance. These approaches significantly reduce the water demanded by cooling and process operations, an essential advantage in regions facing water shortage [1–3].

In many cases, carbon capture facilities can use alternative water sources instead of depending on drinking-quality water. Depending on what is available in the region and suitable for use, options may include treated wastewater, recycled industrial water, or seawater. By utilizing these sources along with appropriate treatment systems, such as filtration, desalination, and purification, plants can operate more sustainably and avoid competing with communities for clean freshwater [45].

An effective strategy for minimizing water loss and enabling water reuse within a system is to select appropriate equipment and plant layouts. For instance, evaporation losses can occur when water used for cooling is exposed to the air, leading to gradual evaporation. Designs that limit open water surfaces to air can reduce these losses [46]. Closed-loop cooling systems are another effective option, as they recirculate the same water repeatedly rather than releasing it, significantly reducing overall water usage. Additionally, improved condensers can recover water vapor by cooling it and converting it back to liquid form, allowing reuse rather than loss to the atmosphere [47].

Well-designed storage tanks also play a crucial role in minimizing leakage and evaporation, as they often use sealed covers and insulation [48]. By implementing these approaches, the demand for fresh water can be significantly reduced over time, thereby improving efficiency and sustainability in carbon capture systems. Finally, continuous monitoring and smart control systems are essential for maintaining efficiency in carbon capture plants. Modern software tools allow operators to track water and energy usage in real time, identify waste, and automatically adjust the system to enhance performance by reducing these wastes. Over time, this data can be utilized to optimize equipment, forecast maintenance requirements, and ensure that the system operates as efficiently as possible. This kind of smart engineering and digital optimization helps make sure that carbon capture supports climate goals while respecting local resource limits [49].

Figure 11.7 illustrates the role of digital monitoring and advanced control systems in the operation of modern carbon capture facilities. An operator sits in front of multiple high-resolution screens displaying real-time process data, flow diagrams, and performance indicators for key units such as absorbers, regenerators, compressors, and storage systems.

Figure 11.7 An operator monitoring real-time process data in a control room, illustrating the role of smart automation and continuous performance tracking in modern carbon capture systems (Courtesy of Shutterstock.com).

This setup reflects how modern carbon capture plants rely on integrated software platforms to track water and energy consumption, detect process inefficiencies, and maintain optimal operating conditions. Continuous data collection and automated adjustments not only improve process stability and safety but also enhance resource efficiency by minimizing waste. Such digital tools are essential for ensuring that carbon capture installations operate reliably, sustainably, and in alignment with long-term climate and resource-management goals.

11.6 Challenges Facing CCS Projects

Analyzing real-world examples reveals how CCS projects can be successfully implemented in areas with limited water resources. These case studies highlight both the challenges of operating in dry regions and the practical solutions that can be employed to minimize water use and protect local water supplies. By exploring how various projects recycle water, use alternative water sources, and implement low-water technologies, we can gain valuable insights into the responsible design of CCS systems.

These examples emphasize a crucial lesson: carbon capture can be effective in water-stressed regions if careful planning is undertaken, local conditions are considered and accepted, and efficient technologies are utilized [50]. Real-world experiences from various locations offer valuable understandings for future projects, ensuring that climate solutions do not place additional pressure on already vulnerable water systems.

In the Middle East, countries like the United Arab Emirates and Saudi Arabia are implementing carbon capture projects aimed at reducing emissions from heavy industry and the oil and gas sector. Given the region's extreme water scarcity, water-intensive systems would not be sustainable. To overcome this challenge, many projects are utilizing seawater, desalinated water, or recycled industrial water for cooling and processing instead of relying on freshwater [51].

Furthermore, many facilities utilize dry-cooling systems, hybrid cooling methods, and energy-efficient capture technologies to further decrease water consumption. Some plants also incorporate renewable energy sources or waste-heat recovery systems, which capture heat that would otherwise be lost during industrial processes. This captured heat can be reused for tasks such as heating solvents for regeneration or powering parts of the carbon capture process. By reusing this heat instead of generating new energy through water-intensive cooling or power production, these systems help reduce both energy demand and the overall water footprint of the facility [52].

These examples from the Middle East demonstrate how technology choices, careful planning, and alternative water sourcing strategies can make carbon capture feasible in desert environments. They also emphasize the importance of adapting CCS solutions to local resource conditions [51].

In the western United States, especially in states such as Texas and New Mexico, limited freshwater supplies make carbon capture more challenging for power plants and industrial facilities. These regions often face long-term droughts and competing demands for water from farmers, cities, and ecosystems [53]. To address this, some carbon capture projects use treated wastewater from municipal systems or produced water from oil and gas operations, helping reduce the need for fresh drinking water [54]. Many facilities also rely on renewable energy integration and waste-heat recovery to lower the additional energy required for carbon capture [29]. By reusing heat already generated at industrial sites and supplementing power with solar or wind energy, these projects can reduce environmental impact.

In North Africa, both agriculture and daily life depend significantly on limited freshwater resources, which makes water management a critical concern for industrial development. As a result, early research and pilot projects on CCS in the region have focused on careful site selection, water-efficient capture technologies, and thorough environmental impact assessments [32]. These studies aim to ensure that carbon capture activities do not interfere with irrigation, drinking water needs, or the preservation of fragile natural ecosystems.

In addition, in North Africa, local planning efforts have emphasized the significance of involving communities and stakeholders early in the process. By understanding local water priorities, consulting with farmers and residents, and assessing long-term water availability, these efforts help prevent conflicts and build trust before larger projects are implemented. These efforts emphasize the importance of planning and community consultation before scaling up projects [55].

These examples emphasize an important lesson: carbon capture can be achieved in water-stressed regions, but it requires careful planning and design tailored to local conditions. It is essential to use alternative water sources, enhance efficiency, and choose technologies that minimize water consumption. Projects cannot simply copy designs from other areas; they must be adapted to suit the specific environments and needs of the local communities. By considering local resource limits and applying smart engineering solutions, CCS projects can support climate goals without creating new environmental pressures.

11.7 Key Highlights (Chapter Summary)

◇ Chapter 11 – Key Highlights:

1) This chapter explains how carbon capture systems are closely linked to both water use and energy demand, forming part of the water–energy nexus.
2) Carbon capture technologies often require large amounts of water for cooling, solvent preparation, regeneration, and cleaning, especially in industrial and power plant settings.
3) The energy needed to run carbon capture systems can significantly increase total power consumption and environmental pollution, making energy efficiency an essential consideration.
4) If carbon capture plants are powered by fossil fuels, additional energy use from CCS can reduce climate benefits, so integrating renewable energy with CCS is important.
5) Water scarcity and limited energy infrastructure in certain regions can make carbon capture deployment more challenging and expensive.
6) Geographical factors such as weather conditions, water availability, electricity access, and the proximity of CO_2 storage sites significantly influence the success of carbon capture projects.

7) Resource-efficient design strategies include the use of dry-cooling systems, water-recycling and reuse schemes, high-efficiency capture technologies, and integration of renewable energy sources with CCS to minimize energy and water demand.
8) Careful site selection and thoughtful infrastructure design help place carbon capture facilities in locations where water and energy are already available, reducing the need for long-distance pipelines, lowering water consumption, and avoiding extra pressure on local services and communities.
9) Real-world case studies from regions such as the Middle East, North Africa, and the western United States show that carbon capture can operate in water-stressed areas if water-saving technologies and alternative water sources are used.
10) Ultimately, sustainable carbon capture must balance climate gains with careful water and energy use to avoid creating new negative environmental impacts.

Keywords: Water–Energy Nexus, Carbon Capture and Storage (CCS), Water Consumption in CCS, Energy Demand, Energy Penalty, Cooling Water Requirements, Steam Generation, Solvent-Based Capture, Solid Sorbents, Membrane Technologies, Direct Air Capture (DAC), Water Scarcity, Resource Constraints, Water Recycling, Dry-Cooling Systems, Hybrid Cooling Technologies, Waste-Heat Recovery, Renewable Energy Integration, CO_2 Compression, CO_2 Transport, Pipeline Transport, Ship-Based CO_2 Transport, Geological Storage, Site Selection, Regional Suitability, Arid and Water-Stressed Regions, Energy Infrastructure, Environmental Sustainability, Process Optimization, Digital Monitoring, Smart Control Systems, Industrial Efficiency, Climate Adaptation, Resource Management

References

1 J. Wang, Z. Yu, X. Zeng, Y. Wang, K. Li, and S. Deng, "Water-energy-carbon nexus: A life cycle assessment of post-combustion carbon capture technology from power plant level," *Journal of Cleaner Production*, vol. 312, p. 127727, Aug., 2021. DOI: 10.1016/j.jclepro.2021.127727.

2 S. A. Nihalani, Y. D. Mishra, R. Singh, P. Shukla, and P. Singh, "Water consumption management for thermal power plant," in *Environmental Processes and Management*, R. Singh, P. Shukla, and P. Singh, Eds, Vol. 91. Cham: Springer, 2020, pp. 145–162, DOI: 10.1007/978-3-030-38152-3_9.

3 L. Rosa, D. L. Sanchez, G. Realmonte, D. Baldocchi, and P. D'Odorico, "The water footprint of carbon capture and storage technologies," *Renewable and Sustainable Energy Reviews*, vol. 138, p. 110511, Mar., 2021. DOI: 10.1016/j.rser.2020.110511.

4 A. Al-Mamoori, A. Krishnamurthy, A. A. Rownaghi, and F. Rezaei, "Carbon capture and utilization update," *Energy Technology*, vol. 5, no. 6, pp. 834–849, Jan., 2017. DOI: 10.1002/ente.201600747.

5 L. Liu *et al.*, "Water demands for electricity generation in the U.S.: Modeling different scenarios for the water–energy nexus, "*Technological Forecasting and Social Change*, vol. 94, pp. 318–334, May 2015. 10.1016/j.techfore.2014.11.004.

6 H. Eldardiry and E. Habib, "Carbon capture and sequestration in power generation: Review of impacts and opportunities for water sustainability," *Energy, Sustainability and Society*, vol. 8, no. 6, pp. 1–13, 2018. DOI: 10.1186/s13705-018-0146-3.

7 N. Ding, J. Liu, J. Yang, and B. Lu, "Water footprints of energy sources in China: Exploring options to improve water efficiency," *Journal of Cleaner Production*, vol. 174, pp. 1021–1031, Feb., 2018. DOI: 10.1016/j.jclepro.2017.10.273.

8 S. K. Jain and V. P. Singh, *Water Resources Systems Planning and Management*, 2nd ed., Vol. 51. Amsterdam, Netherlands: Elsevier Science, Nov. 25, 2023, p. 954.

9 J. Lv, Y. Li, B. Shan, S. Jin, and C. Suo, "Planning energy-water nexus system under multiple uncertainties—a case study of Hebei province," *Applied Energy*, vol. 229, pp. 389–403, Nov., 2018. DOI: 10.1016/j.apenergy.2018.08.010.

10 K. L. Lam, P. A. Lant, K. R. O'Brien, and S. J. Kenway, "Comparison of water–energy trajectories of two major regions experiencing water shortage," *Journal of Environmental Management*, vol. 181, pp. 403–412, Oct., 2016. DOI: 10.1016/j.jenvman.2016.06.068.

11 M. C. Oliveira, M. Iten, and H. A. Matos, "Review on water and energy integration in process industry: Water-heat nexus," *Sustainability*, vol. 14, no. 13, p. 7954, Jun., 2022. DOI: 10.3390/su14137954.

12 C. Ingrao, R. Strippoli, G. Lagiola, and D. Huisingh, "Water scarcity in agriculture: An overview of causes, impacts and approaches for reducing the risks," *Heliyon*, vol. 9, no. 8, p. e18507, Aug., 2023. DOI: 10.1016/j.heliyon.2023.e18507.

13 N. Mancosu, R. L. Snyder, G. Kyriakakis, and D. Spano, "Water scarcity and future challenges for food production," *Water*, vol. 7, no. 3, pp. 975–992, Mar., 2015. DOI: 10.3390/w7030975.

14 A. Lohrmann, C. Lohrmann, and P. Luukka, "Global and regional models for identification of cooling technology in thermal power generation for water demand estimations in water–energy nexus studies," *Journal of Cleaner Production*, vol. 380, p. 134842, Dec., 2022. DOI: 10.1016/j.jclepro.2022.134842.

15 N. Sharma and S. S. Mahapatra, "A preliminary analysis of increase in water use with carbon capture and storage for Indian coal-fired power plants," *Environmental Technology & Innovation*, vol. 9, pp. 51–62 ,Feb., 2018. DOI: 10.1016/j.eti.2017.10.002.

16 Y. Jiang *et al.*, "Energy-effective and low-cost carbon capture from point-sources enabled by water-lean solvents," *Journal of Cleaner Production*, vol. 388, p. 135696, Feb., 2023. DOI: 10.1016/j.jclepro.2022.135696.

17 W. Zhang, H. Liu, Y. Sun, J. Cakstins, C. Sun, and C. E. Snape, "Parametric study on the regeneration heat requirement of an amine-based solid adsorbent process for post-combustion carbon capture," *Applied Energy*, vol. 168, pp. 394–405, Apr., 2016. DOI: 10.1016/j.apenergy.2016.01.049.

18 O. A. Ibigbami, O. D. Onilera, and R. O. Akinyeye, "Post-combustion capture and other carbon capture and sequestration (CCS) technologies: A review," *Environmental Quality Management*, vol. 34, no. 1, p. e22180, Jan., 2024. DOI: 10.1002/tqem.22180.

19 A. S. El Nasr, T. Nelson, A. Kataria, and M. R. M. Abu-Zahra, "Benchmarking of a novel solid sorbent CO_2 capture process for NGCC power generation," *International Journal of Greenhouse Gas Control*, vol. 42, pp. 583–592, Nov., 2015. DOI: 10.1016/j.ijggc.2015.09.014.

20 C. Drechsler and D. W. Agar, "Investigation of water co-adsorption on the energy balance of solid sorbent based direct air capture processes," *Energy*, vol. 192, p. 116587, Feb., 2020. DOI: 10.1016/j.energy.2019.116587.

21 M. Webber, A. Aliyu, S. H. Jin, M. M. Sadiq, B. Sohani, and A. Elseragy, "A techno-economic review of direct air capture of moisture processes: sustainable versus energy-intensive methods," *International journal of Environmental Science and Technology*, vol. 22, pp. 579–613, 2025. DOI: 10.1007/s13762-024-05720-7.

22 P. Udmale, Y. Ichikawa, T. Nakamura, N. Shaowei, H. Ishidaira, and F. Kazama, "Rural drinking water issues in India's drought-prone area: A case of Maharashtra state," *Environmental Research Letters*, vol. 11, no. 7, p. 074013, Jul., 2016. DOI: 10.1088/1748-9326/11/7/074013.

23 V. G. Gude, "Desalination and water reuse to address global water scarcity," *Reviews in Environmental Science and Biotechnology*, vol. 16, pp. 591–609, 2017. DOI: 10.1007/s11157-017-9449-7.

24 M. Capocelli and M. De Falco, "Generalized penalties and standard efficiencies of carbon capture and storage processes," *International Journal of Energy Research* , Nov.17, 2021. DOI: 10.1002/er.7474.

25 A. El-Suleiman, N. B. Anosike, and P. Pilidis, "A preliminary assessment of the initial compression power requirement in CO_2 pipeline 'Carbon Capture and Storage (CCS) Technologies,'," *Technologies*, vol. 4, no. 2, p. 15 , May., 2016. DOI: 10.3390/technologies4020015.

26 Ș. A. Șerban, S. G. Tomescu, I. Vlăducă, and S. Voicu, "Energy improvement of an oil injected screw compressor skid," *EMERG: Energy. Environment. Efficiency. Resources. Globalization*, vol. 2021, no. 1, p. 51, 2021.

27 S. Jackson and E. Brodal, "Optimization of the energy consumption of a carbon capture and sequestration related carbon dioxide compression processes," *Energies*, vol. 12, no. 9, p. 1603, Apr., 2019. DOI: 10.3390/en12091603.

28 Y. Zheng, L. Gao, S. He, and H. Jin, "Reduction potential of the energy penalty for CO_2 capture in CCS," *Frontiers in Energy*, vol. 17, pp. 390–399, 2023. DOI: 10.1007/s11708-023-0864-x.

29 Q. Zhao, "A review of pathways to carbon neutrality from renewable energy and carbon capture," in *E3S Web of Conferences*, vol. 245, p. 01018, EDP Sciences, Mar. 2021, presented at the 2021 5th International Conference on Advances in Energy, Environment and Chemical Science (AEECS 2021), DOI: 10.1051/e3sconf/202124501018.

30 Y. Zhang, J. Gao, M. He, D. Feng, Q. Du, and S. Wu, "Simulation optimization of a new ammonia-based carbon capture process coupled with low-temperature waste heat utilization," *Energy & Fuels*, vol. 31, no. 4, pp. 4219–4225, Mar., 2017. DOI: 10.1021/acs.energyfuels.6b03361.

31 S.-Y. Oh, S. Yun, and J.-K. Kim, "Process integration and design for maximizing energy efficiency of a coal-fired power plant integrated with amine-based CO_2 capture process," *Applied Energy*, vol. 216, pp. 311–322, Apr., 2018. DOI: 10.1016/j.apenergy.2018.02.100.

32 K. F. Bou-Hamdan, F. Sufyan, and A. H. Abbas, "Carbon capture, utilization, and storage in the MENA region: A regional review of projects and challenges," *Arabian Journal for Science and Engineering*, vol. 50, pp. 4529–4549, 2025. DOI: 10.1007/s13369-025-09999-7.

33 S. Selosse and O. Ricci, "Carbon capture and storage: Lessons from a storage potential and localization analysis," *Applied Energy*, vol. 188, pp. 32–44, Feb., 2017. DOI: 10.1016/j.apenergy.2016.11.117.

34 F. B. Ernst, A. I. Karabulut, and M. İ. Yeşilnacar, "Policy planning and implementation for carbon capture technologies," in *Decarbonization Strategies and Drivers to Achieve Carbon Neutrality for Sustainability*, A. K. Haghi and J. G. Speight, Eds. Elsevier, 2024, pp. 259–281, DOI: 10.1016/B978-0-443-13607-8.00001-8.

35 M. Gholami-Shabani and K. Nematpour, "Reuse of wastewater as non-conventional water: A way to reduce water scarcity crisis," in *Wastewater Treatment and Sludge Management Systems—The Gutter-to-Good Approaches*, A. A. Kazmi, Ed. IntechOpen, 2024.

36 H. Rashid and A. Rafey, "Solid adsorbents for carbon dioxide capture: A review," *Chemistry and Ecology*, vol. 39, no. 7, pp. 775–791, 2023. DOI: 10.1080/02757540.2023.2227622.

37 D. S. Shalaeva, O. I. Kukartseva, V. S. Tynchenko, V. V. Kukartsev, S. V. Aponasenko, and E. V. Stepanova, "Analysis of the development of global energy production and consumption by fuel type in various regions of the world," *IOP Conference Series: Materials Science and Engineering*, vol. 952, p. 012025, 2020. DOI: 10.1088/1757-899X/952/1/012025.

38 J. M. Nelson, *Access to Power: Politics and the Urban Poor in Developing Nations*. Princeton, NJ: Princeton University Press, 2023.

39 N. Jain, A. Srivastava, T. N. Singh, V. Vishal, and T. Singh, "Carbon capture, transport and geologic storage: A brief introduction," in *Geologic Carbon Sequestration*, V. Vishal and T. Singh, Eds. Cham: Springer, 2016, DOI: 10.1007/978-3-319-27019-7_1.

40 T. Grant *et al.*, "Comparative analysis of transport and storage options from a CO_2 source perspective," *Int. J. Greenh. Gas Control*, vol. 72, pp. 175–191, May., 2018. DOI: 10.1016/j.ijggc.2018.03.012.

41 A. Raza, R. Gholami, R. Rezaee, V. Rasouli, and M. Rabiei, "Significant aspects of carbon capture and storage—a review," *Petroleum*, vol. 5, no. 4, pp. 335–340, Dec., 2019. DOI: 10.1016/j.petlm.2018.12.007.

42 X. Yu, S. Zhang, L. Liu, and J. Du, "Optimal design and scheduling of carbon capture power plant based on uncertainty decision-making methods," *Journal of Cleaner Production*, vol. 380, no. 2, p. 134852, Dec., 2022. DOI: 10.1016/j.jclepro.2022.134852.

43 A. Saxena *et al.*, "Innovative pathways in carbon capture: Advancements and strategic approaches for effective carbon capture, utilization, and storage," *Sustainability*, vol. 16, no. 22, p. 10132, 2024. DOI: 10.3390/su162210132.

44 T. Tu, X. Yang, Q. Cui, Y. Shang, and S. Yan, "CO_2 regeneration energy requirement of carbon capture process with an enhanced waste heat recovery from stripped gas by advanced transport membrane condenser," *Applied Energy*, vol. 323, p. 119593, Oct., 2022. DOI: 10.1016/j.apenergy.2022.119593.

45 A. Dickinson-Cove *et al.*, "Reactive carbon capture using saline water: Evaluation of prospective sources, processes, and products," *Chemical Society Reviews*, vol. 54, pp. 116–151, 2025. DOI: 10.1039/D4CS00701H.

46 Y. W. Youssef and A. Khodzinskaya, "A review of evaporation reduction methods from water surfaces," *E3S Web of Conferences*, vol. 97, p. 05044, 2019. DOI: 10.1051/e3sconf/20199705044.

47 J. M. Averina, G. E. Kaliakina, D. Y. Zhukov, A. Y. Kurbatov, and V. S. Shumova, "Development and design of a closed water use cycle," *International Multidisciplinary Scientific Geo Conference: SGEM*, vol. 19, no. 3.1, 2019. DOI: 10.5593/sgem2019/3.1/S12.019.

48 K. Sotoodeh, *Storage Tanks Selection, Design, Testing, Inspection, and Maintenance: Emission Management and Environmental Protection*. Elsevier, 2024.

49 A. Chawla, Y. Arellano, M. V. Johansson, H. Darvishi, K. Shaneen, and M. Vitali, "IoT-based monitoring in carbon capture and storage systems," *IEEE Internet of Things Magazine*, vol. 5, no. 4, pp. 34–39, 2022. DOI: 10.1109/IOTM.2022.10012404.

50 N. H. Malik, F. Shaikh, L. Kumar, and M. S. Hossain, "Quantification of the water-energy-carbon nexus of the coal fired powerplant in water stressed area of Pakistan," *Energy*

Sources, Part A: Recovery, Utilization, and Environmental Effects, vol. 47, no. 1, pp. 2094–2116, 2020. DOI: 10.1080/15567036.2020.1840671.

51 N. Darraj, M. M. Abdelrahman, and M. Y. Alkih, "Regional review for large-scale deployment of carbon capture and storage (CCS) in the Middle East," *in Mediterranean Offshore Conference*, SPE-223358-MS, Oct. 2024, DOI: 10.2118/223358-MS.

52 Y. Kong, W. Wang, X. Huang, L. Yang, X. Du, and Y. Yang, "Direct dry cooling system through hybrid ventilation for improving cooling efficiency in power plants," *Applied Thermal Engineering*, vol. 115, pp. 936–944, 2017. DOI: 10.1016/j.applthermaleng.2017.03.067.

53 P. C. Buotte, B. E. Law, W. J. Ripple, and L. T. Berner, "Carbon sequestration and biodiversity co-benefits of preserving forests in the western United States," *Ecological Applications*, vol. 30, no. 2, p. e02039, Mar., 2020. DOI: 10.1002/eap.2039.

54 L. Lu, J. S. Guest, C. A. Peters, X. Zhu, G. H. Rau, and Z. J. Ren, "Wastewater treatment for carbon capture and utilization," *Nature Sustainability*, vol. 1, pp. 750–758, 2018. DOI: 10.1038/s41893-018-0187-9.

55 F. Mulyasari, A. K. Harahap, A. O. Rio, R. Sule, and W. G. A. Kadir, "Potentials of the public engagement strategy for public acceptance and social license to operate: Case study of carbon capture, utilisation, and storage Gundih pilot project in Indonesia," *International Journal of Greenhouse Gas Control*, vol. 108, 103312, 2021. DOI: 10.1016/j.ijggc.2021.103312.

12

Integration of Carbon Capture with Renewable Energy Systems

12.1 Why Integration with Renewables Matters

Integrating carbon capture with renewable energy systems is crucial for ensuring that carbon capture and storage (CCS) technology is really climate-friendly. If the carbon capture equipment is powered by fossil fuels, some of the benefits of reducing emissions can be diminished, as additional fuel may need to be burned to operate the CCS system. This ultimately reduces the overall effectiveness of the process. When carbon capture systems are powered by renewable sources such as solar, wind, hydropower, geothermal energy, or sustainably sourced biomass, the process becomes significantly cleaner. Using renewable energy enables CO_2 to be captured by CCS and prevented from entering the atmosphere without increasing reliance on fossil fuels, thereby strengthening the overall environmental benefits of carbon capture [1, 2].

Another key concept of integrating carbon capture with renewable energy is the opportunity to build a more flexible and resilient energy system. In this context, flexibility refers to the ability to adjust to changing energy supply and demand by using multiple clean energy sources. At the same time, resilience means maintaining reliable operation even when individual energy sources fluctuate or are temporarily unavailable. However, these outcomes are not automatic, and they depend on proper engineering design and supporting infrastructure such as energy storage and grid management [1, 2].

Renewable electricity can power key components of the carbon capture process, such as compressors, pumps, and direct air capture (DAC) units, without increasing emissions. When engineered correctly, this integrated approach helps ensure that carbon capture operates cleanly, efficiently, and reliably within a low-carbon energy system [3].

Carbon capture systems can play a significant role in supporting electricity grids that rely heavily on renewable energy sources. The electricity grid is a network that transmits power from power plants and renewable sources to homes, businesses, and industries [4]. When wind and solar power generation fluctuates, the grid needs to maintain a balance between electricity supply and demand. Many carbon capture systems are designed to adjust their power consumption based on grid conditions [5].

For instance, during times of abundant wind or solar power, these systems can increase their electricity usage to operate fans, pumps, and CO_2 compressors. Conversely, when renewable energy generation is low or when there is high demand on the grid, the carbon capture system can reduce its electricity consumption by slowing down certain processes or temporarily decreasing capture rates. This flexibility helps stabilize the grid, relieves energy consumption stress during peak demand periods, and facilitates the integration of renewable energy into the carbon capture systems [5].

Figure 12.1 illustrates an integrated energy system in which renewable power sources, specifically wind turbines and solar photovoltaic panels, supply clean electricity to support CCS operations at an industrial facility. The image highlights how renewables can provide the substantial electrical energy needed to operate compressors, pumps, capture units, and auxiliary equipment without relying on fossil fuels. By powering CCS with renewable energy, the overall carbon footprint of the system is significantly reduced, making CO_2 capture more sustainable and environmentally responsible. This integration also demonstrates a future-oriented approach in which industrial processes, emissions management, and clean energy generation work together as part of a unified low-carbon energy infrastructure.

In short, combining carbon capture with renewable energy creates a stronger and more effective climate strategy. This approach lowers emissions and supports future options for sustainable negative-emission technologies. Negative emissions refer to the net removal of

Figure 12.1 Integrated renewable energy system combining wind turbines and solar panels to power carbon capture operations at an industrial facility. Using clean electricity reduces the emissions associated with CCS due to burning fossil fuel and supports a more sustainable, low-carbon energy system.

greenhouse gases, especially carbon dioxide (CO_2), from the atmosphere. This happens when more CO_2 is taken out of the air than is released, resulting in an overall reduction in atmospheric CO_2 levels. Rather than treating clean energy and carbon removal as separate solutions, integrating them allows each one to strengthen the other and deliver greater overall benefits [1, 2].

12.2 Opportunities for Bioenergy with CCS (BECCS)

Bioenergy with Carbon Capture and Storage (BECCS) is one of the most promising technologies for achieving net-negative emissions, meaning it can remove more CO_2 from the atmosphere than it releases [6]. Unlike traditional power plants that burn fossil fuels and emit CO_2, BECCS uses biomass as its fuel source. Biomass can come from agricultural residues, forestry by-products, dedicated energy crops, or even organic waste materials [7].

As plants grow, they naturally absorb CO_2 from the atmosphere through photosynthesis, a process in which plants use sunlight to convert carbon dioxide and water into the energy they need to grow [8]. This plant material, known as biomass, can later be used to generate electricity, heat, or fuels [9, 10]. When this process is combined with, or coupled to, CCS, the CO_2 released during biomass use is captured and stored deep underground instead of returning to the atmosphere. This approach or technology, called BECCS, ensures that the carbon absorbed by plants is permanently removed from the atmosphere [1, 2].

There are many ways to use BECCS across the energy system, because biomass can replace fossil fuels in several types of facilities. One option is to apply BECCS in power plants that burn biomass, where the biomass is used like coal or natural gas to produce heat, which then creates steam to generate electricity. In some cases, the heat can also be used directly to supply district heating or industrial steam, not just electricity [9, 10].

BECCS can also be used in biofuel plants that produce renewable fuels such as ethanol, biodiesel, or sustainable aviation fuel. In these facilities, CO_2 is released naturally during fermentation and fuel-processing steps [11]. For example, in a biofuel plant, when biomass such as corn or sugarcane is fermented by bacteria to produce ethanol, the process produces a pure stream of CO_2 as a by-product. Because this CO_2 stream is already concentrated and clean, it is relatively easy and low-cost to capture and compress it for storage [12]. Similar processes exist in biodiesel and aviation fuel production, where CO_2 is generated during feedstock processing and refining [13]. Finally, many industries that naturally use biomass, such as paper mills, pulp plants, and sugar factories, produce CO_2 as part of their operations, making them good candidates for BECCS as well [14]. By capturing and storing these emissions, BECCS can significantly lower the carbon footprint of renewable fuel production and even create negative emissions.

Figure 12.2 illustrates an integrated BECCS system, combining a biomass production and processing facility with a dedicated carbon capture plant. On the left side of the image, the biomass facility includes feedstock storage, anaerobic digesters, and agricultural infrastructure, representing the biological source of carbon that is naturally absorbed from the atmosphere through plant growth. On the right side, the industrial section depicts the carbon capture plant, where CO_2 generated during biomass conversion is captured, purified, and prepared for compression and long-term geological storage. The spatial separation

Figure 12.2 Integrated BECCS system combining biomass production and processing with industrial carbon capture. Biomass-produced CO_2 is captured and prepared for long-term storage, allowing the plant to generate energy while achieving net-negative emissions.

between the two areas reflects real engineering practice, as BECCS facilities require dedicated zones for biomass handling, energy conversion, and high-pressure carbon capture operations. Together, the combined layout demonstrates how BECCS integrates renewable biological resources with advanced industrial processes to produce energy while achieving net-negative emissions.

Countries like the United States, Brazil, Canada, and several nations in Europe and Asia possess substantial agricultural and forestry resources, making them well-suited for large-scale deployment of BECCS. These regions generate significant quantities of crop residues, wood by-products, and organic waste, all of which can be used as fuel for BECCS systems [15].

In addition to its climate benefits, BECCS can create important economic and social benefits, especially in rural regions. Building biomass plants can generate jobs in farming, harvesting, collecting forestry residues, and transporting biomass to energy facilities. BECCS also supports employment in power plants and biofuel facilities by creating new roles in plant operation, maintenance, CO_2 capture, and storage systems. At the same time, it can help expand renewable energy industries by adding value to agricultural and forestry resources. Biomass from farms, forests, and organic waste is used to produce renewable electricity, renewable heat, biogas, and low-carbon fuels such as ethanol and sustainable aviation fuel. These activities strengthen rural economies, provide new income opportunities for farmers and forest owners, and encourage investment in clean technologies [14]. Socially, BECCS can support community development by creating long-term local jobs, improving skills and training, and helping keep economic activity in rural areas rather than moving to cities [16].

BECCS can also help improve energy security. Unlike wind or solar power, which depend on weather conditions and sunlight, biomass systems can generate energy whenever needed. This means they can provide reliable and controllable power continuously, even when there is no wind or sunshine. As a result, BECCS can support the electricity system by providing power during periods when energy output from renewable sources like wind and solar is low, helping to maintain a stable and reliable grid. In this way, BECCS works as a valuable partner to intermittent renewable energy sources, ensuring a steady supply of clean energy [17].

However, BECCS must be planned very carefully to avoid creating new environmental or social challenges. Although it has strong potential to reduce emissions, large-scale biomass production can place pressure on land, water resources, biodiversity, and food systems if not managed responsibly. For instance, growing energy crops on land that is needed for food production, or clearing forests to produce biomass, could damage ecosystems and compete with agriculture [18].

To reduce these risks, BECCS projects should prioritize sustainably sourced biomass, meaning materials that do not compete with food production, damage ecosystems, or place excessive pressure on land and water. This includes agricultural residues (leftover materials from farming, such as straw, which can be collected without affecting food supply), forestry by-products like waste wood, and organic waste from food systems, industry, or agriculture that would otherwise be discarded. In addition, carefully selected energy crops can be used when they are grown on marginal or degraded land that cannot support food production. Protecting natural ecosystems is also essential, meaning biomass supply chains should avoid clearing forests, draining wetlands, or damaging other sensitive habitats. Additionally, water use must be carefully managed so that BECCS facilities operate sustainably, particularly in regions where water resources are already scarce [18, 19].

Overall, BECCS provides a unique opportunity to produce renewable energy while removing CO_2 from the atmosphere, offering a powerful path for achieving significant and long-term emissions reductions. Because it delivers both clean energy and permanent carbon removal, BECCS has become an important element in many global net-zero and even net-negative emissions strategies [6, 7]. When developed responsibly and based on sustainable biomass sources, BECCS can significantly contribute to a cleaner and stronger energy system for the future [18, 19].

12.3 Solar- and Wind-Powered Capture Systems

Solar- and wind-powered carbon capture systems enable the operation of CO_2 capture and removal technologies using clean, renewable energy rather than fossil fuels. By relying on energy from the sun and wind, these systems avoid the additional emissions that would occur if carbon capture were powered by coal or natural gas. This approach makes the process far more sustainable and greatly enhances its overall climate benefits. In simple terms, renewable energy supplies the power needed to capture CO_2, while carbon capture removes emissions from the atmosphere, creating a cleaner system in which energy generation and carbon removal reinforce one another rather than work in opposition [20].

Solar energy can support carbon capture in several important ways. Solar thermal systems can generate heat that is used to regenerate solvents in post-combustion capture or to power certain types of DAC systems. Meanwhile, solar photovoltaic (PV) panels can provide electricity for running pumps, fans, CO_2 compressors, and control systems. By supplying both heat and power, solar energy can help reduce the reliance on fossil fuels and lower the operating emissions of carbon capture systems [21].

Figure 12.3 shows a solar-powered DAC facility operating in an arid desert environment, demonstrating how large-scale CO_2 removal from the air can be achieved using clean and renewable energy. The DAC units, arranged in modular rows, continuously pull CO_2 from the atmosphere, while an expansive solar farm in the foreground supplies the electricity needed to run fans, sorbent regeneration systems, pumps, and CO_2 compression units. The harsh, dry landscape highlights an important advantage of solar-powered DAC: it can operate in regions with abundant sunlight but limited water resources, without relying on fossil fuels or placing additional pressure on local water supplies. By combining DAC with solar photovoltaics, the system enables carbon removal with minimal operational emissions, supporting sustainable, and climate-neutral CO_2 capture in sun-rich regions. Solar photovoltaics (PV) refer to technology that converts sunlight directly into electricity [22].

This approach (solar energy) is especially effective in regions with strong sunlight and plenty of available land, such as the Middle East, Australia, parts of Africa, and the southwestern United States. In these areas, solar power can be used to run carbon capture systems reliably and at a lower cost over time. Although installing solar systems requires an

Figure 12.3 Solar-powered direct air capture (DAC) facility in a desert. Electricity from the solar farm drives DAC units and CO_2 compression systems. This setup enables low-emission carbon removal in water-scarce, high-sunlight regions.

upfront investment, using sunlight as the main energy source reduces the need to buy fuel or electricity later, which lowers operating expenses in the long run. As solar technology continues to become cheaper and more widely used, its role in powering carbon capture is expected to increase, helping support clean energy growth and large-scale emissions reductions [22].

Wind energy can also play a major role in powering carbon capture systems, particularly in regions with strong wind resources such as northern Europe, coastal areas, and offshore zones. Wind turbines can supply clean electricity directly to capture units, providing power for fans, pumps, compressors, and other system components. By using wind energy instead of fossil fuels, the carbon footprint of the capture process is reduced, making the overall system far more climate-friendly [23].

In many windy regions, wind power is available throughout the year and often produces large amounts of electricity, especially at night or during cooler seasons when solar generation is lower. This makes wind energy a valuable combination with solar power [24]. When combined, the two can support nearly continuous clean energy supply to carbon capture facilities, helping maintain stable operation and minimizing reliance on fossil-based backup power [25].

However, both solar and wind power are intermittent, meaning they do not always produce electricity at a steady level throughout the day. Carbon capture systems, on the other hand, normally work best when they have a constant and reliable energy supply, because the capture and compression equipment is designed to run continuously. This makes it challenging to power carbon capture directly with renewables that frequently fluctuate [26, 27].

To solve this challenge, renewable-powered carbon capture systems can use several strategies. One option is flexible operating modes, where the carbon capture unit increases its operation when renewable energy is plentiful and reduces or pauses when renewable supply is low. For example, the capture unit may run at full capacity during sunny or windy hours and operate at a lower level during cloudy or calm periods [27].

Figure 12.4 illustrates how renewable-powered carbon capture systems adjust their operating rate based on the availability of solar and wind energy. On the left side of the image, strong sunlight and steady wind allow solar panels and wind turbines to generate abundant, clean electricity for the carbon capture plant. During these periods, the carbon capture unit shown in the center can operate at full capacity, using the surplus renewable energy to power compressors, fans, and pumps. On the right side of the image, solar and wind output drops—representing nighttime, cloudy conditions, or low-wind periods. When renewable energy is limited, the carbon capture system automatically shifts into a reduced-capacity operating mode, conserving energy while maintaining essential functions.

The visual contrast between the two halves of Figure 12.4 reflects changes in renewable power availability over time, not differences in geography or climate. This highlights the key operating principle of flexible, renewable-powered carbon capture: the system ramps up when renewable energy is abundant and scales down when it is scarce, ensuring stable, efficient performance without relying on fossil-based backup power.

Another approach is the use of hybrid systems, which combine multiple energy sources to ensure stable operation. In some cases, the carbon capture plant may draw power from a mix of renewable sources, such as solar during the day and wind at night, helping smooth

Figure 12.4 Flexible operation of a renewable-powered carbon capture system. The carbon capture unit increases output when solar and wind energy are abundant (left) and reduces operation during periods of low renewable supply (right), ensuring stable performance.

out the natural variability of each resource. In other cases, the system may blend renewable energy with grid electricity or waste heat from industrial processes to fill in power gaps when renewable output is low. For example, the capture system might run mostly on solar and wind, but switch to the grid or use excess heat from nearby industrial equipment when needed. By combining different energy inputs, whether multiple renewables or renewables with supplemental power, hybrid systems help carbon capture plants remain steady and reliable, even when renewable energy supply fluctuates [28].

New technologies are making this easier. Battery storage can supply electricity when the sun is not shining or the wind is not blowing [29]. Thermal energy storage, on the other hand, can store heat generated during sunny periods and release it later when sunlight is limited, providing the heat needed for solvent regeneration or other processes [30]. In addition, smart control and automation systems can monitor energy supply and plant conditions in real time and automatically adjust the carbon capture system to use energy more efficiently. For example, when renewable power is abundant, the system can increase capture rates, and when supply is lower, it can temporarily reduce energy use. This flexibility allows the carbon capture process to remain stable, efficient, and cost-effective without wasting energy or interrupting operation [31].

In summary, solar- and wind-powered carbon capture systems combine two key climate strategies, using clean renewable energy and removing CO_2, into one integrated approach. By operating capture technologies with clean electricity and heat, they avoid additional emissions while removing carbon from the atmosphere. This strengthens the overall climate benefit of CCS and supports a more sustainable path to decarbonization [20]. Although challenges remain, especially related to the variability of renewable power, rapid advances in energy storage, using hybrid systems, and smart control systems are making this approach more practical and reliable.

12.4 Hybrid Plant Designs and Case Examples

Hybrid plant designs integrate multiple energy sources, including renewable electricity, industrial waste heat, and grid power, to enhance the reliability and efficiency of carbon capture systems. Carbon capture technology operates most effectively when supplied with a steady and predictable energy source. However, depending solely on solar or wind energy can be problematic due to their inherent variability. Hybrid systems address this issue by enabling the plant to switch between different energy sources or use them simultaneously, depending on each energy's availability [21, 28].

This flexibility helps carbon capture systems keep running smoothly even when renewable power changes throughout the day. For example, the system can operate on solar power during daylight hours and switch to wind energy when wind speeds are strong enough to generate excess electricity [21, 28]. It can also draw on waste heat from nearby industrial processes, such as refineries or chemical plants, using heat that would otherwise be wasted to support solvent regeneration or other thermal needs [32]. When none of these sources are sufficient, the system can rely on the electricity grid, meaning the local or regional power network, to provide stable backup power. By intelligently blending these different energy sources, hybrid systems avoid sudden shutdowns, prevent large swings in energy demand, and maintain a steady capture process [21, 28]. At the same time, they reduce emissions by maximizing the use of clean energy and minimizing reliance on fossil-based electricity, making carbon capture more reliable and practical across a wide range of regions and energy conditions.

Another hybrid model uses battery storage or thermal energy storage, often working alongside solar and wind power. Energy storage plays a crucial role by allowing carbon capture units to keep running even when renewable output drops, for example, at night, during cloudy weather when solar energy is limited, or when wind speeds are too low to generate sufficient electricity. Battery storage provides immediate electrical power to keep pumps, fans, compressors, and control systems operating. Thermal storage captures excess heat during sunny periods. It is released later to support heat-intensive steps such as solvent regeneration in amine systems or specific processes in DAC [29, 30]. Maintaining stable power and heat levels is essential because sudden changes in temperature or pressure can stress equipment, create thermal fatigue, and accelerate wear on pumps, valves, and heat exchangers. By smoothing out these fluctuations, storage technologies help protect equipment, reduce the risk of shutdowns, and improve overall system efficiency. Storage technologies make renewable-powered carbon capture more reliable, durable, and practical under real-world operating conditions.

Figure 12.5 illustrates a modern hybrid carbon capture facility that is fully integrated with renewable energy sources, energy-storage systems, and advanced automation for efficient load management. In the background, large wind turbines and solar photovoltaic

Figure 12.5 Hybrid renewable-powered carbon capture system with integrated battery storage and smart automation. Wind and solar energy supply the primary power for CCS operations, while energy-storage and advanced control systems maintain adjustable and stable operation performance during fluctuating renewable output.

panels generate clean electricity for the plant, thereby reducing reliance on fossil fuels. At the center of the image is the carbon capture unit, which is supported by a dedicated battery storage system in addition to the wind and solar plant. The battery storage system stores excess renewable energy and releases it during periods of low sunlight or weak wind, ensuring stable and continuous CCS operation. The two cabinets on the left represent automation, control, and digital monitoring systems, not electrical power cabinets. These two cabinets containing an array of sensors, control cabinets, and digital monitoring systems highlight the importance of automation and intelligent control in balancing the fluctuating inputs from renewable sources with the energy demands of compressors, pumps, and other capture equipment. By integrating renewable power, carbon capture technology, energy storage, and smart control systems, the image exemplifies a next-generation CCS configuration that is flexible, low-carbon, and optimized for grid-friendly operation.

Real-world examples of hybrid energy systems for carbon capture are already emerging. In Europe, several major industrial hubs are developing designs that connect offshore wind farms to CO_2 capture systems at steel, cement, and chemical plants [33]. Offshore wind in the North Sea is attractive because it provides strong, steady, and relatively predictable winds, making it one of the most reliable renewable resources in the region. These projects aim to use this clean power to operate large-scale carbon capture units, reducing the need for fossil-based electricity. At the same time, the systems are supported by the electricity grid, which provides backup power when wind output drops [34]. Flexible plant controls and smart automation allow the capture units to adjust their energy use in real time, ensuring stable operation [31]. By combining offshore wind, grid support, and advanced control systems, these industrial clusters demonstrate how hybrid energy models can power carbon capture reliably while lowering emissions across hard-to-abate industries.

In the United States, several pilot projects are linking solar farms directly to DAC units, with battery storage added to smooth out changes in sunlight. During sunny periods, solar panels provide clean electricity to power fans, sorbent regeneration systems, and CO_2 compression. When sunlight drops, such as during cloudy weather or in the late afternoon, batteries supply the necessary power to keep the DAC unit running without interruption. This hybrid setup allows DAC facilities to operate more consistently and efficiently, even in areas where solar conditions can vary widely from hour to hour [35].

In countries such as Japan and South Korea, hybrid systems combine renewable energy and industrial waste heat as primary inputs to keep carbon capture operating reliably throughout the year [36]. Because renewable output can change quickly with weather conditions, waste heat from nearby industrial plants is often used to support the system during periods of lower renewable supply [37]. In situations where neither renewables nor waste heat are sufficient, some facilities may temporarily rely on limited backup power from existing natural gas units, though this is generally considered a transitional solution and not ideal for long-term decarbonization.

Hybrid plant designs demonstrate how flexible and adaptable carbon capture can be when multiple energy sources are combined in a thoughtful way. By blending renewables, waste heat, and grid electricity, these systems can adjust to changing conditions while maintaining stable CO_2 capture performance [28]. This flexibility makes carbon capture more reliable across different climates, seasons, and industrial settings.

12.5 Energy Storage, Load Management, and Flexibility

As carbon capture technology becomes increasingly integrated with renewable energy systems, the ability of CCS units to operate flexibly is becoming essential. Rather than operating at a constant capacity, next-generation CCS facilities are being designed to adapt their operations to changing grid conditions, varying outputs from renewable energy sources, and dynamic energy markets, where energy prices and availability fluctuate throughout the day in response to demand and renewable supply. To achieve this flexibility, CCS systems must utilize more than just batteries or thermal storage [38]. They also require effective load management (the capability to adjust their energy consumption based on available power) and process flexibility, which enables different parts of the capture system to speed up, slow down, or temporarily pause operations without damaging the equipment [39].

12.5.1 Long-Duration Energy Storage (LDES)

While conventional batteries are helpful for smoothing short-term changes in renewable energy, such as passing clouds or brief drops in wind, they cannot supply power for many hours or days. Many carbon capture systems, however, need long, steady periods of energy to run equipment like solvent regenerators, CO_2 compressors, and DAC units.

LDES technologies fill this gap by storing large amounts of energy and releasing it slowly over extended periods [40].

LDES includes several technologies that can hold energy for many hours or even days [40]. The main types are flow batteries, compressed air energy storage (CAES), pumped hydro storage, and hydrogen-based energy storage. Flow batteries store energy in large liquid-filled tanks and can hold power for many hours. When renewable electricity is available, the energy is stored as chemical energy in these liquids. Later, when power is needed, the liquids flow through a cell stack to generate electricity. They are easy to expand by increasing tank size, which makes them suitable for industrial and long-duration applications. CAES uses surplus renewable electricity to compress air into underground caverns or tanks, releasing it later to generate electricity when the renewable supply is low. Pumped hydro storage works like a large rechargeable battery. Extra renewable electricity is used to pump water uphill into a higher reservoir, storing the energy in the water as potential energy for the future. When electricity is needed, the water is released downhill through turbines to generate power. The purpose is not to store water, but to store energy in the water in a simple and reliable way. Emerging hydrogen-based storage converts extra renewable electricity into hydrogen, which can be stored for long periods and later used to produce clean electricity or heat. Together, these LDES technologies provide steady, reliable energy for carbon capture systems during extended periods of low wind or solar output [41].

Figure 12.6 illustrates a renewable-powered LDES setup designed to support carbon capture operations when solar and wind output is low. In the foreground, the single

Figure 12.6 Long-duration energy storage (LDES) supporting renewable-powered carbon capture. Flow-battery tanks store excess solar and wind energy and release it during low-renewable energy periods. This ensures a steady and reliable power supply for energy-intensive CCS operations.

large cylindrical tank represents a flow-battery electrolyte reservoir, a key component of modern LDES systems that store renewable electricity as chemical energy for long periods. Behind the tank, solar photovoltaic panels capture sunlight during the day, while wind turbines generate electricity whenever wind conditions are favorable. When renewable energy is abundant, excess power is stored in the tank as chemical energy. During periods of low solar or wind availability, the stored energy is discharged and supplied to CCS units, helping power CO_2 compressors, solvent regeneration, or DAC equipment. This configuration demonstrates how LDES acts as a long-term storage buffer, allowing carbon capture systems to operate reliably even when renewable generation is variable.

12.5.2 Demand Response and Grid Services

Modern CCS plants can support the electricity grid by adjusting how much power they use in response to changing grid conditions, a strategy known as demand response. When renewable energy is abundant, such as during windy periods or strong solar output, or when the grid needs extra electrical load to stay stable, CCS units can increase their activity. When we say that the grid needs extra electrical load to stay stable, it means that sometimes the electricity supply is higher than the electricity demand, especially when there is a lot of solar or wind power being produced at once. If too much electricity is fed into the grid without enough consumers using it, the grid's frequency and voltage can rise and become unstable. This may involve temporarily raising CO_2 capture rates or running solvent regeneration and compression cycles more intensively to absorb the surplus renewable power [42].

During times when electricity supply is limited, such as low-wind evenings or cloudy winter days, CCS plants can reduce nonessential energy use. They may slow down regeneration cycles, shift certain operations to a later time when renewable electricity becomes available again. This helps avoid stressing the grid during peak demand. Some advanced CCS designs can even act as controllable loads, meaning they can quickly increase or decrease their electricity consumption to help maintain grid stability [39].

12.5.3 Curtailment Utilization and Power-to-X Integration

Curtailment happens when renewable energy (wind or solar) produces more electricity than the grid can accept, forcing operators to waste that extra clean energy [43]. Power-to-X refers to technologies that convert surplus renewable electricity into other useful products, such as hydrogen, synthetic fuels, heat, or chemicals [44].

In many regions, large amounts of renewable electricity are wasted during curtailment, which occurs when wind or solar power generation exceeds what the grid can handle. Instead of letting this clean energy go unused, flexible CCS plants can use it [43]. They can increase energy-intensive operations in CCS, such as CO_2 compression or solvent regeneration, precisely during these periods of surplus renewable supply.

More advanced systems go a step further by linking CCS with Power-to-X technologies. These technologies use excess renewable electricity to produce hydrogen, synthetic fuels, industrial heat, or other valuable products. When CCS is combined with Power-to-X, each technology supports the other. Surplus renewable electricity produces hydrogen through electrolysis, which can then be combined with captured CO_2 to make synthetic fuels. Heat released during hydrogen production or fuel synthesis can be reused in CCS solvent regeneration, reducing the system's energy demand. CCS also captures CO_2 from industrial and fuel-production processes, providing a clean feedstock for Power-to-X and reducing total emissions [44]. This creates an integrated low-carbon system where excess renewable electricity is never wasted, and both fuel production and CCS become more efficient.

Figure 12.7 shows an integrated low-carbon energy and carbon-management hub where point-source carbon capture DAC, renewable power, hydrogen production, and Power-to-X processes operate together on a single industrial site. The image illustrates how surplus renewable electricity from the wind turbine and solar panel, which would normally be curtailed when production exceeds grid capacity, can instead be redirected into productive low-carbon paths. In the background, solar panels and wind turbines generate more electricity than the grid can absorb. Rather than wasting this excess energy, the system channels it into a hydrogen electrolysis facility, visible on the right, where water is split into hydrogen using clean power. This hydrogen can then be combined with captured CO_2 from the adjacent carbon capture facility (center-right) to produce synthetic fuels, chemicals, or heat. The captured CO_2 is shown being processed and conditioned before entering Power-to-X conversion units. The integration of CCS with Power-to-X ensures that excess

Figure 12.7 Integrated CCS and Power-to-X system utilizing surplus renewable electricity for hydrogen production and synthetic fuel synthesis, preventing curtailment and enhancing overall energy efficiency.

renewable energy is fully utilized, CO_2 is recycled into valuable products, and the overall system minimizes emissions while improving energy efficiency. The picture captures the essential idea that curtailment need not be wasted energy but can instead drive a circular, low-carbon industrial ecosystem.

Although DAC and CCS perform different functions, co-locating them is increasingly common because both technologies require similar supporting infrastructure. DAC units benefit from being placed near CCS facilities, since they can use the same compression systems, CO_2 conditioning (purifying) equipment, transport pipelines, and geological storage sites. In addition, colocation allows DAC to take advantage of surplus renewable electricity and excess heat released from hydrogen production or synthetic-fuel synthesis, improving overall efficiency. By placing DAC and CCS next to one another, the site can operate as a fully integrated carbon-management system: industrial sources are decarbonized through CCS, atmospheric CO_2 is removed through DAC, and the combined CO_2 streams can be stored underground or used in Power-to-X processes.

12.5.4 Flexible Operation and Process Optimization

Emerging CCS systems are being designed with flexible solvent regeneration, which allows them to operate efficiently even when steam or electricity supply fluctuates. In traditional amine-based systems, steam provides the heat needed to release CO_2 from the solvent, while electricity powers pumps, fans, and control systems. Because these systems normally require steady energy input, they are difficult to run alongside variable renewable energy. New designs solve this by making regeneration more adaptable [45]. For example, the system can delay solvent (e.g. amine) regeneration cycles, meaning it allows more CO_2-rich solvent to accumulate and waits to heat it until renewable power is available. The modern CCS system can also adjust the solvent concentration, using more solvent for CO_2 capture when energy is abundant and less when energy is limited. In addition, the capture unit can operate at partial load, reducing steam and electricity demand during low renewable output. These flexible strategies reduce energy use when renewable energy is not available, improve energy efficiency, and reduce the overall energy penalty.

12.5.5 Hybrid Operation with Industrial Processes

In industrial environments, carbon capture systems can be connected to nearby facilities that produce waste heat or generate intermittent surplus electricity during their operations. By utilizing these otherwise wasted energy resources, carbon capture systems can operate more efficiently, enhancing regeneration or compression activities when additional heat or power is available. This opportunistic approach reduces the reliance on dedicated steam or electricity, and improves the efficiency and cost-effectiveness of the entire industrial setup. For example, a steel mill produces large amounts of high-temperature waste heat that can be used to support solvent regeneration in a CCS system. By using these available energy streams, CCS systems can operate more efficiently and reduce their reliance on dedicated external energy sources [46].

12.6 Technical and Policy Challenges of Integration

Integrating carbon capture systems with renewable energy offers significant benefits, but it also introduces important technical and policy challenges that must be resolved before large-scale deployment becomes feasible. A key technical difficulty is that most carbon capture technologies were originally designed to operate on a steady, uninterrupted power supply, while renewable sources such as solar and wind generate electricity that fluctuates throughout the day. As a result, adapting carbon capture units to function reliably under variable power conditions requires new system designs, more advanced control strategies, and greater operational flexibility. Many existing capture technologies are not built to ramp up and down quickly, and modifying them to handle changing power levels can be complex and expensive [2, 18].

Another major challenge is the lack of supporting infrastructure. In many regions, the basic systems needed to link renewable energy with carbon capture simply do not exist. These include transmission lines to carry clean electricity over long distances, energy-storage facilities to balance variable renewable output, and CO_2 pipelines or transport networks to move captured carbon to storage sites. Without this infrastructure, it becomes difficult for a renewable-powered carbon capture project to operate efficiently or at a large scale [2].

Building these systems requires long-term planning, strong coordination between governments and industry, and significant financial investment. In addition, facilities for renewable-powered carbon capture equipment, such as high-efficiency compressors, smart control systems, or advanced DAC components, are still being developed. This makes it harder for some countries, especially those where clean-energy technologies are not yet widely available, to deploy renewable-powered carbon capture quickly. Supplying these facilities will be essential for making such projects more possible, reliable, and widely accessible [47].

From a policy perspective, one of the biggest barriers is that most existing regulations and government incentive programs were designed for traditional, fossil-powered carbon capture systems, not for capture systems that run on renewable energy [48]. Many national systems provide financial support for CCS in general, but they do not recognize or reward the additional climate benefits that come from pairing CCS with clean power. As a result, renewable-integrated carbon capture projects often face higher upfront costs while receiving no extra political-economic credit compared with fossil-powered systems. This creates a financial imbalance and obstacles that discourage companies from choosing cleaner, renewable-driven options [49]. To accelerate renewable-integrated carbon capture deployment, policymakers need to develop clearer rules, updated standards, and incentives that specifically support integrated renewable-powered carbon capture systems [48, 49].

There are permitting and regulatory delays that can significantly prolong the time it takes for project developers to obtain the necessary approvals and licenses from government authorities. This challenge becomes even greater for projects that integrate multiple technologies into a single system. Many policymakers and regulatory agencies lack clear guidelines for hybrid plants, flexible operating modes, or facilities that combine renewable energy with CO_2 capture and storage at the same location. Since these integrated systems are relatively new, the approval process can be slow and uncertain, leaving developers unsure about how long it will take to obtain approval or what specific requirements they need to meet. Furthermore, the process is often inconsistent, as different agencies or regions may apply various standards or make different decisions for similar projects. These regulatory gaps pose additional barriers for developers attempting to establish renewable-powered carbon capture systems. Therefore, it is essential to create clear and modern permitting frameworks to accelerate deployment and minimize these regulatory risks [50].

In summary, integrating carbon capture with renewable energy requires both technical adjustments and modernized policy frameworks. Addressing these challenges will make it easier for renewable-powered carbon capture systems to expand and operate reliably, allowing them to play a more effective role in achieving long-term climate and decarbonization goals.

12.7 Key Highlights (Chapter Summary)

◇ **Chapter 12—Key Highlights:**

1) Integrating carbon capture with renewable energy systems helps reduce reliance on fossil fuels for energy production and improves the overall sustainability and climate benefits of carbon removal.
2) Renewable-powered capture systems avoid the additional emissions that would be produced if carbon capture relied on electricity or heat generated from fossil fuels.
3) BECCS provides one of the strongest pathways to net-negative emissions by combining biomass energy production with long-term CO_2 storage.
4) Solar and wind energy can directly power carbon capture units, making the carbon capture process cleaner and more climate-friendly.
5) Hybrid plant designs that mix renewables, waste heat, storage, or grid power help maintain stable operation even when renewable output varies.
6) Energy storage and flexible load management allow carbon capture systems to adapt to fluctuating solar and wind supply without interrupting operations.
7) Renewable-powered carbon capture systems support grid stability by absorbing surplus electricity from solar and wind during periods of high renewable output. This extra energy is then used to power the carbon capture process, helping balance supply and demand on the grid.
8) Technical challenges remain regarding upgrading CCS with renewable energy, such as redesigning carbon capture systems so they can operate under fluctuating power from solar and wind, and upgrading the electricity grid, energy-storage systems, and CO_2 transport infrastructure needed to support renewable-powered operation.
9) Policy gaps persist, as many current regulations and incentives were designed for fossil-powered carbon capture and do not adequately support renewable-driven projects.

10) Successful integration of CCS with renewable energy requires coordinated planning, updated regulatory frameworks, and targeted incentives to encourage investment and ensure long-term sustainability.

Keywords: Renewable Integration, BECCS, Bioenergy Systems, Direct Air Capture, Solar-Powered Capture, Wind-Powered Capture, Hybrid Plant Designs, Energy Storage, Load Management, Grid Flexibility, Waste Heat Utilization, Variable Renewable Energy, Carbon Removal Technologies, Sustainable Decarbonization, Net-Negative Emissions, Infrastructure Planning, Renewable Deployment, CO_2 Transport Networks, Energy-Resource Optimization, Policy Incentives

References

1 A. N. Arnette, "Renewable energy and carbon capture and sequestration for a reduced carbon energy plan: An optimization model," *Renewable and Sustainable Energy Reviews*, vol. 70, pp. 254–265, Apr., 2017. DOI: 10.1016/j.rser.2016.11.218.

2 M. S. Zantye, A. Arora, and M. F. Hasan, "Renewable-integrated flexible carbon capture: A synergistic path forward to clean energy future," *Energy & Environmental Science*, vol. 14, no. 7, pp. 3986–4008, 2021.

3 M. Sharifzadeh, H. Lubiano-Walochik, and N. Shah, "Integrated renewable electricity generation considering uncertainties: The UK roadmap to 50% power generation from wind and solar energies," *Renewable and Sustainable Energy Reviews*, vol. 72, pp. 385–398, May, 2017. DOI: 10.1016/j.rser.2017.01.069.

4 S. Sgouridis, M. Carbajales-Dale, and D. Csala, "Comparative net energy analysis of renewable electricity and carbon capture and storage," *Nature Energy*, vol. 4, pp. 456–465, 2019. DOI: 10.1038/s41560-019-0365-7.

5 M. Anvari *et al.*, "Short term fluctuations of wind and solar power systems," *New Journal of Physics*, vol. 18 , Jun., 063027, 2016. DOI: 10.1088/1367-2630/18/6/063027.

6 M. Bui, M. Fajardy, and N. Mac Dowell, "Bio-energy with carbon capture and storage (BECCS): Opportunities for performance improvement," *Fuel*, vol. 213, pp. 164–175, Feb., 2018. DOI: 10.1016/j.fuel.2017.10.100.

7 A. C. Opia, M. K. B. A. Hamid, S. Syahrullail, A. B. A. Rahim, and C. A. N. Johnson, "Biomass as a potential source of sustainable fuel, chemical and tribological materials—overview," *Materials Today: Proceedings*, vol. 39, no. 2, pp. 922–928, 2021. DOI: 10.1016/j.matpr.2020.04.045.

8 A. Stirbet, D. Lazár, Y. Guo, and G. Govindjee, "Photosynthesis: Basics, history and modelling," *Annals of Botany*, vol. 126, no. 4, pp. 511–537, Sep., 2020. DOI: 10.1093/aob/mcz171.

9 H. Alidrisi and A. Demirbas, "Enhanced electricity generation using biomass materials," *Energy Sources, Part A: Recovery, Utilization, and Environmental Effects*, vol. 38, no. 10, pp. 1419–1427, 2016. DOI: 10.1080/15567036.2014.948647.

10 Z. Luo and J. Zhou, "Thermal conversion of biomass," in *Handbook of Climate Change Mitigation and Adaptation*, M. Lackner, B. Sajjadi, and W. Y. Chen, Eds. Cham: Springer, 2025, DOI: 10.1007/978-3-031-84483-6_27.

11 S. V. Vassilev and C. G. Vassileva, “Composition, properties and challenges of algae biomass for biofuel application: An overview,” *Fuel*, vol. 181, pp. 1–33, Oct., 2016. DOI: 10.1016/j.fuel.2016.04.106.

12 O. Pardo-Planas, H. K. Atiyeh, J. R. Phillips, C. P. Aichele, and S. Mohammad, “Process simulation of ethanol production from biomass gasification and syngas fermentation,” *Bioresource Technology*, vol. 245, no. pt. A, pp. 925–932, Dec., 2017. DOI: 10.1016/j.biortech.2017.08.193.

13 E. Emmanouilidou, S. Mitkidou, A. Agapiou, and N. C. Kokkinos, “Solid waste biomass as a potential feedstock for producing sustainable aviation fuel: A systematic review,” *Renewable Energy*, vol. 206, pp. 897–907, Apr., 2023. DOI: 10.1016/j.renene.2023.02.113.

14 K. Kuparinen, E. Vakkilainen, and T. Tynjälä, “Biomass-based carbon capture and utilization in Kraft pulp mills,” *Mitigation and Adaptation Strategies for Global Change*, vol. 24, pp. 1213–1230, 2019. DOI: 10.1007/s11027-018-9833-9.

15 M. Fridahl and M. Lehtveer, “Bioenergy with carbon capture and storage (BECCS): Global potential, investment preferences, and deployment barriers,” *Energy Research & Social Science*, vol. 42, pp. 155–165, Aug., 2018. DOI: 10.1016/j.erss.2018.03.019.

16 C. Gough and S. Mander, “Beyond social acceptability: Applying lessons from CCS social science to support deployment of BECCS,” *Current Sustainable/Renewable Energy Reports*, vol. 6, pp. 116–123, 2019. DOI: 10.1007/s40518-019-00137-0.

17 M. Fajardy and N. Mac Dowell, “The energy return on investment of BECCS: Is BECCS a threat to energy security?,” *Energy & Environmental Science*, vol. 11, pp. 1581–1594, 2018. DOI: 10.1039/C7EE03610H.

18 H. J. Buck, “Challenges and opportunities of bioenergy with carbon capture and storage (BECCS) for communities,” *Current Sustainable/Renewable Energy Reports*, vol. 6, pp. 124–130, 2019. DOI: 10.1007/s40518-019-00139-y.

19 C. Gough, L. Mabon, and S. Mander, “Social and ethical dimensions of BECCS,” in *Biomass Energy with Carbon Capture and Storage (BECCS): Unlocking Negative Emissions*, C. Gough, P. Thornley, S. Mander, N. Vaughan, and T. Falano, Eds. Hoboken, NJ: Wiley, 2018, pp. 185–202 , ch. 12, DOI: 10.1002/9781119237716.ch12.

20 H. Ishaq, O. Siddiqui, G. Chehade, and I. Dincer, “A solar and wind driven energy system for hydrogen and urea production with CO_2 capturing,” *International Journal of Hydrogen Energy*, vol. 46, no. 6, pp. 4749–4760, Jan., 2021. DOI: 10.1016/j.ijhydene.2020.01.208.

21 H. Fu, X. Zhao, L. Ma, T. Zhang, Q. Wu, and H. Sun, “A comparative study on three types of solar utilization technologies for buildings: Photovoltaic, solar thermal and hybrid photovoltaic/thermal systems,” *Energy Conversion and Management*, vol. 140, pp. 1–13, May., 2017. DOI: 10.1016/j.enconman.2017.02.059.

22 E. Prats-Salvado, N. Jagtap, N. Monnerie, and C. Sattler, “Solar-powered direct air capture: Techno-economic and environmental assessment,” *Environmental Science & Technology*, vol. 58, no. 5, pp. 2282–2292, Jan., 2024. DOI: 10.1021/acs.est.3c08269.

23 J. Song *et al.*, “Enhancing environmental sustainability via interval optimization for low-carbon economic dispatch in renewable energy power systems: Leveraging the flexible cooperation of wind energy and carbon capture power plants,” *Journal of Cleaner Production*, vol. 442, p. 140937, Feb., 2024. DOI: 10.1016/j.jclepro.2024.140937.

24 B. Alonzo, H.-K. Ringkjøb, B. Jourdier, P. Drobinski, R. Plougonven, and P. Tankov, “Modelling the variability of the wind energy resource on monthly and seasonal timescales," *Renewable Energy*, vol. 113, pp. 1434–1446, Dec., 2017. DOI: 10.1016/j.renene.2017.07.019.

25 F. J. Santos-Alamillos, D. Pozo-Vázquez, J. A. Ruiz-Arias, L. Von Bremen, and J. Tovar-Pescador, "Combining wind farms with concentrating solar plants to provide stable renewable power," *Renewable Energy*, vol. 76, pp. 539–550, Apr., 2015. DOI: 10.1016/j.renene.2014.11.055.

26 S. Zhou, Y. Wang, Y. Zhou, L. E. Clarke, and J. A. Edmonds, "Roles of wind and solar energy in China's power sector: Implications of intermittency constraints," *Applied Energy*, vol. 213, pp. 22–30, Mar., 2018. DOI: 10.1016/j.apenergy.2018.01.025.

27 X. Fu, Y. Zhang, X. Liu, and Z. Liu, "Stable power supply system consisting of solar, wind and liquid carbon dioxide energy storage," *Renewable Energy*, vol. 221, p. 119730, Feb., 2024. DOI: 10.1016/j.renene.2023.119730.

28 D. J. Arent, S. M. Bragg-Sitton, D. C. Miller, M. F. Ruth, and J. Cox, "Multi-input, multi-output hybrid energy systems," *Joule*, vol. 5, no. 1, pp. 47–58, Jan., 2021. DOI: 10.1016/j.joule.2020.12.006.

29 Z. Ma, Z. Zhao, Y. Wang, and Z. Hou, "Study on the optimal configuration of battery energy storage system in distribution networks considering carbon capture units," *Recent Advances in Electrical & Electronic Engineering*, vol. 17, no. 6, pp. 582–601, Jul., 2024. DOI: 10.2174/2352096516666230714154857.

30 B. J. Limb *et al.*, "Economic viability of using thermal energy storage for flexible carbon capture on natural gas power plants," *Journal of Energy Storage*, vol. 55, no. Part D, Nov., 105836, 2022. DOI: 10.1016/j.est.2022.105836.

31 A. P. Anyebe, O. K. K. Yeboah, O. I. Bakinson, T. Y. Adeyinka, and F. C. Okafor, "Optimizing carbon capture efficiency through AI-driven process automation for enhancing predictive maintenance and CO_2 sequestration in oil and gas facilities," *American Journal of Environment and Climate*, vol. 3, no. 3, pp. 44–58, 2024.

32 J. Beiron, F. Normann, and F. Johnsson, "Carbon capture from combined heat and power plants: Impact on the supply and cost of electricity and district heating in cities," *International Journal of Greenhouse Gas Control*, vol. 129, Oct., 103973, 2023. DOI: 10.1016/j.ijggc.2023.103973.

33 S. Roussanaly *et al.*, "Offshore power generation with carbon capture and storage to decarbonise mainland electricity and offshore oil and gas installations: A techno-economic analysis," *Applied Energy*, vol. 233–234, pp. 234–251, Jan., 2019. DOI: 10.1016/j.apenergy.2018.10.020.

34 G. Durakovic, "*Hydrogen production in the North Sea: Decarbonizing European energy with offshore wind, natural gas and carbon capture and storage*," 2024.

35 M. Gutsch and J. Leker, "Co-assessment of costs and environmental impacts for off-grid direct air carbon capture and storage systems," *Communications Engineering*, vol. 3, p. 14, 2024. DOI: 10.1038/s44172-023-00152-6.

36 P. Fragkos, "Energy system transitions and low-carbon pathways in Australia, Brazil, Canada, China, EU-28, India, Indonesia, Japan, Republic of Korea, Russia and the United States," *Energy*, vol. 216, p. 119385, Feb., 2021. DOI: 10.1016/j.energy.2020.119385.

37 X. Ye, Z. Dong, J. Lu, and C. Li, "Thermoeconomic evaluation of double-reheat coal-fired power units with carbon capture and storage and waste heat recovery using organic Rankine cycle," *International Journal of Greenhouse Gas Control*, vol. 105, p. 103247, Feb., 2021. DOI: 10.1016/j.ijggc.2020.103247.

38 A. Qadir, M. Sharma, F. Parvaresh, R. Khalilpour, and A. Abbas, "Flexible dynamic operation of solar-integrated power plant with solvent based post-combustion carbon capture (PCC) process," *Energy Conversion and Management*, vol. 97, pp. 7–19, Jun., 2015. DOI: 10.1016/j.enconman.2015.02.074.

39 N. A. Manaf, A. Qadir, and A. Abbas, "Efficient energy management of CO_2 capture plant using control-based optimization approach under plant and market uncertainties," *Journal of Process Control*, vol. 74, pp. 2–12, Feb., 2019. DOI: 10.1016/j.jprocont.2018.07.015.

40 J. W. McNamara, V. DeAngelis, and R. H. Byrne, "Long-duration energy storage in a decarbonized future: Policy gaps, needs, and opportunities," *MRS Energy & Sustainability*, vol. 9, pp. 142–170, 2022. DOI: 10.1557/s43581-022-00037-9.

41 Y. Zeng, T. Zhou, T. Wang, M. Zhang, S. Zhang, and H. Yang, "Long-duration energy storage: A critical enabler for renewable integration and decarbonization," *Energies*, vol. 18, no. 3, p. 466, Jan., 2025. DOI: 10.3390/en18030466.

42 C. F. Heuberger, I. Staffell, N. Shah, and N. Mac Dowell, "Quantifying the value of CCS for the future electricity system," *Energy & Environmental Science*, vol. 9, pp. 2497–2510, 2016. DOI: 10.1039/C6EE01120A.

43 M. Laimon, "Renewable energy curtailment: a problem or an opportunity?," *Results in Engineering*, vol. 26, p. 104925, Jun., 2025. DOI: 10.1016/j.rineng.2025.104925.

44 V. Battaglia and L. Vanoli, "Optimizing renewable energy integration in new districts: Power-to-X strategies for improved efficiency and sustainability," *Energy*, vol. 305, p. 132312, Oct., 2024. DOI: 10.1016/j.energy.2024.132312.

45 A. M. Abdillahi, M. W. Mustafa, S. Y. Abujarad, and M. Mustapha, "Harnessing flexibility potential of flexible carbon capture power plants for future low carbon power systems: Review," *Renewable and Sustainable Energy Reviews*, vol. 81, pp. 3101–3110, Jan., 2018. DOI: 10.1016/j.rser.2017.08.085.

46 J. Asadi and P. Kazempoor, "Advancing power plant decarbonization with a flexible hybrid carbon capture system," *Energy Conversion and Management*, vol. 299, Jan., 117821, 2024. DOI: 10.1016/j.enconman.2023.117821.

47 E. Smirnova, S. Kot, E. Kolpak, and V. Shestak, "Governmental support and renewable energy production: A cross-country review," *Energy*, vol. 230, Sept., 120903, 2021. DOI: 10.1016/j.energy.2021.120903.

48 P. Oduro, P. Simpa, and D. E. Ekechukwu, "Renewable energy expansion: Legal strategies for overcoming regulatory barriers and promoting innovation," *World Journal of Advanced Engineering Technology and Sciences*, vol. 12, no. 1, pp. 168–186, 2024. DOI: 10.30574/wjaets.2024.12.1.0208.

49 J. Kim and K. Park, "Financial development and deployment of renewable energy technologies," *Energy Economics*, vol. 59, pp. 238–250, Sep., 2016. DOI: 10.1016/j.eneco.2016.08.012.

50 I. Michael, "Regulatory hurdles and permitting delays," 2025.

13

Carbon Capture, Circular Economy, and Resource Efficiency

13.1 Circular Economy Structure

The circular economy is a model that aims to reduce waste, keep materials in use for as long as possible, and design systems that regenerate instead of harming natural resources. It challenges the traditional "take, make, use, and dispose" approach, where products are used once and then thrown away [1]. In a circular system, materials are reused, repaired, recycled, or recovered so they stay in the economy rather than ending up as waste. The circular economy model encourages companies to design products that last longer, can be easily repaired, and can be broken down into useful parts and recycled at the end of their life. It also promotes the use of renewable resources and the reduction of dependence on raw materials that are costly, limited, or environmentally damaging to extract [1, 2].

By keeping materials circulating within the system, the circular economy helps conserve natural resources, reduce pollution, and lower greenhouse gas emissions. It also creates more robust and efficient industrial systems, in which one industry's waste can become another's input. This way of thinking supports both environmental protection and economic innovation, making it a robust framework for building a more sustainable future [1, 3].

When the circular economy is applied to carbon management, it asks us to view CO_2 in a new way. Instead of seeing it only as pollution that needs to be removed, CO_2 can be treated as a useful resource that can be captured, reused, and turned into valuable products. In this approach, carbon does not simply move in a straight line from emission to atmosphere. Instead, it becomes part of a continuous loop, where captured CO_2 is fed back into industrial processes, transformed into new materials, or used to create cleaner fuels [4].

This shift in thinking fits naturally with global climate goals. By keeping CO_2 in circulation rather than releasing it, we reduce emissions while opening the door to new industries and economic opportunities. Turning CO_2 into fuels, chemicals, building materials, and other products helps create a system in which environmental protection and economic development can happen at the same time. In simple terms, circular carbon systems help us manage emissions more intelligently, making the entire industrial system cleaner, more efficient, and more innovative [3, 4].

In a circular economy framework, industries try to design processes that avoid waste from the very beginning. Instead of allowing valuable resources to escape into the environment, the goal is to keep them circulating within production cycles for as long as possible.

For carbon-intensive sectors such as cement, steel, and power generation, this means capturing CO_2 before it leaves the smokestack and directing it into new manufacturing pathways for reuse [5].

The circular approach also encourages companies to use energy and materials more efficiently, reducing unnecessary losses throughout the system [6]. This includes improving energy efficiency, recovering heat, recycling solvents, and choosing materials that are easier to reuse or repurpose. Products are designed so that when they reach the end of their useful life, their components can be repaired, reused, or transformed into new products rather than discarded [7].

Figure 13.1 shows a visual overview of the circular carbon economy, highlighting how CO_2 can move through a continuous loop of removal, reduction from the atmosphere, reuse, and recycling rather than being released into the atmosphere [3]. The diagram begins at the top with renewable energy sources, such as solar and wind. These technologies provide clean electricity to power industrial systems without adding new emissions [8]. This renewable energy drives the central processes that enable circular carbon systems by supplying the electricity and heat needed to power every significant step of the circular carbon loop, including carbon capture, utilization, and transportation, without adding new emissions [9]. When fossil fuels power these energy-intensive steps,

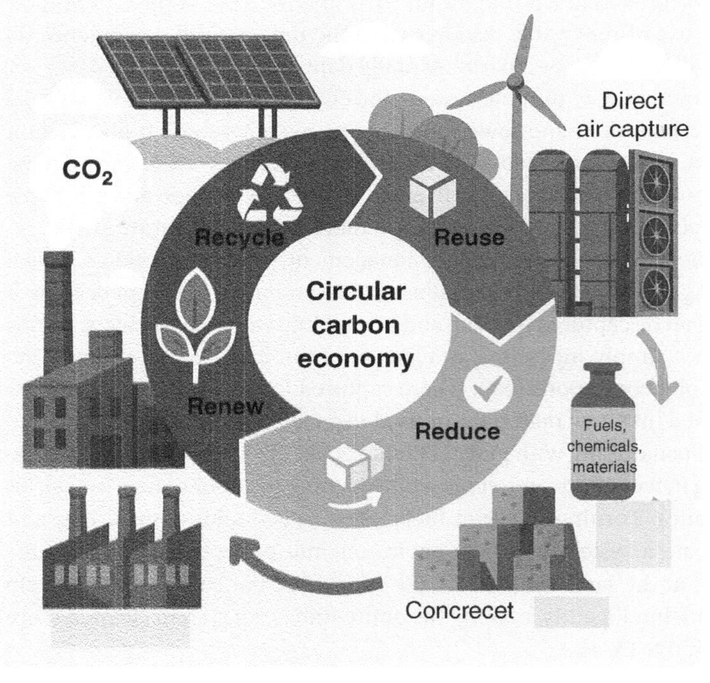

Figure 13.1 Circular carbon system showing how CO_2 is reduced, reused, recycled, and powered by renewable energies through integrated industrial processes powered by renewable energy. The closed-loop structure keeps carbon circulating within the economy instead of being released to the atmosphere, supporting sustainability and resource efficiency.

additional CO_2 is created, reducing the net benefit of the system. But when solar, wind, hydropower, or other renewable sources supply that energy, the entire cycle becomes clean [8].

At the top right of the figure, carbon dioxide removal technologies collect CO_2 from industrial facilities, power plants, or directly from the air (DAC) [10]. Instead of allowing these emissions to escape, the system redirects them into new pathways. The captured CO_2 then moves into the next phase, represented on the bottom right side, where it can be converted into useful products such as fuels, chemicals, and concrete, and other products through carbon utilization. By transforming CO_2 into valuable goods, the system keeps carbon circulating within the economy rather than contributing to atmospheric buildup [11].

The circular loop at the center of the figure illustrates the core principles of a circular carbon system, showing how carbon can be continuously reused rather than released into the atmosphere. **Reduce** refers to minimizing emissions and resource use at the industrial plants by improving efficiency, lowering energy consumption, and optimizing industrial processes [12]. For instance, a cement plant can reduce emissions by improving the efficiency of its kilns, using alternative low-carbon fuels, or recovering waste heat so that less energy is required during production. **Reuse** represents the direct use of captured CO_2 in industrial operations, where CO_2 becomes a feedstock rather than a waste product. **Recycle** involves transforming CO_2 into new materials or fuels such as methanol, polymers, synthetic fuels, or mineralized products, turning carbon into a valuable input [13]. **Renew** highlights the role of renewable energy in powering these processes, supplying clean electricity and heat for carbon capture, conversion, and product manufacturing [8]. Together, these four principles create a closed-loop system in which carbon circulates repeatedly through industrial and energy pathways, reducing waste, lowering emissions, and supporting long-term sustainability.

In summary, by integrating carbon capture with circular economy principles, industries can shift toward regenerative systems in which resources continually cycle rather than moving in a one-way, use-and-discard direction. In this approach, captured CO_2 becomes part of a closed loop, feeding back into new products, materials, and industrial processes instead of being released into the atmosphere.

13.2 From Waste to Value: CO_2 as a Resource

Traditionally, carbon dioxide has been treated purely as a waste product, something harmful that escapes from industrial processes and contributes to climate change [14]. The main focus has been on capturing it and storing it away so it can no longer damage the atmosphere. In a circular economy, however, CO_2 is reimagined as a resource rather than a substance that must be thrown away [4]. Instead of viewing carbon only as a problem, this approach encourages us to see its potential as a raw material that can be captured and put to productive use.

This shift in mindset opens the door to many new possibilities. Captured CO_2 can be transformed into fuels, chemicals, plastics, building materials, fertilizers, and other products that support industry while lowering emissions. By turning CO_2 into something useful, industries can create new economic opportunities, reduce reliance on fossil-based materials, and contribute to a cleaner and more efficient industrial system [11].

Using captured CO_2 as a raw material also helps reduce dependence on fossil-based resources such as crude oil, natural gas, and mined minerals. When CO_2 is transformed into industrial feedstocks, it can directly replace chemicals and materials that are normally produced from petroleum or gas [15]. For example, CO_2 can be converted into methanol, which is widely used to make plastics, resins, and fuels. Traditionally, methanol is produced from natural gas; using CO_2 instead avoids both extraction and combustion emissions of the natural gas [16].

Similarly, CO_2 can be used to produce synthetic fuels such as e-diesel, e-kerosene, and e-gasoline. These fuels can substitute for fossil-based transport fuels in aviation, shipping, and heavy industry [17]. CO_2 is also used to create polycarbonates and polyurethanes, which are the building blocks of many plastics and foams, again replacing materials normally derived from fossil hydrocarbons. In the construction sector, captured CO_2 can be mineralized and used to make concrete blocks, aggregates , and cement additives. These products not only avoid emissions associated with producing traditional cement and aggregates but also store CO_2 permanently within the material [11]. Usage of carbon supports more efficient supply chains and helps industries move toward cleaner production methods.

Figure 13.2 illustrates the core idea of transforming CO_2 from an unwanted emission into a valuable product within a circular carbon economy. At the center of the diagram is the concept of a circular economy, emphasizing continuous reuse of carbon rather than its one-way release into the atmosphere. The image shows CO_2 being captured from an industrial facility and redirected into several value-creating pathways. On the right, one major route is the synthesis of low-carbon fuels, such as e-diesel, e-kerosene, and methanol, where captured CO_2 becomes a feedstock for energy carriers that can directly replace

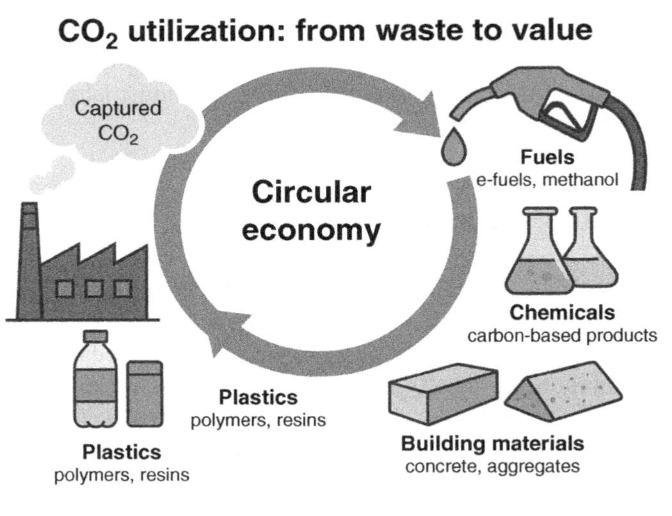

Figure 13.2 Conceptual pathways for CO_2 utilization within a circular carbon economy, illustrating how captured CO_2 can be converted into low-carbon fuels, chemicals, plastics, and mineralized building materials, transforming emissions into valuable products rather than atmospheric waste.

fossil-derived fuels. Adjacent to this, the illustration highlights the production of chemicals and polymers through processes like methanol-to-olefins, enabling CO_2 to serve as a building block for resins, plastics, and a broad range of carbon-based products traditionally derived from petroleum. At the bottom, the diagram presents the mineralization route, where CO_2 is chemically bound into solid building materials such as aggregates, concrete, and cement alternatives, effectively storing carbon in long-lived infrastructure. Together, these pathways visually reinforce the shift from treating CO_2 as waste to integrating it into industrial value chains, reducing emissions while generating new economic opportunities.

At the same time, CO_2 utilization helps create more efficient and better-connected industrial systems. For example, a cement plant that typically releases large amounts of CO_2 can capture that gas and supply it to a nearby company that mineralizes CO_2 into concrete blocks or aggregates. In this case, a waste emission becomes a building material [18]. Similarly, a carbon capture plant can capture CO_2 from refinery or power plant and sell it to a chemical manufacturer that converts it into methanol, synthetic fuels, or plastics [11]. These exchanges reduce emissions while creating practical uses for captured carbon. These relationships form networks where the output of one industry becomes the input of another, strengthening overall system efficiency. Instead of allowing resources to flow in a one-way direction toward waste, carbon, and materials circulate through productive loops.

These emerging technologies make emissions reduction far more attractive to industry by introducing new ways to generate revenue from captured carbon. When CO_2 can be transformed into marketable products such as fuels, polymers, construction materials, or chemicals, it becomes a value-creating asset rather than a cost [11]. This shift changes the economics of carbon capture: companies no longer face only the expenses of installing capture equipment and complying with regulations. Instead, they have the potential to earn income from selling CO_2-based products, licensing new technologies, or entering growing carbon-to-materials markets [19].

The financial advantages of utilizing CO_2 are increasing as more individuals and companies seek low-carbon products. Additionally, many governments are providing support through grants, tax credits, and carbon markets to encourage the use of CO_2. Companies that adopt CO_2 utilization early can gain advantages, offering innovative products, accessing new markets, and reducing their reliance on fossil fuels, which often have volatile prices. Because of these advantages, CO_2 utilization changes the economics of carbon capture. Managing emissions is no longer just an extra cost for industries. Instead, it becomes a chance to create valuable products, earn new revenue, and innovate. What used to be seen as a problem, carbon emissions, can now help drive industrial growth [20].

Figure 13.3 highlights the economic potential of CO_2 utilization by showing how carbon that would normally be emitted from industrial facilities can become a valuable economic resource. On the left side, a factory releases CO_2, represented by a cloud labeled "CO_2," symbolizing traditional industrial emissions. A set of consumer products, packaging, and bottles marked with green leaf symbols, representing the transformation of captured CO_2 into marketable low-carbon goods such as fuels, chemicals, and plastics. To the right, a rising bar chart and bold upward arrow symbolize economic growth and increasing revenue, showing how companies benefit financially when they convert captured CO_2 into useful products. The stacked gold coins emphasize profit generation and the emergence of new carbon-based

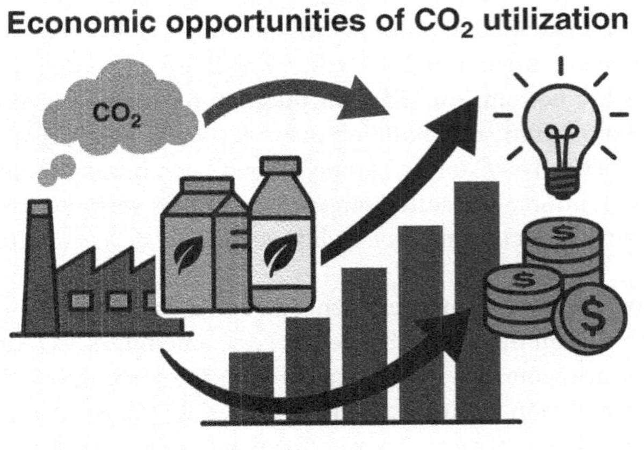

Figure 13.3 Economic opportunities created by CO_2 utilization, showing how captured carbon can be transformed into low-carbon products, innovation, and new revenue streams.

markets. The light bulb, placed above the financial symbols, represents innovation, signaling that CO_2 utilization drives technological development and creates space for new ideas, new products, and new business models. Overall, the picture communicates that CO_2 utilization is not only an environmental strategy but also a powerful economic opportunity.

This transformation from "waste to value" is a core idea of the circular economy. It encourages industries to move away from the old linear model of using materials once and then throwing them away. Instead, companies are motivated to close loops, keep resources in circulation, and design processes that make smarter use of materials [21]. When CO_2 is treated as a valuable input rather than something to be discarded, it changes the entire role of carbon capture. Carbon capture is no longer just a technology for cleaning up emissions at the end of a process. It becomes part of a larger strategy that supports new products, new business opportunities, and more efficient use of resources. In this way, CO_2 utilization turns carbon capture into a tool for sustainable industrial growth, helping industries innovate while reducing environmental impact [22].

13.3 Utilization Pathways: Fuels, Chemicals, and Building Materials

Captured CO_2 can be directed into multiple utilization pathways to produce valuable outputs. Instead of simply focusing on preventing emissions, these pathways turn CO_2 into a functional ingredient within industrial processes. Fuels, chemicals, and building materials are still the main areas where CO_2 can be used, but each of these categories now includes many new and creative technologies. Some methods use CO_2 in its current form, while others mix it with renewable hydrogen or special catalysts to make useful products. For example, CO_2 and hydrogen can be combined to make clean synthetic fuels for planes and ships, or converted into methanol, which is used to produce everyday items like plastics, paints, and adhesives [11, 16–18].

One major pathway is converting CO_2 into synthetic fuels. These include methanol, methane, aviation fuels, and other liquid fuels made by combining captured CO_2 with hydrogen produced from renewable energy. These fuels can replace fossil fuels in transportation, especially in sectors that are hard to electrify, such as aviation and shipping. While producing synthetic fuels requires energy, the overall process can become low-carbon when powered by renewables. The result is a fuel system that reuses carbon instead of adding new carbon to the atmosphere [17].

In the construction sector, CO_2 can react with minerals to create strong building materials, such as carbon-cured concrete blocks or CO_2-based cement additives. These products lock the carbon into solid forms, keeping it out of the atmosphere [18]. New research is pushing CO_2 utilization even further, opening the door to products that were not possible only a few years ago. One promising area is the development of advanced materials made partly or entirely from CO_2. For example, some companies can turn CO_2 into lightweight carbon-based fibers, which are strong materials used in applications such as bicycles, car parts, sports equipment, airplane components, and wind turbine blades. These fibers are normally made from petroleum, so using CO_2 instead reduces the need for fossil fuels while creating high-performance materials [23].

Biotechnology is also expanding the range of CO_2-based products. Some companies now use microorganisms, such as bacteria or algae, to convert CO_2 into valuable biological substances. These microorganisms act as biological conversion systems: they take in CO_2 and transform it into proteins, enzymes, or specialized chemicals. Proteins produced this way can be used for animal feed, food ingredients, or sustainable fertilizers, offering an alternative to land- and water-intensive agricultural sources. Enzymes made from CO_2 are used in detergents, food processing, medicine, and biofuel production, acting as natural "tools" that speed up chemical reactions. These biological and material innovations show that CO_2 is far more useful than once assumed [24].

Figure 13.4 presents a modern and visually engaging overview of how carbon dioxide can be transformed into eatable protein using renewable electricity. In the background, wind turbines and an industrial facility symbolize both the source of CO_2 emissions and the clean energy used to power the CO_2 conversion process. On the left side of the image, CO_2 molecules appear near the industrial plant, emphasizing that the system begins with captured carbon dioxide. A lightning-bolt icon represents renewable electricity, highlighting the essential role of clean energy in driving the biological and electrochemical reactions that follow [25].

The central portion of the diagram shows two interconnected pathways for turning CO_2 into useful biological products. In the first route, CO_2 and renewable electricity contribute to reactions that generate oxygen, which then supports microbial processes that ultimately produce protein. In the second route, the diagram illustrates electro-methanogenesis, a process in which specialized microorganisms use electricity to convert CO_2 into methane (CH_4). In simple terms, these microbes receive electrons from renewable electricity and combine them with CO_2 and water to produce methane. This methane then serves as the main energy and carbon source for methanotrophic microorganisms, a special group of bacteria that naturally grow by consuming methane. As these methanotrophs feed on the methane, they multiply and build cellular biomass [26]. Because microbial cells are rich in proteins, amino acids, and nutrients, this growing biomass can be harvested and processed into microbial protein, a sustainable, low-carbon food ingredient [27].

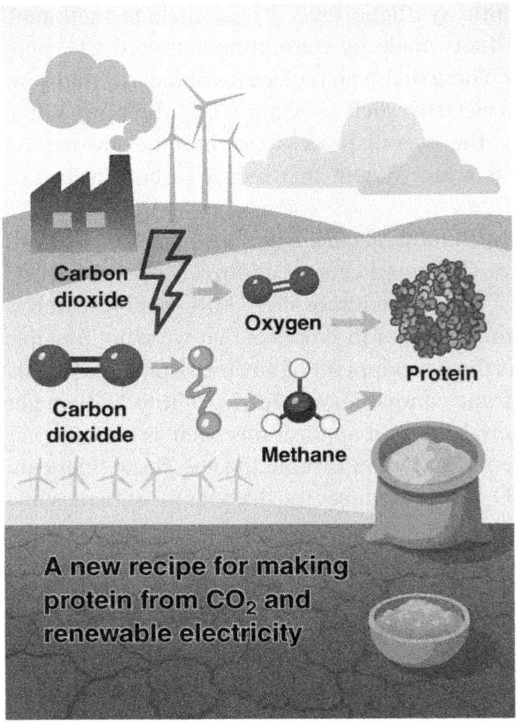

Figure 13.4 Schematic representation of converting CO_2 indirectly into microbial protein through renewable-electricity-driven pathways, including electro-methanogenesis and methanotrophic biomass production.

At the bottom right, bowls and bags filled with powder represent the final protein product, illustrating how this process can yield a nutritious and scalable food source. The cracked earth beneath the scene evokes the global challenges of climate change, food security, and resource stress, reinforcing the relevance of sustainable protein production. Overall, the image illustrates a circular, low-carbon system in which waste CO_2 and renewable electricity are transformed into valuable protein, showcasing an innovative pathway for future food and climate resilience.

In addition to choosing the right utilization pathway, it is important to consider the purity of the captured CO_2. Different applications require different quality levels, which can influence energy use and processing costs. For example, mineralization processes can often use low-purity CO_2 directly from industrial exhaust streams, making them relatively easy to integrate. In contrast, producing synthetic fuels and chemicals usually requires high-purity CO_2, which must be carefully separated and cleaned before use. Biotechnology sits somewhere in between: microorganisms can tolerate some impurities but still require a consistent and stable CO_2 stream. Understanding these purity needs helps determine which utilization options are most suitable for specific industries [28].

Many CO_2 conversion pathways, especially those that produce fuels or chemical feedstocks, require significant amounts of energy. This is because breaking and reforming chemical bonds takes heat, electricity, or hydrogen produced from clean power sources [29]. As a result, the climate benefits of CO_2 utilization depend heavily on using renewable or low-carbon energy. When powered by solar, wind, hydropower, or waste heat, these

Figure 13.5 Many CO_2 utilization technologies need large amounts of energy, both electrical and thermal, to convert captured CO_2 into fuels or chemicals, showing why carbon utilization technology is not yet widely used at an industrial scale.

systems can achieve meaningful reductions in lifecycle emissions. But if fossil-based energy is used, the overall environmental gains can be limited. Ensuring access to affordable renewable energy is therefore a key requirement for CO_2 utilization technologies.

Although CO_2 utilization offers many exciting possibilities, not all of these technologies are ready for large-scale industrial use yet. Many of the processes still need a lot of energy, depend on advanced or expensive catalysts, or require CO_2 that is extremely pure, which can be difficult or costly to produce [28, 29]. Because of these limitations, CO_2 utilization should be viewed as a helpful addition to long-term storage, not a full replacement for it. In other words, utilization can handle part of the captured CO_2, but deep emissions reduction will still depend on storing CO_2 safely underground. As research continues, costs come down, and supportive government policies are developed, more CO_2 utilization pathways will become practical and widely used in the future [30].

Figure 13.5 shows a conceptual illustration of CO_2 utilization systems and the significant energy inputs they require. Captured CO_2 is directed toward reactors that convert it into fuels, chemicals, or other products. The top of the illustration highlights the need for substantial electricity, shown through power lines and energy-flow symbols. At the bottom, the image also includes a large heat source, representing the thermal energy required for many CO_2 conversion processes, such as catalytic reactions and electrolysis processes. By combining both electricity and heat requirements, the figure emphasizes the core message: although CO_2 utilization offers promising opportunities, many pathways remain highly energy-intensive and depend on reliable, large-scale energy supply in both electrical and thermal forms [29]. As a result, energy demand remains a major barrier to deploying CO_2 utilization at an industrial scale.

Several industrial projects around the world demonstrate how CO_2 utilization is moving from laboratory research to commercial reality. Companies such as CarbonCure and Carbfix are incorporating CO_2 into concrete and mineralized products at growing scales [31]. In the chemicals sector, firms like Covestro use captured CO_2 to produce plastics, reducing the need for petroleum-derived ingredients [32]. Biotechnology leaders such as LanzaTech convert CO_2-rich gas streams into fuels and specialty chemicals using microbial fermentation [33]. Specialty chemicals are high-value chemicals made for very specific uses, often in small amounts. They are used in products like medicines, cosmetics, electronics, and coatings to give them special properties or make them work better. Making these chemicals from CO_2 offers a cleaner alternative to using fossil fuels. These examples show that CO_2 utilization is not just a theoretical idea but an emerging industry that continues to expand as technologies mature.

13.4 Resource Efficiency in Capture Technologies

Resource efficiency is an important part of making carbon capture technologies more sustainable. These systems can use large amounts of energy, water, heat, and chemical solvents, so improving efficiency helps reduce energy usage, environmental impact, and operating costs. In simple terms, resource efficiency means getting the same amount of CO_2 captured while using fewer resources. This is especially important as carbon capture becomes more common in industries such as power generation, cement, steel, and chemicals [34].

One of the main areas of improvement is energy efficiency. Traditional solvent-based capture systems need heat to regenerate the solvent and electricity to run pumps, fans, and compressors. Newer technologies focus on reducing these energy needs by using advanced solvents, solid sorbents, membranes, and better heat-recovery systems [35, 36]. For example many plants recover waste heat from industrial processes and use it to power solvent regeneration instead of burning extra fuel [36]. Using renewable electricity also helps reduce the carbon footprint of the capture process.

Another key area is solvent and material efficiency. Carbon capture systems can lose small amounts of solvent during operation, which must be replaced. Engineers try to reduce these losses by improving solvent stability and minimizing solvent evaporation [37]. Solid sorbent systems and membrane technologies are also being developed to reduce chemical consumption and lower overall maintenance requirements.

Water use is also a critical resource consideration. Some carbon capture technologies, especially those in power or industrial plants, need water for cooling, cleaning, or steam generation. To improve water efficiency, facilities can recycle process water, use dry-cooling technologies, or use alternative water sources such as treated wastewater. These approaches help ensure that carbon capture does not create extra pressure on local water supplies, particularly in dry or water-stressed regions [38].

Digital tools and automation are becoming increasingly important for making carbon capture systems more efficient. Modern plants use sensors and real-time monitoring to constantly measure how much energy, water, and chemicals the system is using. If something is not working well, for example, if a pump is using too much electricity or a cooling system is wasting water, smart software as a part of digital tool can quickly detect the

Figure 13.6 Engineer using digital monitoring and automation tools to optimize the performance of a carbon capture system through real-time data analysis.

problem. After recognizing the problem, these digital systems can also adjust operations automatically. For instance, they can change airflow, control temperatures, or fine-tune solvent circulation without waiting for a human operator. This helps the plant run smoothly and avoids unnecessary resource use. By using data and automation, carbon capture units can operate at their highest possible efficiency, reduce waste, and lower costs [39].

Figure 13.6 shows an engineer using a digital monitoring interface in an industrial facility, symbolizing the growing role of automation and smart technologies in modern carbon capture operations. The engineer is analyzing a real-time process schematic displayed on a laptop. Modern sensors and digital monitoring systems allow operators to continuously track key performance indicators, such as energy consumption, solvent flow, and cooling efficiency, and to identify inefficiencies instantly [40]. By combining digital control platforms with the presence of experienced field engineers, plants can make rapid operational adjustments, ensuring higher reliability, reduced emissions, and more sustainable use of resources.

Improving resource efficiency also makes carbon capture much easier to integrate into existing plants and industrial sites. When a capture system uses less energy and water, it places less pressure on the host facility's resources. This means it can fit more easily into power plants, cement factories, steel mills, or chemical plants without pressurizing their infrastructure. Efficient carbon capture units can share heat, electricity, and water systems with the main plant, making a more compact and simpler layout and installation and reducing the need for expensive upgrades. In other words, the more resource-efficient a carbon capture system is, the more smoothly it can connect with other industries and become part of their everyday operations.

Overall, improving resource efficiency makes carbon capture more efficient, cleaner, cheaper, and easier to integrate into different industries. By using less energy, fewer chemicals, and smarter water management, capture technologies can better support long-term sustainability and help industries meet climate goals without placing unnecessary strain on resources.

13.5 Designing Closed-Loop Carbon Systems

Designing closed-loop carbon systems means building industries where carbon does not escape into the atmosphere but stays circulating within the system for as long as possible. In a closed loop, CO_2 becomes a useful resource instead of a waste product. This idea fits within the broader concept of a circular economy, where materials are reused, recycled, and recovered rather than thrown away. The goal is not only to reduce carbon emissions from industrial systems and prevent pollution, but also to utilize carbon dioxide for producing valuable products that benefit industries, economies, and societies [41].

A closed-loop carbon system begins when CO_2 is captured from factories, power plants, bioenergy facilities, or even from the air. Instead of sending all this CO_2 for permanent storage, some of it can be transformed into valuable products, such as synthetic fuels, chemical feedstocks, plastics, fertilizers, or cement-like building materials. When these products eventually reach the end of their life, the carbon inside them can be recovered again, either through recycling or by capturing CO_2 during combustion or breakdown. This creates a repeating cycle where carbon keeps flowing from one useful application to another [41].

To make these loops work effectively, products must be designed with future carbon recapture in mind. For example, if CO_2-based fuels are burned in a plant equipped with carbon capture, the released CO_2 can immediately reenter the cycle. CO_2-based building materials can hold carbon safely for decades, and CO_2-derived plastics can be mechanically or chemically recycled several times. In this way, industries begin thinking about carbon not just at the point of use, but across the entire lifespan of a product [41].

Closed-loop carbon systems depend on cooperation between different industries, because no single facility can create a full carbon cycle on its own. Each sector contributes something different to keep carbon circulating [41]. For instance, a power plant equipped with a carbon capture system can collect CO_2 from its flue gas and send that captured CO_2 to a nearby chemical plant, where it can be converted into synthetic fuels (such as e-methanol or e-kerosene), plastics, fertilizers, or other industrial products [42]. Cement plants, which operate at very high temperatures, often generate large amounts of waste heat. This waste heat can be used to support solvent regeneration in CCS units or to power heat-intensive steps in CO_2 conversion processes [43]. Cement facilities can also direct their captured CO_2 to industrial hubs where it becomes a feedstock for fuels, chemicals, or mineralized materials [44].

Figure 13.7 shows how a closed-loop carbon system works through collaboration between a power plant, a chemical plant, and a cement plant. The power plant, equipped with a carbon capture unit, collects CO_2 from its flue gas and sends it to the chemical plant, where the captured CO_2 is converted into valuable products such as synthetic fuels, fertilizers, or plastics. The cement plant, which operates at very high temperatures, provides waste heat that can be reused to support solvent regeneration in the capture unit or drive CO_2 conversion reactions in the chemical plant. Electricity from renewable sources, such as wind or solar power, helps to run the capture and conversion processes with minimal additional emissions. The flow of materials, heat, and CO_2 between these facilities demonstrates how industries can connect their processes to create a circular carbon loop, where CO_2 is captured, reused, and kept in industrial circulation instead of being released into the atmosphere [45].

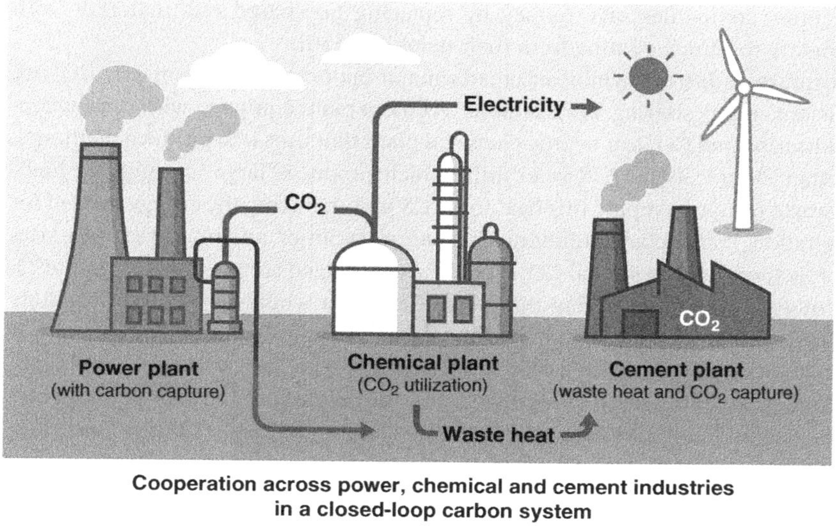

Figure 13.7 Cooperation across power, chemical, and cement industries in a closed-loop carbon system.

Waste-to-energy plants can also support the cycle. When CO_2-based fuels or plastics reach the end of their life, these facilities can burn them for energy and capture the resulting CO_2 again, keeping the carbon in circulation instead of releasing it [46, 47]. Meanwhile, renewable electricity from wind, solar, or hydro can supply the clean power needed for electrolysis, CO_2 conversion, and other low-carbon industrial steps [48]. When these industries exchange CO_2, share waste heat, and use renewable electricity together, they form an integrated network where carbon, energy, and materials move efficiently across the system. This cooperation allows carbon to stay in use for much longer, forming a regenerative, closed-loop carbon cycle [41].

13.6 Industrial Symbiosis and Innovation

Industrial symbiosis or cooperation is a core idea in the circular economy. It describes situations where different industries work together by sharing resources such as energy, heat, materials, and even waste streams. Instead of operating separately, these industries form connected networks in which the by-products from one facility become useful inputs for another. This cooperation helps reduce emissions, lower operating costs, and spark new technological ideas across entire industrial clusters [49].

Industrial symbiosis helps reduce emissions, lower costs, and encourage innovation by turning waste streams into shared resources across connected industries. When one facility sends its excess heat, captured CO_2, wastewater, or by-products to another facility that can use them as inputs, both sides avoid the environmental impact and expense of producing those resources from scratch. This exchange reduces overall energy consumption and cuts emissions that would have occurred from burning additional fuel or disposing of waste.

At the same time, companies save money by replacing purchased raw materials with low-cost secondary resources coming from their neighbors [50].

In carbon capture, industrial symbiosis creates major opportunities for smarter CO_2 use and more efficient energy sharing. For example, a cement plant equipped with carbon capture can send its captured CO_2 to a nearby chemical plant that uses it to produce methanol, plastics, or other chemicals [51]. A steel mill, which produces large amounts of high-temperature waste heat, can supply this heat to a CCS unit, reducing the energy needed for solvent regeneration. Power plants, refineries, bioenergy facilities, and manufacturing sites can also link together through shared CO_2 pipelines, centralized storage hubs, or joint CO_2 utilization centers [52]. By connecting these facilities, what would normally be separate projects becomes an integrated carbon network.

Industrial networks make it much easier to develop and scale new carbon technologies, meaning technologies that capture CO_2, convert CO_2 into useful products, or permanently store CO_2, because different facilities are physically close and already sharing materials. When a cement plant, steel mill, chemical plant, and recycling facility are connected, they can supply each other with real industrial inputs such as captured CO_2, waste heat, or by-products without needing expensive transportation or special arrangements. This enables a wide range of new carbon technologies, such as CO_2-to-fuel reactors or CO_2-to-chemicals processes, to be tested under real factory conditions instead of only in small laboratory settings [49, 50]. For example, if a chemical plant wants to test a new CO_2-to-fuel conversion process, it can receive a stable, continuous flow of captured CO_2 directly from the cement plant next door, allowing researchers to evaluate how the system performs during real, constant industrial operation. This close cooperation helps companies experiment faster, solve technical challenges earlier, and refine their technologies based on real performance data. As a result, innovative carbon technologies can progress from pilot scale to full commercial deployment much more quickly and successfully.

Shared infrastructure also plays a major role. Building CO_2 pipelines, storage sites, or processing hubs is very expensive for one company alone. But when multiple industries use the same CO_2 pipeline network, the cost is divided among them, making participation affordable even for smaller companies. The same is true for shared storage sites, where CO_2 from several factories can be stored safely underground. This collective approach lowers costs, increases efficiency, and encourages more companies to take part in decarbonization efforts [53].

Industrial symbiosis helps make entire regions more sustainable by encouraging industries to work together instead of operating separately. This cooperation reduces the total amount of energy, water, and raw materials that each company needs to use. By sharing resources and reusing what would normally be wasted, industries can lower costs, cut emissions, and reduce pressure on the environment. It also helps create local networks where companies depend on each other, building a sense of collaboration rather than competition. Over time, this approach can lead to low-carbon industrial ecosystems, where carbon capture becomes a key connecting link, collecting emissions from one site and supplying useful CO_2 or heat to another [49, 50].

In this way, industrial symbiosis not only supports cleaner production but also strengthens the overall sustainability and resilience of regional economies [49, 50]. It shows how cooperation and smart resource use can turn waste into value while helping multiple industries reach their climate goals together.

13.7 Key Global CCS Initiatives and Projects

◇ Chapter 13—Key Highlights:

1) This chapter connects carbon capture and storage (CCS) with the broader goals of the circular economy, emphasizing the reuse of carbon rather than simple disposal.
2) This chapter explains how captured CO_2 can be treated as a valuable resource for new industrial products, not just a waste material to be stored underground.
3) The circular economy structure encourages coordination between industries to create carbon closed-loop systems, where emissions are reduced, resources are reused, and waste is minimized across the value chain.
4) CO_2 utilization technologies can convert captured carbon into useful products, such as synthetic fuels, chemicals, and construction materials, supporting both decarbonization and economic growth.
5) Using CO_2 in concrete, plastics, and fuels demonstrates how carbon capture can contribute to low-carbon manufacturing and material innovation.
6) This chapter highlights resource efficiency in carbon capture technologies through better water and energy use, waste heat recovery, and using digital technologies.
7) Designing closed-loop carbon systems helps integrate capture, utilization, and reusing and recycling processes across multiple industries.
8) Industrial symbiosis, where waste streams from one sector become inputs for another, reduces costs by sharing infrastructure and reducing the need for new resources, while creating new economic and environmental opportunities.
9) CO_2 reuse technologies must be supported by strong policy frameworks, market incentives, and lifecycle assessment tools to ensure real environmental benefits.
10) Integrating carbon capture into a circular economy model can transform industries from being major emitters into active participants in climate solutions.

Keywords: Circular Economy, Carbon Capture, Carbon Utilization, Resource Efficiency, CO_2 Recycling, Industrial Symbiosis, Synthetic Fuels, Building Materials, Carbon Innovation, Closed-Loop Systems, Sustainable Manufacturing, Green Economy, Waste-to-Value, Carbon Neutrality, Regenerative Systems

References

1 S. Daga, K. Yadav, P. Singh, and V. Mishra, "Beyond the take-make-dispose model—unlocking the power of circular economy for an environmentally resilient future," in *Circular Economy and Environmental Resilience*, P. Singh, S. Daga, K. Yadav, and V. Mishra, Eds. Cham: Springer, 2025, DOI: 10.1007/978-3-031-93091-1_1.

2 M. Geissdoerfer, P. Savaget, N. M. P. Bocken, and E. J. Hultink, "The circular economy: A new sustainability paradigm?," *Journal of Cleaner Production*, vol. 143, pp. 757–768, Feb., 2017.

3 M. Yang *et al.*, "Circular economy strategies for combating climate change and other environmental issues," *Environmental Chemistry Letters*, vol. 21, pp. 55–80, 2023. DOI: 10.1007/s10311-022-01499-6.

4 L. M. Alsarhan, A. S. Alayyar, N. B. Alqahtani, and N. H. Khdary, "Circular carbon economy (CCE): A way to invest CO_2 and protect the environment, a review," *Sustainability*, vol. 13, no. 21, p. 11625, 2021. DOI: 10.3390/su132111625.

5 R. Karim, A. Waaje, M. Mustaqim Roshid, and M. B. Yeamin, "Turning the waste into wealth: Progressing toward global sustainability through the circular economy in waste management," in *Sustainable Waste Management in the Tourism and Hospitality Sectors*, A. Kumar, S. Sharma, and R. Gupta, Eds. IGI Global, 2025, p. 46, DOI: 10.4018/978-8-3693-6110-8.ch019.

6 F. Di Maio, P. C. Rem, K. Baldé, and M. Polder, "Measuring resource efficiency and circular economy: A market value approach," *Resources, Conservation and Recycling*, vol. 122, pp. 163–171, Jul., 2017. DOI: 10.1016/j.resconrec.2017.02.009.

7 Y. Cao, H. Lu, and C. Zhu, "Consumer preference for end-of-life scenarios and recycled products in circular economy," *Sustainability*, vol. 14, no. 19, p. 12129, 2022. DOI: 10.3390/su141912129.

8 S. Algarni, V. Tirth, T. Alqahtani, S. Alshehery, and P. Kshirsagar, "Contribution of renewable energy sources to the environmental impacts and economic benefits for sustainable development," *Sustainable Energy Technologies and Assessments*, vol. 56, p. 103098, Mar., 2023. DOI: 10.1016/j.seta.2023.103098.

9 A. G. Olabi, "Circular economy and renewable energy," *Energy*, vol. 181, pp. 450–454, Aug., 2019. DOI: 10.1016/j.energy.2019.05.196.

10 M. Z. Jacobson, "The health and climate impacts of carbon capture and direct air capture," *Energy & Environmental Science*, vol. 12, no. 12, pp. 3567–3574, 2019.

11 Q. Zhu, "Developments on CO_2-utilization technologies," *Clean Energy*, vol. 3, no. 2, pp. 85–100, Jun., 2019. DOI: 10.1093/ce/zkz008.

12 S. Walker, N. Coleman, P. Hodgson, N. Collins, and L. Brimacombe, "Evaluating the environmental dimension of material efficiency strategies relating to the circular economy," *Sustainability*, vol. 10, no. 3, p. 666, 2018. DOI: 10.3390/su10030666.

13 R. Minunno, T. O'Grady, G. M. Morrison, and R. L. Gruner, "Exploring environmental benefits of reuse and recycle practices: A circular economy case study of a modular building," *Resources, Conservation and Recycling*, vol. 160, p. 104855, Sep., 2020. DOI: 10.1016/j.resconrec.2020.104855.

14 T. A. Jacobson, J. S. Kler, M. T. Hernke, *et al.*, "Direct human health risks of increased atmospheric carbon dioxide," *Nature Sustainability*, vol. 2, pp. 691–701, 2019. DOI: 10.1038/s41893-019-0323-1.

15 P. Styring, "Carbon dioxide utilization as a mitigation tool," in *Managing Global Warming: An Interface of Technology and Human Issues*, I. Pioro, Ed, 1st ed.. Academic Press, 2019, pp. 527–551, DOI: 10.1016/B978-0-12-814104-5.00018-1.

16 L. P. M. Costa *et al.*, "Capture and reuse of carbon dioxide (CO_2) for a plastics circular economy: A review," *Processes*, vol. 9, no. 5, p. 759, 2021. DOI: 10.3390/pr9050759.

17 S. S. Ravi, J. Mazumder, J. Sun, C. Brace, and J. W. G. Turner, "Techno-economic assessment of synthetic e-fuels derived from atmospheric CO_2 and green hydrogen," *Energy Conversion and Management*, vol. 291, p. 117271, 2023. DOI: 10.1016/j.enconman.2023.117271.

18 S. Hughes, *et al.*, Analysis Of Carbon Capture Retrofits For Cement Plants, U.S. Department of Energy, Technical Report, Mar. 2023. DOI: 10.2172/1970135

19 F. Ganda, "The effect of carbon performance on corporate financial performance in a growing economy," *Social Responsibility Journal*, vol. 14, no. 4, pp. 895–916, 2018. DOI: 10.1108/SRJ-12-2016-0212.

20 C. Hepburn, E. Adlen, J. Beddington, *et al.*, "The technological and economic prospects for CO_2 utilization and removal," *Nature*, vol. 575, pp. 87–97, 2019. DOI: 10.1038/s41586-019-1681-6.

21 O. Romero-Hernández and S. Romero, "Maximizing the value of waste: From waste management to the circular economy," *Thunderbird International Business Review*, vol. 60, no. 5, pp. 757–764, Sep./Oct., 2018. DOI: 10.1002/tie.21968.

22 M. Aresta, "Carbon dioxide utilization: The way to the circular economy," *Greenhouse Gases: Science and Technology*, vol. 9, no. 4, pp. 610–612, 2019.

23 N. Hsan, S. Kumar, Y. Cho, *et al.*, "Advancements in carbon-based materials for enhanced carbon dioxide capture and conversion: A comprehensive review," *Fibers and Polymers*, vol. 26, pp. 947–974, 2025. DOI: 10.1007/s12221-025-00871-x.

24 I. S. Thakur, M. Kumar, S. J. Varjani, Y. Wu, E. Gnansounou, and S. Ravindran, "Sequestration and utilization of carbon dioxide by chemical and biological methods for biofuels and biomaterials by chemoautotrophs: Opportunities and challenges," *Bioresource Technology*, vol. 256, pp. 478–490, May., 2018. DOI: 10.1016/j.biortech.2018.02.038.

25 B. Molitor, A. Mishra, and L. T. Angenent, "Power-to-protein: Converting renewable electric power and carbon dioxide into single cell protein with a two-stage bioprocess," *Energy & Environmental Science*, vol. 12, no. 12, pp. 3515–3521, 2019. DOI: 10.1039/C9EE02381J.

26 K. K. Sahoo, G. Goswami, and D. Das, "Biotransformation of methane and carbon dioxide into high-value products by methanotrophs: Current state of art and future prospects," *Frontiers in Microbiology*, vol. 12, Mar., 2021. DOI: 10.3389/fmicb.2021.636486.

27 S. Matassa, N. Boon, I. Pikaar, and W. Verstraete, "Microbial protein: Future sustainable food supply route with low environmental footprint," *Microbial Biotechnology*, vol. 9, no. 5, pp. 568–575, Sept., 2016. DOI: 10.1111/1751-7915.12369.

28 A. Murugan *et al.*, "Purity requirements of carbon dioxide for carbon capture and storage," *Environmental Science: Processes & Impacts*, vol. 21, no. 3, pp. 404–420, 2019. DOI: 10.1039/C8EM00498F.

29 A. Al-Mamoori, A. Krishnamurthy, A. A. Rownaghi, and F. Rezaei, "Carbon capture and utilization update," *Energy Technology*, vol. 5, no. 6, pp. 834–849, June, 2017. DOI: 10.1002/ente.201600747.

30 Global CO_2 Initiative, "Global roadmap for implementing co_2 utilization," Technical Report, Global CO_2 Initiative, Ann Arbor, MI, USA, 2016. [Online]. Available: https://hdl.handle.net/2027.42/150624

31 C. Rausch, R. Ferron, C. Dubay, A. Baral, L. Rogers, and L. Loftus-Otway, "Synthesis of carbon capture and repurposing by-products," *Environmental Science & Technology*, vol. 59, no. 2, pp. 1234–1242, 2025.

32 C. Gürtler, "R&D&I and industry examples: Covestro's dream production," in *CO_2 and CO as Feedstock. Circular Economy and Sustainability*, M. Kircher and T. Schwarz, Eds. Cham, Switzerland: Springer, 2023, pp. 317–326, DOI: 10.1007/978-3-031-27811-2_26.

33 C. Mihalcea, F. Burton, R. Conrado, and S. Simpson, "R&D&I and industry examples: Lanzatech's gas fermentation," in *CO_2 and CO as Feedstock. Circular Economy and Sustainability*, M. Kircher and T. Schwarz, Eds. Cham, Switzerland: Springer, 2023, pp. 327–336, DOI: 10.1007/978-3-031-27811-2_27.

34 M. He, Y. Sun, and B. Han, "Green carbon science: Efficient carbon resource processing, utilization, and recycling towards carbon neutrality," *Angewandte Chemie International Edition*, vol. 61, no. 15, p. e202112835, Apr., 2022. DOI: 10.1002/ange.202112835.

35 H. Guo, C. Li, X. Shi, H. Li, and S. Shen, "Nonaqueous amine-based absorbents for energy efficient CO_2 capture," *Applied Energy*, vol. 239, pp. 725–734, Apr., 2019. DOI: 10.1016/j.apenergy.2019.02.019.

36 Z. Zhang, D.-N. Vo, J. Kum, S.-H. Hong, and C.-H. Lee, "Enhancing energy efficiency of chemical absorption-based CO_2 capture process with advanced waste-heat recovery modules at a high capture rate," *Chemical Engineering Journal*, vol. 472, p. 144918, Sept., 2023. DOI: 10.1016/j.cej.2023.144918.

37 D. Loachamin, J. Casierra, V. Calva, A. Palma-Cando, E. E. Ávila and M. Ricaurte, "Amine-based solvents and additives to improve the CO_2 capture processes: A review," *ChemEngineering*, vol. 8, no. 6, art. no. 129, 2024. 10.3390/chemengineering8060129

38 L. Rosa, D. L. Sanchez, G. Realmonte, D. Baldocchi, and P. D'Odorico, "The water footprint of carbon capture and storage technologies," *Renewable and Sustainable Energy Reviews*, vol. 138, p. 110511, Mar., 2021. DOI: 10.1016/j.rser.2020.110511.

39 K. Puskarskiy, R. Loftager, J. Hoffner, P. Haastrup, S. Kugendran, and J. Moody, "Applying digitalization for energy efficiency insight and advancing it towards carbon emission reduction targets," in Proc. IADC/SPE Int. Drilling Conf. and Exhibition, Galveston, TX, USA, Mar. 2022, Paper SPE-208742-MS. DOI: 10.2118/208742-MS

40 A. Sivasuriyan *et al.*, "Emerging trends in the integration of smart sensor technologies in structural health monitoring: A contemporary perspective," *Sensors*, vol. 24, no. 24, p. 8161, 2024. DOI: 10.3390/s24248161.

41 H. Soleimani, M. Mohammadi, M. Fadaki, and S. M. J. Mirzapour al-e-hashem, "Carbon-efficient closed-loop supply chain network: An integrated modeling approach under uncertainty," *Environmental Science and Pollution Research*, 2021. DOI: 10.1007/s11356-021-15100-0.

42 C. Ampelli, S. Perathoner, and G. Centi, "CO_2 utilization: An enabling element to move to a resource- and energy-efficient chemical and fuel production," *Philosophical Transactions of the Royal Society A—Mathematical Physical and Engineering Sciences*, vol. 373, no. 2037, Mar., 2015. DOI: 10.1098/rsta.2014.0177.

43 J. J. Fierro, A. Escudero-Atehortua, C. Nieto-Londoño, M. Giraldo, H. Jouhara, and L. C. Wrobel, "Evaluation of waste heat recovery technologies for the cement industry," *International Journal of Thermofluids*, vol. 7–8, p. 100040, Nov., 2020. DOI: 10.1016/j.ijft.2020.100040.

44 M. Meijssen, V. Becattini, and M. Mazzotti, "Integrated carbon capture and utilization in the cement industry: A comparative study," *ACS Sustainable Chem. Eng.*, vol. 12, no. 7, pp. 2709–2718, Feb., 2024. DOI: 10.1021/acssuschemeng.3c07081.

45 P. Majumder and A. Dutta, "Closed-loop carbon management strategies for climate and energy-resilient India," *iScience*, vol. 28, no. 10, p. 113509, Oct., 2025. DOI: 10.1016/j.isci.2025.113509.

46 A. R. Kalair, M. Seyedmahmoudian, A. Stojcevski, N. Abas, and N. Khan, "Waste to energy conversion for a sustainable future," *Heliyon*, vol. 7, no. 10, p. e08155, Oct., 2021. DOI: 10.1016/j.heliyon.2021.e08155.

47 P. Wienchol, A. Szklek, and M. Ditaranto, "Waste-to-energy technology integrated with carbon capture—challenges and opportunities," *Energy*, vol. 198, p. 117352, May, 2020. DOI: 10.1016/j.energy.2020.117352.

48 M. Kim, K. Kim, T.-h. Kim, and J. Kim, "Economic and environmental benefit analysis of a renewable energy supply system integrated with carbon capture and utilization framework," *Chemical Engineering Research and Design*, vol. 147, pp. 200–213, Jul., 2019. DOI: 10.1016/j.cherd.2019.05.008.

49 Y. Zhang, H. Zheng, B. Chen, M. Su, and G. Liu, "A review of industrial symbiosis research: Theory and methodology," *Frontiers in Earth Science*, vol. 9, pp. 91–104, 2015. DOI: 10.1007/s11707-014-0445-8.

50 A. P. M. Velenturf, "Promoting industrial symbiosis: empirical observations of low-carbon innovations in the Humber region, UK," *Journal of Cleaner Production*, vol. 128, pp. 116–130, Aug., 2016. DOI: 10.1016/j.jclepro.2015.06.027.

51 J. Ammenberg, L. Baas, M. Eklund, R. Feiz, A. Helgstrand, and R. Marshall, "Improving the CO_2 performance of cement, part III: The relevance of industrial symbiosis and how to measure its impact," *Journal of Cleaner Production*, vol. 98, pp. 145–155, Jul., 2015. DOI: 10.1016/j.jclepro.2014.01.086.

52 M. Biermann, H. Ali, M. Sundqvist, M. Larsson, F. Normann, and F. Johnsson, "Excess heat-driven carbon capture at an integrated steel mill—considerations for capture cost optimization," *Int. J. Greenh. Gas Control*, vol. 91, p. 102833, Dec., 2019. DOI: 10.1016/j.ijggc.2019.102833.

53 R. Molinier and P. da Costa, "Infrastructure sharing synergies and industrial symbiosis: Optimal capacity oversizing and pricing," *Journal of Industrial and Intelligent Information*, vol. 7, no. 1, pp. 24–32, 2019. DOI: 10.18178/jiii.7.1.24-32.

14

Economic Considerations and Financial Models

14.1 Overview of Global Climate Targets and the Role of CCS

Understanding the cost structure of carbon capture projects is essential for assessing their financial feasibility and long-term sustainability. Every carbon capture, utilization, and storage (CCUS) project involves large investments, and breaking down these costs helps identify where savings can be achieved and how to make projects more attractive to investors [1]. The total cost of a carbon capture project is usually divided into two main categories: capital expenditure (CAPEX) and operating expenditure (OPEX) [2].

Capital costs (CAPEX) represent the initial investment needed to design and build the project. This includes purchasing and installing major equipment such as absorbers, compressors, heat exchangers, and storage tanks. It also covers pipelines for CO_2 transport, site preparation, and civil construction works. For large power plants or industrial facilities, CAPEX is often the largest cost component, typically more than 50% of the total project cost [3].

Additional capital expenditure (CAPEX) in CCUS projects can arise from several non-technical factors such as permitting, environmental studies, and contingency allowances for design or cost variations [4, 5]. Permitting refers to the process of obtaining formal government approvals to capture, transport, and store CO_2. This step often involves extensive documentation, legal reviews of the project, and coordination with multiple regulatory agencies to ensure compliance with safety and environmental standards [4].

Legal reviews involve checking that all project plans, contracts, and permit applications comply with national laws and safety regulations. It can be time-consuming and expensive, especially when public consultations and risk assessments are required [6–8]. Public consultations and risk assessments make the permitting stage more expensive because they require additional work, experts, and time [7, 8]. A risk assessment is a technical study that identifies and quantifies potential hazards (like CO_2 leaks, explosions, or groundwater contamination). It often requires simulations, safety modeling software, and specialist engineers to evaluate possible accident scenarios [7].

Environmental studies or Environmental Impact Assessments (EIAs) evaluate how the CCUS project might affect air, water, soil, ecosystems, and local communities. These studies involve field data collection, modeling of potential CO_2 releases, groundwater analysis, and biodiversity assessments. All these activities require specialized consultants and long review processes, which significantly increase project costs [9]. Contingency allowances are extra funds (typically 10–20% of the total CAPEX) added to the project budget to cover unforeseen costs such as design modifications, inflation during construction, or new safety requirements [10].

Operating costs (OPEX) refer to the ongoing expenses of running and maintaining the carbon capture plant after it is built. The largest portion of OPEX usually comes from energy consumption, as capture systems often require large amounts of electricity or steam to separate and compress CO_2. Other key OPEX elements include the replacement of solvents or sorbents, maintenance and repair of equipment, labor costs, and process monitoring. Efficient process design and the use of renewable or low-cost energy sources can greatly reduce OPEX over time [11].

As a major part of the operating expenditure (OPEX) in a CCUS project, CO_2 transportation and storage contribute significantly to the total cost [12]. After the CO_2 is captured, it must be transported, most commonly through high-pressure pipelines, and in some cases by ship, to an appropriate storage site such as a depleted oil or gas reservoir or a deep saline formation. The transportation cost depends on several factors, including the distance to the storage site, pipeline diameter and material, terrain complexity, and the safety and compression systems required to maintain CO_2 in dense phase [13]. The storage phase includes drilling and completing injection wells, installing surface facilities, and long-term monitoring to ensure the CO_2 remains safely contained [14]. Together, these transportation and storage activities form one of the largest components of the project's OPEX throughout its operating lifetime.

Figure 14.1 illustrates the transportation and storage stages of a CCUS project, which together make up one of the largest components of operating expenditure (OPEX). Once

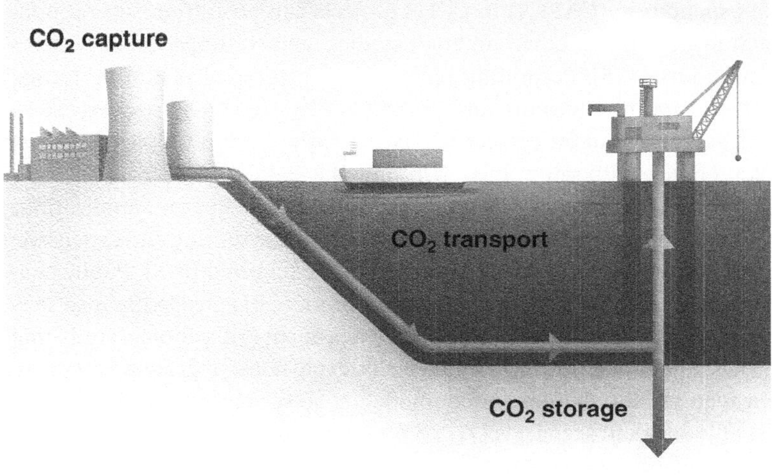

Figure 14.1 Illustration of CO_2 transportation and geological storage, showing pipeline or ship transport to an offshore injection site. These stages require significant ongoing energy and monitoring activities, making them major contributors to CCUS operating expenditure (OPEX).

CO_2 is captured at the industrial facility, it must be moved to a suitable storage site, most commonly through high-pressure pipelines or, in some cases, by specialized ships. CO_2 transport requires continuous compression and pressure management, both of which significantly increase energy consumption and operating costs. At the storage site, injection wells, surface equipment, and long-term monitoring systems must operate reliably over decades to ensure safe CO_2 storage, further adding to OPEX. The combined need for pumping and compressor power, shipping, well maintenance, and monitoring makes the transport-and-storage chain one of the most resource-intensive parts of the CCUS system, increasing OPEX.

Understanding these cost components allows stakeholders, including engineers, investors, and policymakers, to understand the project cost structures and identify where cost reductions can be achieved [15]. Technological improvements such as advanced capture materials, improved heat recovery systems, and automation mainly help reduce operating expenditure (OPEX) by lowering energy use, maintenance, and labor requirements, while their impact on capital expenditure (CAPEX) is generally minor [16]. Similarly, integrating carbon capture with existing industrial plants or renewable energy systems can lower overall operational costs and improve economic performance.

In summary, the cost structure of a carbon capture project is shaped by two major factors: the initial construction (CAPEX) and the day-to-day operation (OPEX), including the CO_2 transport and storage expenses [1]. By focusing on innovation, efficiency, and smart design, future projects can become more cost-effective, making carbon capture a more practical tool for achieving global climate goals.

14.2 Levelized Cost of CO_2 Abatement (LCCA)

14.2.1 What Is LCCA and Why Is It Important?

The Levelized Cost of CO_2 Abatement (LCCA) is a way of answering a very simple question: "If we build and operate this carbon capture project for its whole life, how much does it really cost to stop one tonne of CO_2 from reaching the atmosphere?" Instead of looking only at first-year costs or the investment, LCCA adds up all relevant expenses over the project's entire lifetime and compares them to the total amount of CO_2 successfully captured and stored. Because of this, LCCA is one of the most important economic indicators for carbon capture projects. It helps engineers, companies, governments, and investors compare different technologies on a fair and consistent basis. If two projects use other equipment, run for various lengths of time, and capture different amounts of CO_2, LCCA still allows us to compare them using a standard measure: cost per tonne of CO_2 avoided [17].

14.2.2 How LCCA Is Calculated?

To calculate the LCCA, we first add together all of the project's costs. This includes capital expenditure (CAPEX), the money spent to design, purchase, and install the plant and its equipment. It also includes operating expenses (OPEX), which cover day-to-day running costs such as energy, chemicals, labor, maintenance, and spare parts. When we talk about

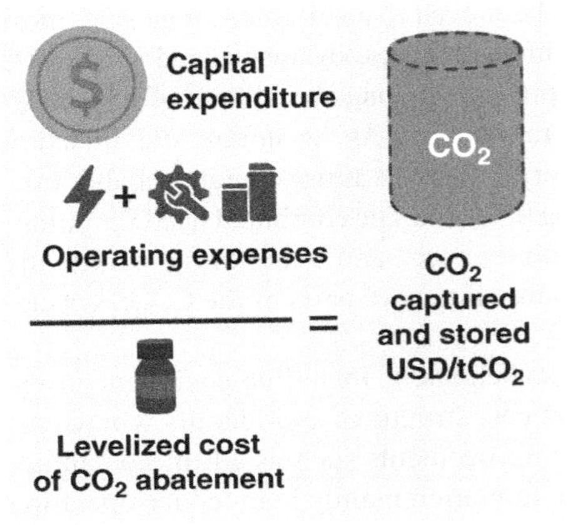

Figure 14.2 Simplified illustration of how the Levelized Cost of CO_2 Abatement (LCCA) is calculated by combining capital expenses, operating costs, and total CO_2 captured and stored.

carbon capture and particularly OPEX, energy use is significant because these systems often require substantial steam (heat), electricity, or both. Once we have the total cost over the project's lifetime, we divide it by the total amount of CO_2 that is captured and permanently stored during the same period. The result is usually expressed in currency per tonne of CO_2, for example USD/tCO_2 or EUR/tCO_2. In simple words, the LCCA tells us the "cost" for preventing one tonne of CO_2 emissions using a particular project or technology [17].

Figure 14.2 illustrates the Levelized Cost of CO_2 Abatement (LCCA) by showing how all key cost components of a carbon capture project come together. On the left, capital expenditure (CAPEX) represents the upfront investment needed for major equipment and construction. Below it, operating expenses (OPEX), such as electricity for compressors and pumps and labor, are shown as ongoing costs throughout the project's lifetime. On the right, the graphic displays the total amount of CO_2 captured and permanently stored. The LCCA is calculated by dividing all lifetime costs (CAPEX + OPEX) by the total tonnes of CO_2 avoided, giving a final value in USD per tonne of CO_2. This visualization highlights that LCCA is a full-lifecycle measure used to compare different carbon capture technologies on equal terms [17, 18].

14.2.3 Key Factors Affecting LCCA

14.2.3.1 Capital and Operating Costs

Many different factors can influence the LCCA, and understanding them helps us see why some projects are cheaper and some are more expensive. The first obvious factor is the level of capital and operating costs. If a project needs costly equipment, complex construction, or high-priced materials, the CAPEX will be high, which tends to increase the LCCA. Similarly, if the plant consumes a lot of energy, uses costly chemicals, or requires frequent repairs and maintenance, OPEX will be high, thereby raising the cost per tonne of CO_2. On the other hand, if engineers reduce energy consumption, simplify the design, or improve integration with existing processes (e.g. by using waste heat from another unit), both CAPEX and OPEX can be reduced, and the LCCA will decrease [17, 18].

14.2.3.2 Capture Efficiency

Another critical factor is capture efficiency, which is the percentage of CO_2 that the system is able to remove from the gas stream. If we can capture more CO_2 using roughly the same equipment and similar energy input, the cost per tonne of CO_2 will fall. For example, imagine a plant that captures 85% of the CO_2. If modifications can increase capture efficiency to

95% without significantly increasing energy consumption or other costs, the same plant will now prevent more CO_2 emissions over its lifetime. Because the total cost is divided by a larger number of tonnes of CO_2 captured, the LCCA becomes lower. Of course, in real projects, there is usually a trade-off: pushing efficiency very high may require more energy or more complex equipment, thereby increasing costs. Engineers must therefore balance efficiency gains against the extra costs they create [17, 18].

14.2.3.3 Project Size (Economies of Scale)

Project size, or scale, strongly influences Life-Cycle Cost Assessment. In general, larger carbon capture projects can take advantage of economies of scale, meaning that some costs do not increase as quickly as the plant's size. For example, if a plant doubles its capacity, the cost of engineering design, project management, specific control systems, and monitoring equipment does not automatically double. Many of these expenses include fixed or semi-fixed components, so they can be shared across a larger amount of captured CO_2. In contrast, most other costs, such as detailed engineering, control system hardware, major equipment, construction labor, and OPEX, do increase with project size. Therefore, the term "fixed costs" in LCCA does not mean the cost stays the same, but rather that it increases more slowly than the plant's output, allowing larger plants to spread these semifixed expenses over more tonnes of CO_2 captured. Because of this, a large facility that captures several million tonnes of CO_2 per year may achieve a lower cost per tonne compared with a small pilot or demonstration plant [19].

Figure 14.3 illustrates the general concept of economies of scale in carbon capture projects. As plant size increases, certain costs, particularly engineering design, project management, and parts of the monitoring and control systems, do not rise as quickly as the total

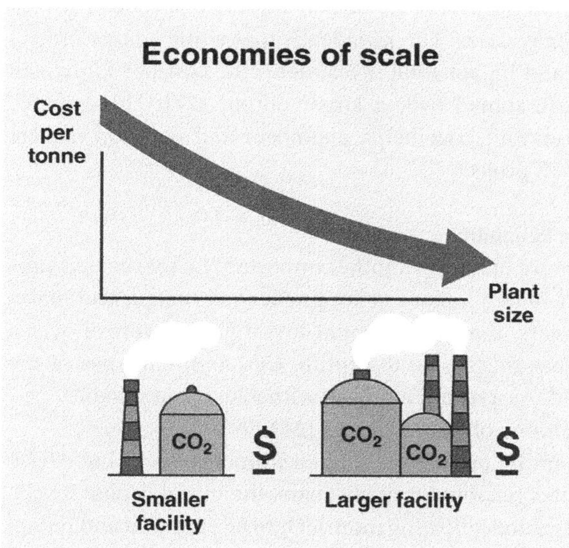

Figure 14.3 Typical economies of scale in carbon capture projects. Larger facilities can often reduce the cost per tonne of CO_2 by spreading semifixed costs across higher capture volumes.

amount of CO_2 captured. This means that larger facilities can often spread their semifixed costs over a greater volume of captured CO_2, resulting in a lower cost per tonne. The downward-sloping arrow in the diagram represents this typical trend, showing how unit costs tend to fall as plant capacity grows. However, the figure should be interpreted as a simplified illustration of the general relationship rather than a strict rule.

However, it is important to understand that larger projects are not always better in every situation. When the capacity increases, the project becomes more complex. Bigger plants usually require more land, more equipment, more construction work, and greater coordination among contractors and suppliers. Financing can also be more challenging because larger projects need higher capital investment and must manage greater financial risk. In addition, the construction schedule may lengthen, and delays may have a greater impact [20].

To illustrate how economies of scale work on the fixed-cost part of a project, imagine two CO_2 capture plants: Plant A (a smaller plant) captures 0.3 million tonnes of CO_2 per year, while Plant B (a larger plant) captures 1.5 million tonnes per year. Suppose both plants require roughly the same spending on the engineering team, parts of the control system, and monitoring equipment, for example, USD 20 million in fixed costs. If we look only at this fixed-cost component (and not the full LCCA), the cost per tonne of CO_2 becomes very different: for Plant A, dividing USD 20 million by 0.3 million tonnes gives about USD 67 per tonne, whereas for Plant B, the same fixed cost divided by 1.5 million tonnes is only about USD 13 per tonne. In reality, the larger plant will have a higher total capital cost, because major equipment must be larger or duplicated, and higher operating costs due to greater energy use, maintenance needs, and staffing. However, this simple example shows that some fixed or semifixed costs contribute less per tonne in a larger plant, which is one of the reasons why bigger, well-designed CCS facilities can potentially achieve a lower overall LCCA than very small demonstration plants.

Even with these challenges, increasing the scale of a carbon capture project can still be an effective way to reduce LCCA in many cases. The key idea is that while a large project does require more engineering work and higher total investment, the cost per tonne can still decrease because certain costs are spread over a larger output [21]. This balance between higher complexity and lower unit cost helps engineers and decision-makers choose the most efficient scale for a CCS project.

14.2.3.4 Operating Lifetime and Plant Reliability

The operating lifetime of a carbon capture project is another important factor that strongly affects its Life-Cycle Cost Assessment. When a plant is designed, constructed, and maintained to operate reliably for many years, the initial capital investment is spread over a longer period. This means the plant has more time to capture CO_2 and make use of the expensive equipment that was installed. As a result, a project with a long and reliable operating life can achieve a lower cost per tonne of CO_2 captured [21].

The operating lifetime of a carbon capture project has a major influence on its Life-Cycle Cost Assessment. When a plant operates reliably for many years, the initial capital investment (CAPEX) is spread over a longer period, allowing more CO_2 to be captured and reducing the cost per tonne. Although operating expenditures (OPEX) increase year by year during the plant's lifetime, because equipment must be maintained, repaired, and replaced, OPEX is usually much smaller than CAPEX and does not dominate the total LCCA [22].

For example, a plant that runs successfully for 30 years will normally have a better LCCA than a similar plant that operates for only 10 years. Even though the 30-year plant will face higher **operational expenditures (OPEX)**, such as maintenance costs, repairs, replacement of worn components, staffing, and energy consumption [17]. These ongoing costs grow gradually over time but are still smaller compared to the large capital investment made at the beginning. Because the equipment is used for a much longer period, the initial investment is divided across more years of operation and more tonnes of CO_2 captured.

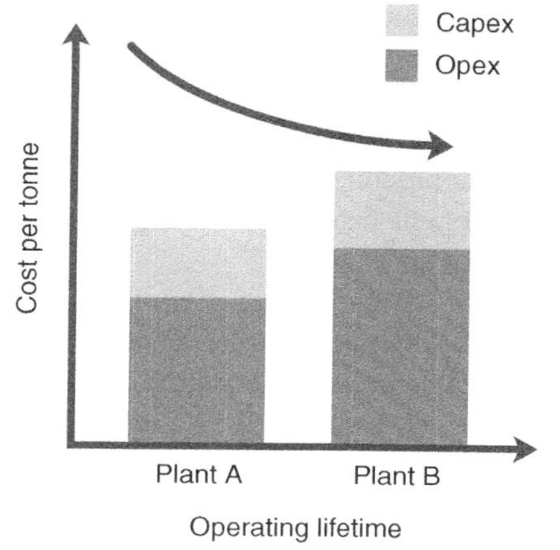

Figure 14.4 Effect of operating lifetime on the cost per tonne of CO_2 captured. Longer-lived plants can spread both CAPEX and OPEX over a larger total amount of CO_2. Because the growth of OPEX over time is slower than the increase in cumulative CO_2 captured, the overall cost per tonne decreases as operating lifetime increases.

Figure 14.4 illustrates how the operating lifetime of a carbon capture plant influences the cost per tonne of CO_2 captured. The bars in the figure show this effect by dividing the total cost per tonne into CAPEX (light color) and OPEX (dark color). Plant A represents a facility with a shorter operational life, while Plant B operates for a much longer period. When the plant runs for more years (Plant B), the initial capital expenditure (CAPEX) is distributed over a larger total amount of CO_2 captured. Because the same upfront investment supports a longer period of CO_2 removal in plant B, the CAPEX cost per tonne becomes lower. In contrast, a short-lived plant captures less CO_2 over its lifespan, so each tonne carries a higher share of the original investment. Although operating costs (OPEX) continue accumulating every year and may eventually exceed the initial capital cost (CAPEX), the key point is that the total amount of CO_2 captured grows much faster when a plant operates for many years. Thus, although OPEX increases over time, it does not increase fast enough to outweigh the benefit of capturing far more CO_2.

However, reliability, high-quality materials, and proper maintenance are not just technical considerations but also strong economic drivers. If a plant experiences frequent breakdowns, long unplanned shutdowns, or is forced to close permanently after a short period, the total amount of CO_2 captured over its lifetime becomes much lower. In such cases, the large CAPEX is spread over fewer tonnes of CO_2, which increases the LCCA. Here, it is OPEX that raises the LCCA because of reduced operating lifetime and poor reliability, which limit the amount of CO_2 abated [23].

14.2.3.5 Transport and Storage Costs

In many carbon capture projects, carbon dioxide (CO_2) is captured, transported, and stored safely. Each of these steps incurs its own costs [12]. For example, if the storage site is located far away, the CO_2 may need to be compressed and transported by pipeline or ship over long distances, which increases expenses [12, 24]. Additionally, storing CO_2 in deep or technically challenging geological formations may require more advanced drilling techniques, compressors, safety systems, and monitoring technologies [25]. All these factors contribute to a higher cost per tonne of CO_2, thereby raising the Levelized Cost of Carbon Abatement (LCCA) [12]. In contrast, projects situated near suitable storage formations, such as depleted gas fields or appropriate saline aquifers, often have a significant advantage. With shorter pipelines, fewer compressor stations, and easier injection operations, these projects can benefit from lower transportation and storage costs, which helps reduce the overall LCCA [12].

14.2.3.6 Energy Prices and Energy Sources

Carbon capture systems often require significant energy, so energy prices directly affect the LCCA. When the prices of electricity, steam, or fuel are high, the capture plant's operational costs increase. This can significantly increase the LCCA, particularly for technologies such as solvent-based amine systems that require substantial heat or power. Conversely, suppose a project can use low-cost renewable energy from sources such as wind turbines and solar panels, harness waste heat from another industrial process, or take advantage of surplus electricity from the grid or renewable energy sources during off-peak hours. In that case, it can significantly reduce its energy use and operating costs (OPEX) [26]. As a result, the LCCA can be reduced, making the project more competitive. This is one reason why integrating carbon capture with renewable energy sources or with industrial plants that generate excess waste heat is often considered a beneficial strategy both technically and financially [26].

Figure 14.5 compares CO_2 transport and storage costs for distant versus nearby storage sites, and the impact of energy prices and energy sources on overall project economics. On the left, when the storage formation is far away, long pipelines and additional compressor stations are required, which increase energy use, infrastructure needs, and the cost per

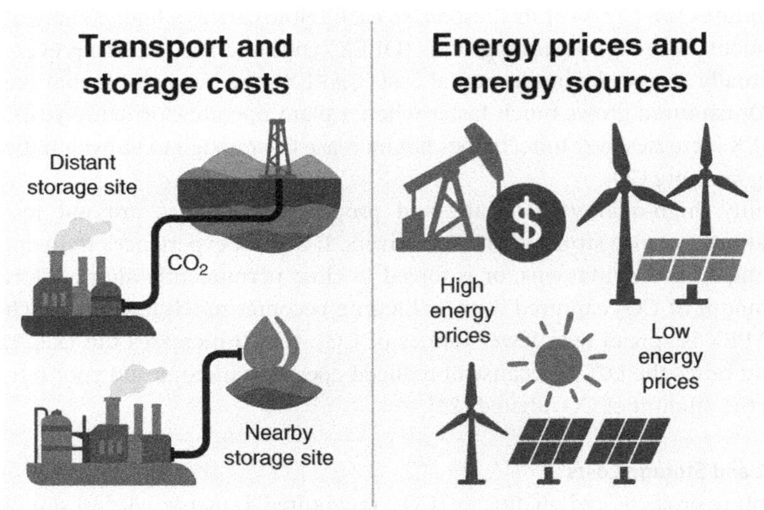

Figure 14.5 Influence of distance to storage site and energy prices on CCUS costs. Distant storage requires longer pipelines and more compression, increasing cost per tonne. In contrast, nearby storage reduces infrastructure needs. Similarly, high fossil-based energy prices raise operating costs, while low-cost renewable energy helps lower the overall LCCA.

tonne of CO_2. In contrast, projects located close to suitable storage formations need shorter pipelines, fewer compressors, and simpler injection systems; nearby storage, therefore, significantly reduces transport and storage costs and lowers the overall LCCA. On the right, the figure contrasts high energy prices associated with conventional fossil-based power with lower effective energy costs when carbon capture is supplied by low-cost renewables such as wind and solar. Using renewable electricity or other low-cost energy sources reduces OPEX and further improves the competitiveness of CCUS projects.

14.2.4 How Stakeholders Use LCCA and Other Indicators

Understanding LCCA is very useful for different groups of stakeholders. Government agencies and regulators use LCCA to decide whether it makes sense to support a project through subsidies, tax credits, or other incentives. If the cost per tonne of CO_2 is very high, they may decide that public money would be better spent elsewhere, or they may design new policies to help reduce costs. Governments have several tools to achieve this. For example, they can simplify permitting procedures to minimize delays and engineering costs, provide low-interest loans or loan guarantees that lower financing costs [27, 28].

Industries and project developers use LCCA to choose between different capture technologies, process designs, and project sizes. They can run scenarios to see how changes in project size, efficiency, energy price, or transport distance affect the cost per tonne. This helps them design smarter, more economical systems. Investors and financial institutions look at LCCA to understand long-term cost stability and risk. A project with a reasonably low and stable LCCA is more likely to attract investment than one with very high or highly uncertain costs per tonne [17].

Figure 14.6 visually summarizes how LCCA sits at the center of decision-making by stakeholders across the CCUS value chain. Each stakeholder group, governments, industries,

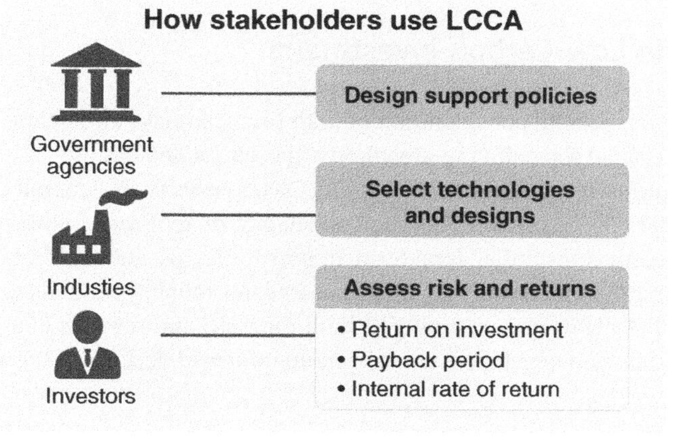

Figure 14.6 Central role of LCCA in stakeholder decision-making. The diagram shows how LCCA affects government policy design, industrial technology choices, and investor evaluations of financial performance and risk.

and investors, is represented with an icon to show its role. The picture demonstrates how LCCA functions as the starting point for each group's evaluation: policymakers use it to determine what level of policy support is needed, industries use it to guide technical and design choices, and investors link it to broader financial indicators to judge project feasibility. Rather than describing what each stakeholder does in detail, the diagram highlights the structure of the decision process itself, showing that LCCA serves as the common reference that connects policy design, technological planning, and financial assessment.

To understand whether a carbon capture project is financially attractive, investors rely on several financial performance indicators in addition to LCCA. The Return on Investment (ROI) measures how much profit a project generates compared to its total investment cost [29]. The payback period shows how long it takes for the project to recover its initial capital through net savings or revenue. The Internal Rate of Return (IRR) reflects the project's overall profitability across its entire lifetime, taking into account both capital and operating costs [30]. These metrics help investors determine whether a project can deliver acceptable returns despite high upfront costs and technology risks, and they often play a decisive role in securing financing.

14.2.5 Summary and Importance of LCCA

In conclusion, the Levelized Cost of CO_2 Abatement is a powerful and transparent tool for evaluating carbon capture projects. It brings together all the major cost elements, capital investment, operating expenses (including energy consumption, transport, and storage), and project lifetime, and relates them to the total amount of CO_2 actually captured and stored. By expressing everything in terms of the cost per tonne of CO_2 avoided, LCCA enables comparisons of very different projects and technologies on a common basis. A **lower LCCA** means that the project is more cost-efficient and more competitive in the market. Such projects are not only more attractive to investors and industries but also more likely to play a significant role in large-scale climate mitigation, as they can reduce emissions at lower cost to society [17].

14.3 Risk and Return in Low-Carbon Investment

Investing in CCUS projects requires careful consideration of both financial risks and potential returns. These projects are crucial for reducing greenhouse gas emissions but are relatively new and often demand significant investments. As a result, CCUS projects are generally viewed as riskier than traditional energy investments such as oil, gas, or renewable power plants. However, they also present substantial long-term opportunities as governments enhance climate policies, carbon prices increase, and industries look for reliable methods to achieve net-zero targets [31, 32]. A clear understanding of both the risks and rewards enables investors, governments, and project developers to make more informed decisions.

14.3.1 Investment Risks in Carbon Capture Technology

A significant source of financial uncertainty originates from policy and regulatory risks. Carbon capture projects rely heavily on government support, which can take several forms, including carbon pricing, subsidies, tax credits, grants, or guaranteed storage payments.

These policies are crucial for making CCUS projects financially feasible, particularly during the initial investment phase and the early years when costs tend to be high. However, if government policies weaken, shift direction, or are eliminated altogether, the profitability of such projects can decline rapidly. For instance, if a country reduces its carbon tax or postpones new climate regulations in favor of adopting carbon capture technologies, companies may no longer receive enough financial incentives to justify the costs of capturing and storing CO_2. As a result, investors prefer stable, long-term policies that minimize uncertainty and signal that CCUS will continue to be a vital component of national climate strategies [33].

Another risk comes from market volatility, especially in carbon credit and emissions trading markets. In many countries, companies that capture and store CO_2 can earn carbon credits if they emit less than their budget, which they can sell to other companies. These credits provide essential revenue for CCUS projects. However, the price of carbon credits does not always stay the same. It can go up or down depending on supply and demand, government rules, and economic conditions. If carbon prices fall, project revenues decrease, increasing investment risk [34]. Investors generally prefer markets that have clear rules, strong enforcement, consistent demand for emissions reductions, and predictable pricing mechanisms. The more stable the carbon market is, the lower the financial risk for a CCUS project.

CCUS projects also face technical and operational risks, especially because many of the technologies involved are still developing. When a new carbon capture plant is built, engineers may not know exactly how the system will perform under all real-world conditions. Capture rates may be lower than expected, energy consumption may be higher, or equipment may need more maintenance than planned. Each of these problems can increase operating costs (OPEX) and reduce CO_2 capture, thereby lowering revenue [35]. These risks are especially common in first-of-a-kind or early-stage projects, where experience is limited. To reduce technical and operational risks, companies must invest in high-quality engineering design, thorough testing, skilled operators, and strong monitoring systems. As technologies mature and more projects are successfully built, technical risks gradually decrease [36].

Figure 14.7 provides a clear overview of the main financial risks that influence the success of carbon capture projects, grouping them into three categories: policy and regulatory uncertainty, carbon market instability, and technical or operational performance issues. Policy risk refers to changes in government incentive schemes, climate regulations, or long-term decarbonization commitments that can alter the financial outlook for industries and investors. Market risk captures the impact of fluctuating carbon prices, especially downward price movements, and shifts in demand for CO_2 reductions, both of which can lower project revenues. Technical risk highlights the possibility that real-world plant performance may fall short of engineering estimates, leading to higher operating costs, lower capture rates, or reduced system reliability.

When all these factors, technology maturity, stable policies, healthy carbon markets, and strong cost control, come together, carbon capture projects can become competitive and attractive investments. Instead of being viewed as risky or experimental, they can function as reliable infrastructure similar to renewable energy projects or traditional utilities. Despite these risks, there are strong reasons why carbon capture can offer attractive long-term returns. As more countries commit to reaching net-zero emissions by 2050, industries such

Figure 14.7 Key categories of financial risk in carbon capture projects. These include policy and regulatory uncertainty, carbon market volatility, and technical and operational performance risks. These interconnected factors shape the financial outlook of CCUS investments.

as cement, steel, chemicals, and power generation will need reliable options for serious decarbonization. In many cases, CCUS is one of the few available solutions for reducing emissions from these sectors. This means that the demand for carbon capture technologies is expected to grow significantly in the coming decades. Early investors who enter the market now may benefit from being part of an industry that is expected to expand rapidly [15].

14.3.2 Potential Revenue Sources in Carbon Capture Technology

Several different sources of revenue can make CCUS projects financially attractive, even though they are often expensive to build and operate. One major source of income comes from carbon credit sales. When a company captures and stores CO_2, it can earn carbon credits that represent verified emission reductions. These credits can be sold to other companies that need to buy extra credits to offset their excess emissions, creating an important revenue stream for the seller. The higher the carbon price, the more valuable these credits become, thereby improving CCUS's financial performance [37].

Another source of revenue is enhanced oil recovery (EOR), which is used in some regions. In EOR, captured CO_2 is injected into older oil reservoirs to help push out remaining oil. Companies purchase CO_2 for this purpose because it increases oil production, and the CCUS project earns money by supplying the CO_2. Although EOR is not used everywhere, it has historically been one of the earliest and most reliable income sources for carbon capture projects [38].

CCUS projects can also generate revenue through long-term CO_2 storage contracts. In this model, industrial emitters, such as cement plants, steel factories, and power stations, pay a storage provider to store the CO_2 they capture permanently. The storage provider, also known as the carbon storage operator or storage site owner, receives a stable, predictable payment for each tonne of CO_2 injected underground for many years. Because these contracts often last for many years, they greatly reduce financial uncertainty, making it easier for storage operators to plan future operations, secure financing, and recover their investment costs over time [39].

Many CCUS projects are now developed as industrial hubs, where several companies share the same pipelines and storage sites. This approach allows the costs of transport infrastructure, monitoring systems, and injection wells to be shared among multiple emitters. Because each company pays only a portion of the total cost, their individual financial burden becomes smaller [40]. This "shared-cost model" encourages collaboration and makes CCUS more affordable for smaller industries that could not build the system on their own.

Finally, government incentives play a major role in supporting CCUS revenues. These incentives can take many forms, including tax credits, direct subsidies, grants, or guaranteed payments for each tonne of CO_2 stored. For example, some governments offer large tax reductions for companies that capture and store CO_2, helping them recover part of their investment. Others provide low-interest loans or guarantee that the project will receive a minimum payment for captured CO_2, even if carbon market prices fall. Such policies reduce financial risk and make CCUS a more secure investment [41].

Figure 14.8 provides a structured overview of the main revenue pathways that strengthen the financial structure for CCUS projects. The diagram positions the CCUS

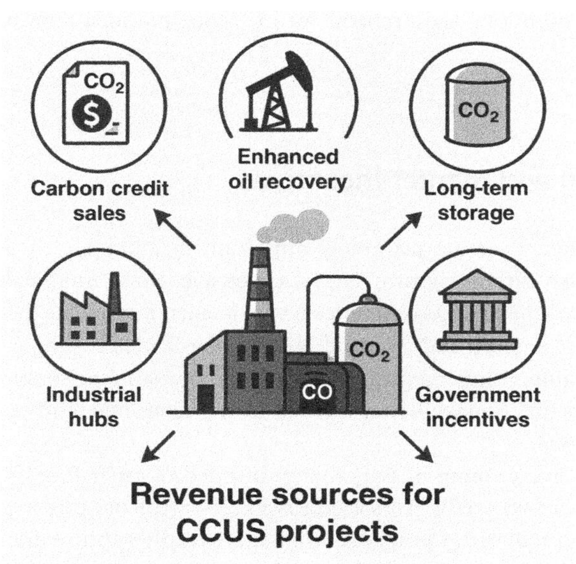

Figure 14.8 Key revenue sources supporting the financial viability of CCUS projects. These include carbon credit sales, EOR-based income, long-term CO_2 storage contracts, cost-sharing in industrial hubs, and government policy.

project at the center and shows four outward revenue flows, each represented with a transparent, technical icon. Carbon credit sales illustrate how verified CO_2 reductions can be sold into compliance or voluntary markets, generating market-based income. EOR reflects revenue from supplying CO_2 to oil operators who inject it into mature reservoirs to increase production. The long-term CO_2 storage icon represents payments from industrial emitters that rely on dedicated storage operators to handle permanent storage over many years. Industrial hub cost-sharing shows how multiple carbon capture companies with shared facilities, such as a pipeline, divide infrastructure and operational expenses, thereby improving the overall financial performance of the system. The figure visually emphasizes that CCUS economics depend not on a single income stream but on a combination of market, contractual, and collaborative mechanisms. Finally, government incentives highlight direct public-sector support, such as tax credits, storage payments, grants, or guaranteed carbon prices, that reduce project risk and strengthen financial performance.

Today, many countries are also developing new and more innovative business models to make returns more stable. Examples include carbon contracts for difference, where governments guarantee a fixed carbon price; storage payment guarantees, which ensure revenue for operators; and long-term CO_2 offtake agreements, where buyers commit to purchasing CO_2 storage services for many years. These mechanisms help reduce uncertainty and make the financial returns from CCUS more predictable [42].

CCUS will deliver two major benefits at the same time: environmental benefits, because it reduces large amounts of CO_2 emissions, and financial benefits, because it becomes a stable and profitable sector within the low-carbon economy. As global climate commitments strengthen and industries seek effective ways to reduce their emissions, carbon capture can play a central role, offering both long-term returns for investors and meaningful progress toward climate goals.

14.4 Carbon Credit Trading and Market Incentives

Carbon credit trading and market incentives are extremely important in making CCUS projects financially attractive. Because carbon capture can be expensive, these financial mechanisms help lower the overall cost by giving companies a way to earn money for the CO_2 they remove from the atmosphere or store underground [43]. The basic idea is simple: if reducing emissions has financial value, companies are more likely to invest in technologies like CCUS. When carbon markets function well, they create a strong economic motivation for industries to cut their emissions.

In a typical carbon credit system, every tonne of CO_2 that is avoided or permanently stored underground can generate a carbon credit. This credit is like a certificate proving that one tonne of emissions has been reduced. Companies that successfully capture and store CO_2 can sell these credits to other companies that need them to offset their excess emissions [44].

There are two main types of carbon markets. The first type is compliance markets, where governments legally require companies to reduce emissions. A well-known example is the EU Emissions Trading System (EU ETS), where companies must either cut their emissions or buy credits to comply with strict environmental laws. The second type is voluntary carbon markets, where companies buy carbon credits even though they are not legally required to do so. In the voluntary market, companies typically purchase credits to meet corporate sustainability targets, achieve carbon-neutral or net-zero commitments, and improve their environmental reputation. Unlike compliance markets, voluntary markets are not usually used for trading credits at a higher price later; instead, they are mainly a tool for companies to show environmental leadership and show progress toward net-zero commitments. In both market types, CCUS projects can earn revenue by generating and selling high-quality carbon credits [45].

Market incentives also include a wide range of government-supported financial tools that make CCUS more predictable and less risky for investors. Some governments offer performance-based payments, which reward projects for every tonne of CO_2 they capture and store. This provides a stable and reliable income stream [46]. Another important mechanism is carbon contracts for difference (CCfDs). These contracts guarantee a fixed carbon price, even if the market price falls. For example, if the guaranteed price is \$100 per tonne but the market price drops to \$60, the government pays the difference. This removes financial uncertainty and gives investors confidence that they will earn enough revenue over time to justify building the project [42].

Figure 14.9 illustrates how carbon credit trading and market incentives create financial value for CO_2 capture from industries, helping support the economics of CCUS. On the left side, the graphic shows an industrial facility producing carbon credits by capturing and

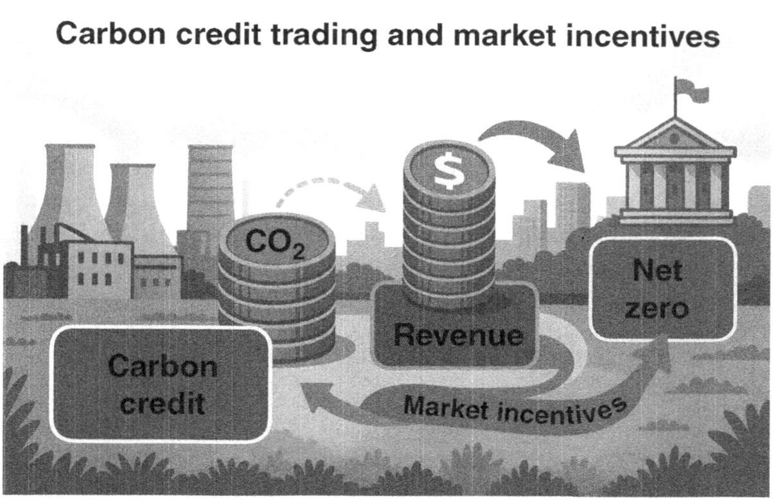

Figure 14.9 Overview of how carbon markets create financial value for captured and stored CO_2. It shows how carbon credits translate into revenue and support progress toward net-zero targets.

permanently storing CO_2. These credits are represented as green CO_2-labeled coins. The center of the image shows how these credits can be sold to generate revenue, which is a key financial driver for CCUS projects in both compliance and voluntary markets. The arrows of "Market Incentives" lead toward a government building and a "Net Zero" sign. The arrow and sign highlight the role of government to achieve Net Zero emissions through national climate policies, carbon pricing, and incentives, such as guaranteed carbon prices or performance-based payments.

Despite these benefits, carbon markets also face several challenges that can affect the financial stability of CCUS projects [47]. In voluntary carbon markets, the price of carbon credits is often low or unstable. One reason is that voluntary markets are not regulated or controlled by governments in the same way as compliance markets. Because there is no legal obligation to buy credits that can reduce the demand. Demand depends heavily on companies' sustainability goals, public pressure, and corporate reputation. These factors can change quickly, leading to fluctuating prices [48].

In voluntary carbon markets, the term low liquidity means that the market is not very active—there are not enough people buying and selling credits regularly. Even if many companies want to buy high-quality carbon credits, there may not be enough projects that can supply the kind of credits buyers are looking for (for example credits from permanent CO_2 storage such as CCUS). As a result, the number of credits available for sale is small, and trades do not happen very often [49, 50].

Another major challenge is trust, meaning that both from investors and the public. For carbon markets to function properly, everyone must trust that each carbon credit represents a real, permanent, and measurable reduction in CO_2 emissions. If buyers start to doubt the quality of credits, the entire market becomes less credible and less useful [51]. To maintain this trust, strict rules and strong verification systems are essential. Reliable monitoring, reporting, and verification (MRV) must demonstrate that the CO_2 captured and stored truly stays underground. This is especially important because leakage from storage sites could undo the climate benefit and make the associated carbon credits invalid [52].

When carbon markets are well-designed and supported by strong incentives, they can transform carbon capture from a costly, hard-to-finance technology into a financially rewarding opportunity. By giving clear economic value to emission reductions, these markets help industries adopt CCUS more quickly and at larger scale. In this way, carbon credit trading and supportive market incentives play a crucial role in making CCUS feasible across different regions and industries [53].

14.5 Public–Private Partnerships and Financing Mechanisms

Public–private partnerships (PPPs) are essential for making CCUS projects financially feasible. These projects are often large, expensive, and involve long-term technical and financial risks, making it challenging for private companies to fund them independently. Additionally, governments face their own challenges, which may result in a lack of technical expertise, operational capacity, or resources to manage large CCUS facilities alone. By collaborating in a PPP, both parties can share costs, risks, responsibilities, and the benefits of the project. This cooperation significantly increases the feasibility of developing large-scale CCUS systems [54].

In a typical PPP arrangement, the government provides support through various means: funding, access to land necessary for building the plant, regulatory approvals, or long-term agreements for purchasing or storing CO_2. This support helps to ensure more financial resources for the private partner. In return, private companies contribute their strengths, including technology, engineering expertise, project delivery skills, and capabilities for long-term operation and maintenance. By combining public support with private expertise, the overall financial burden is reduced, and investors gain greater confidence in the project's likelihood of success. For example, some governments guarantee the long-term feasibility of CO_2 storage, support the construction of single or shared CO_2 transport pipeline networks, or offer minimum CO_2 price guarantees to decrease financial uncertainty in the event of a downturn in the carbon market [54, 55].

Figure 14.10 illustrates how PPPs support the development of CCUS projects. On the left side, the government provides essential forms of support—such as funding, land access, regulatory approvals, long-term CO_2 storage guarantees, or minimum CO_2 price stabilization. On the right side, private companies contribute their technical expertise, engineering capabilities, and operational experience. At the center of the figure, the carbon capture facility represents the shared project outcome, where public support reduces financial risk and private expertise ensures efficient design, construction, and operation. The figure also highlights how PPPs create a balanced and mutually beneficial structure that makes large CCUS projects more reliable and feasible.

Another important financial tool is blended finance, which involves combining public and private capital to make carbon capture projects less risky and more attractive to investors. In this approach, governments or international organizations provide support, such as

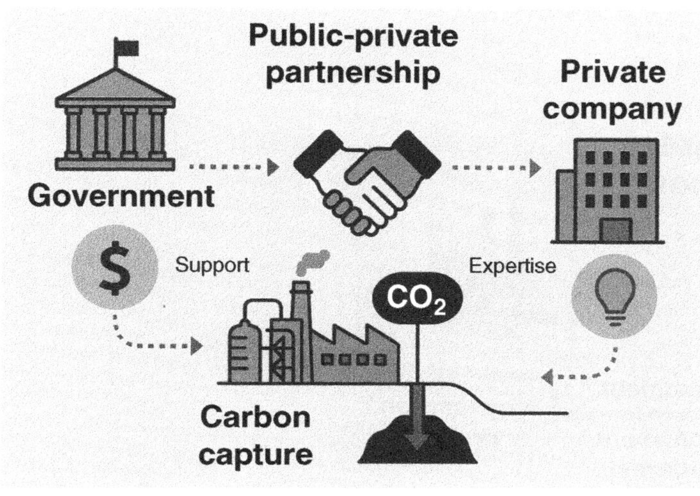

Figure 14.10 Public–private partnership structure for CCUS projects. It shows how government support and private-sector expertise combine to reduce financial risk.

low-interest loans, tax credits, grants, or loan guarantees, to help reduce the high upfront costs that private companies usually hesitate to take on. When public funding takes on part of the risk, banks see the project as safer and are more willing to lend money [56]. This improved "bankability" makes it easier for developers to secure the funding needed to build and operate CCUS facilities [57]. Institutions such as the World Bank, the European Investment Bank, and the Green Climate Fund often use blended finance to support CCUS and other clean-energy projects in developing countries, where private investment alone may be insufficient to launch large projects [58].

Green bonds are a special type of loan used to raise money for environmentally friendly projects, such as carbon capture, renewable energy, or energy-efficient buildings. When a company or government needs money to build a low-carbon project, they can issue (sell) green bonds to investors. Investors who buy these bonds lend their money to the project for a fixed period. In return, they earn money through interest payments, just like with regular bonds. The interest is paid every year or every six months, and at the end of the bond's term, the investor gets back the full amount they originally invested [59].

Figure 14.11 illustrates how green bonds help finance low-carbon projects such as renewable energy and carbon capture systems. On the left, an investor provides capital by purchasing a green bond, shown in the center as an official financial document. The investment flows to the right, where it supports climate-friendly infrastructure like solar power, wind turbines, and CO_2 capture equipment. Over time, the project pays the investor back through scheduled interest payments, shown by the return arrow. The image highlights the full cycle: investors supply funds, the project uses them for environmental purposes, and investors receive steady financial returns. In this way, green bonds link private capital with sustainable development.

For example, if an investor buys a $1,000 green bond with 5% interest, they earn $50 every year until the bond ends, and then they receive their original $1,000 back. The "green" part simply means the money must be used for climate-friendly activities, and the project must report how the funds are spent. Investors like green bonds because they allow them

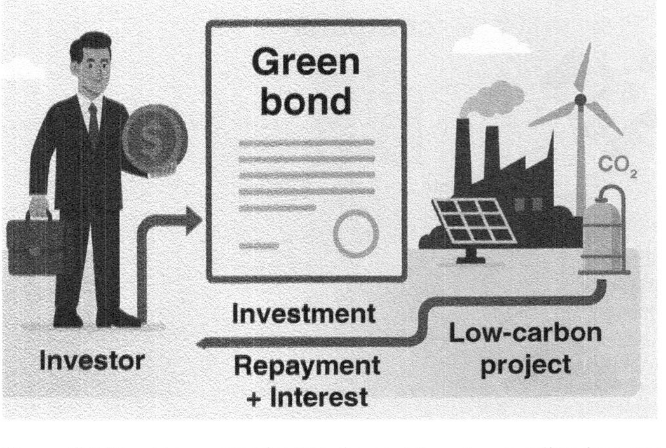

Figure 14.11 How green bonds channel investor funding into renewable energy and CO_2-reducing projects while providing stable financial returns to the investor.

to support climate solutions while still making a profit, and many green bonds are considered low-risk, especially when issued by strong governments or large companies. In some countries, there are even tax benefits for buying them [59].

Governments can also create carbon capture hubs, where multiple companies share common CO_2 transport and storage infrastructure. Instead of each factory building its own pipeline or storage site, several emitters use the same network and a shared geological storage site. Shared infrastructure reduces costs dramatically and makes CCUS possible for smaller industries, such as food processing, district heating, or smaller cement plants, which would not be able to pay for the whole system individually [60].

PPPs are an important form of innovative financing that help make CCUS projects economically realistic, especially for small and medium-sized companies. In a PPP, governments and private companies share the financial risks and responsibilities of developing carbon capture projects. Public support, through grants, low-interest loans, loan guarantees, or tax incentives, reduces the upfront costs that private companies would otherwise struggle to cover alone. At the same time, private partners contribute technical expertise, project management, and operational experience. By combining resources and skills from both sectors, PPPs make CCUS projects more financially secure, more attractive to investors, and easier to implement. These partnerships play a key role in accelerating global CCUS deployment and supporting long-term climate goals [54, 55].

14.6 Case Studies of Financially Feasible Projects

Real-world examples from different countries show that carbon capture projects can become financially successful when the right combination of technology, government support, and market incentives is in place. These case studies help readers understand what makes CCUS projects work in practice and why some projects become strong business opportunities while others struggle.

One of the most well-known examples is the Northern Lights project in Norway. This project is part of the larger "Longship" climate initiative, where the Norwegian government covers a major portion of the initial investment costs. By sharing the financial burden with private companies, the government reduces the risk typically associated with building large CCUS projects. Northern Lights is designed as an "open-access" CO_2 transport and storage system, meaning that industries from Norway and other European countries can ship their captured CO_2 to the facility for permanent storage. This shared model enables multiple companies to use the same infrastructure instead of building their own pipelines and storage sites, lowering overall costs. The project demonstrates how strong government commitment, cost-sharing, and international cooperation can make CCUS commercially attractive and successful [61].

Figure 14.12 shows a coastal CO_2 transport and storage facility inspired by Norway's Northern Lights project. In the foreground, a marine jetty equipped with loading arms allows ships to deliver liquefied CO_2 from industrial plants across Europe. A marine jetty is a specialized structure built along the shoreline that extends out into the sea, fjord, or river. Its main purpose is to allow ships to moor (tie up) and transfer materials safely. The mid-section of the image displays the main onshore processing area, including safety

Figure 14.12 Onshore CO_2 handling and export terminal similar to the Northern Lights project in Norway.

systems, compression units, and operational buildings where CO_2 is measured, conditioned, and prepared for injection. On the right side, large vertical storage tanks temporarily hold the CO_2 before it is transported through underground pipelines to offshore geological formations.

The figure highlights how open-access CO_2 transport and storage hubs reduce costs by allowing many emitters to use the same pipelines, port terminals, and geological storage sites. Instead of each company building its own infrastructure, shared hubs distribute operating expenses (OPEX) such as staffing, maintenance, energy use, and monitoring across many users, making CCUS economically viable even for smaller industries. Large-scale hubs also reduce financial risk because long-term storage monitoring, injection well integrity, and offshore operations are managed collectively rather than by individual firms. This collaborative model, supported by government cost-sharing and private-sector participation, is one of the main reasons projects like Northern Lights can operate reliably and efficiently.

In the United States, numerous successful projects have been supported by the 45Q tax credit, a significant financial incentive that pays companies for every ton of CO_2 they capture and permanently store. This stable, long-term policy has encouraged private investors to support CCUS in industries such as fertilizer production, ethanol manufacturing, and natural gas processing. Since companies can receive the tax credit for multiple years, the financial risk is considerably reduced, allowing projects to recover their costs more easily. These examples illustrate that clear and predictable incentives can quickly accelerate the deployment of CCUS, particularly in sectors where capture costs are lower [62].

Canada provides one of the clearest examples of financially successful CCUS projects. Several large facilities in the power, chemical, and industrial sectors have integrated carbon capture with EOR. In EOR, captured CO_2 is injected into mature oil reservoirs to increase oil production. Because this CO_2 helps boost output, oil companies are willing to pay CCUS operators for the captured CO_2. This creates a stable revenue stream that helps offset major costs such as capture, compression, and transport. The Canadian experience shows that linking CCUS with CO_2-based EOR can significantly reduce financial pressure, improve economic performance, and strengthen the long-term business case for deploying carbon capture technologies [63].

Many newer projects around the world are exploring additional ways to make CCUS profitable by converting captured CO_2 into valuable products. This method is known as carbon utilization. These projects can transform CO_2 into sustainable fuels, chemicals, plastics, and even building materials like cement and concrete. While many of these technologies are still in development, they show great promise in creating new industries, economies, and markets. When CO_2 is used as a raw material for valuable products, companies have multiple avenues to generate income from carbon capture [64].

Together, these case studies show that CCUS can be financially practical when the right conditions are in place. Successful projects are not driven solely by engineering; they also require supportive policies, innovative financing tools, and business models that reduce risk and generate stable revenue. When governments provide clear rules and motivations, and when companies find ways to combine CO_2 capture with profitable activities (such as CO_2-EOR, carbon markets, or carbon utilization), the economics of CCUS become much stronger. In these situations, carbon capture is no longer just an environmental responsibility; it becomes a realistic investment opportunity within the low-carbon economy. As technology continues to improve and more financial support mechanisms appear, CCUS is likely to play an increasingly important and commercially viable role in global climate strategies.

14.7 Key Highlights (Chapter Summary)

◇ Chapter 14—Key Highlights:

1) This chapter explains how capital and operating costs shape the financial feasibility of carbon capture projects.
2) This chapter shows how the Levelized Cost of CO_2 Abatement (LCCA) helps compare CCUS technologies with alternative decarbonization strategies by expressing all costs in terms of cost per tonne of CO_2 avoided.
3) Investors use financial indicators such as payback period, internal rate of return (IRR), and Return on Investment (ROI) to judge project profitability.
4) Carbon capture investments involve several types of risk, including technology risk (when capture or storage systems do not perform as expected or face operational problems), market risk, and policy risk, that must be carefully managed through strong planning, reliable engineering, and proper financing strategies.
5) Stable carbon prices and predictable regulatory frameworks reduce financial uncertainty, build investor confidence, and make it easier to secure long-term investment for carbon capture projects.
6) Carbon credit markets provide financial rewards for each tonne of CO_2 captured and permanently stored, creating an important additional revenue stream for project developers.

7) Market incentives such as carbon taxes, emissions trading systems, and performance-based payments can greatly strengthen project economics by increasing revenue and reducing financial risk.
8) Public–private partnerships (PPPs) help distribute financial risk more evenly by combining government funding and support with private-sector expertise and investment.
9) Successful CCUS projects rely on a mix of strong technology performance, government support, and market incentives, as shown by examples such as Norway's Northern Lights and U.S. projects supported by the 45Q tax credit.
10) Different business models, such as shared infrastructure, CO_2-EOR revenue, and carbon utilization, can significantly improve financial viability, helping companies reduce costs and generate additional income.

Keywords: Carbon Capture Economics, CAPEX, OPEX, Project Financing, Levelized Cost of CO_2 Abatement, Carbon Markets, Green Bonds, Investment Risk, Public–Private Partnerships, Market Incentives, Financial Viability, Climate Finance, Carbon Credits, Cost–Benefit Analysis, Return on Investment (ROI), Payback Period, Internal Rate of Return (IRR)

References

1 T. Pieri and A. Angelis-Dimakis, "Model development for carbon capture cost estimation," *Clean Technologies*, vol. 3, no. 4, pp. 787–803, 2021. DOI: 10.3390/cleantechnol3040046.

2 F. Rowaihy *et al.*, "Decarbonizing Saudi Arabia energy and industrial sectors: Assessment of carbon capture cost," *Carbon Capture Science & Technology*, vol. 14, p. 100375, Mar., 2025. DOI: 10.1016/j.ccst.2025.100375.

3 M. Garcia and N. Berghout, "Toward a common method of cost-review for carbon capture technologies in the industrial sector: cement and iron and steel plants," *Int. J. Greenh. Gas Control*, vol. 87, pp. 142–148, Aug., 2019. DOI: 10.1016/j.ijggc.2019.05.005.

4 C. C. Steincamp, T. R. Birdie, D. Stanley, E. Holubnyak, J. Raney, and W. L. Watney, "Regulation of carbon capture and storage," *Environmental Law*, vol. 51, no. 4, pp. 1149–1219, 2021.

5 G. García-García, M. C. Fernandez, K. Armstrong, S. Woolass, and P. Styring, "Analytical review of life-cycle environmental impacts of carbon capture and utilization technologies," *ChemSusChem*, vol. 14, no. 4, pp. 995–910, Feb., 2021. DOI: 10.1002/cssc.2020002126.

6 H. Zhang, "Regulations for carbon capture, utilization and storage: Comparative analysis of development in Europe, China and the Middle East," *Resources, Conservation and Recycling*, vol. 173, p. 105722, Oct., 2021. DOI: 10.1016/j.resconrec.2021.105722.

7 P. Larkin, W. Leiss, and D. Krewski, "Risk assessment and management frameworks for carbon capture and geological storage: A global perspective," *International Journal of Risk Assessment and Management*, vol. 22, no. 3–4, pp. 254–285, Oct., 2019. DOI: 10.04/IJRAM.2019.103332.

8 F. Mulyasari, A. K. Harahap, A. O. Rio, R. Sule, and W. G. A. Kadir, "Potentials of the public engagement strategy for public acceptance and social license to operate: Case study of carbon capture, utilisation, and storage Gundih pilot project in Indonesia," *International Journal of Greenhouse Gas Control*, vol. 108, p. 103312, June, 2021. DOI: 10.1016/j.ijggc.2021.103312.

9 K. Yanagi, A. Nakamura, and E. Komatsu, "The importance of designing a comprehensive Strategic Environmental Assessment (SEA) & Environmental Impact Assessment (EIA) for carbon capture and storage in Japan," *International Journal of Greenhouse Gas Control*, vol. 91, p. 102823, Dec., 2019. DOI: 10.1016/j.ijggc.2019.102823.

10 M. W. Hammad, A. Abbasi, and M. J. Ryan, "Allocation and management of cost contingency in projects," *Journal of Management in Engineering*, vol. 32, no. 6, Apr., 2016. DOI: 10.1061/(ASCE)ME.1943-5479.0000447.

11 H. Pilorgé *et al.*, "Cost analysis of carbon capture and sequestration of process emissions from the U.S. industrial sector," *Environmental Science & Technology*, vol. 54, no. 12, pp. 7524–7532, May, 2020. DOI: 10.1021/acs.est.9b07930.

12 E. Smith, J. Morris, H. Kheshgi, G. Teletzke, H. Herzog, and S. Paltsev, "The cost of CO_2 transport and storage in global integrated assessment modeling," *International Journal of Greenhouse Gas Control*, vol. 109, Jul., 103367, 2021. DOI: 10.1016/j.ijggc.2021.103367.

13 M. M. J. Knoope, A. Ramírez, and A. P. C. Faaij, "Investing in CO_2 transport infrastructure under uncertainty: A comparison between ships and pipelines," *International Journal of Greenhouse Gas Control*, vol. 41, pp. 174–193, Oct., 20. DOI: 10.1016/j.ijggc.20.07.013.

14 C. Cao, H. Liu, Z. Hou, F. Mehmood, J. Liao, and W. Feng, "A review of CO_2 storage in view of safety and cost-effectiveness," *Energies*, vol. 13, no. 3, p. 600, Jan., 2020. DOI: 10.3390/en13030600.

15 R. Wennersten, Q. Sun, and H. Li, "The future potential for carbon capture and storage in climate change mitigation—an overview from perspectives of technology, economy and risk," *Journal of Cleaner Production*, vol. 103, pp. 724–736, Sep., 20. DOI: 10.1016/j.jclepro.2014.09.023.

16 C. A. Moreira, "Digital monitoring of heavy equipment: Advancing cost optimization and operational efficiency," *Brazilian Journal of Development*, vol. 11, no. 2, 2025. DOI: 10.34117/bjdv11n2-011.

17 J. Florez, J. Dedhia, and I. Karunagaran, "Comparative assessment of levelized costs in carbon abatement strategies," in *Carbon Capture, Utilization, and Storage Conference and Exhibition*, 3–5 Mar., 2025, DOI: 10.530/ccus-2025-4183758.

18 H. E. Ashkanani *et al.*, "Effect of power plant capacity on the CAPEX, OPEX, and LCOC of the CO_2 capture process in pre-combustion applications," *International Journal of Greenhouse Gas Control*, vol. 109, Jul., 103371, 2021. DOI: 10.1016/j.ijggc.2021.103371.

19 R. Jayaraman, "Project cost control: A new method to plan and control costs in large projects," *Business Process Management Journal*, vol. 22, no. 6, pp. 1247–1268, Nov., 2016. DOI: 10.1108/BPMJ-10-2014-0102.

20 Y. Lu, L. Luo, H. Wang, Y. Le, and Q. Shi, "Measurement model of project complexity for large-scale projects from task and organization perspective," *International Journal of Project Management*, vol. 33, no. 3, pp. 610–622, Apr, 20. DOI: 10.1016/j.ijproman.2014.12.005.

21 W. S. Alaloul, M. Altaf, M. A. Musarat, M. F. Javed, and A. Mosavi, "Life cycle assessment and life cycle cost analysis in infrastructure projects: A systematic review," *Journal of Cleaner Production*, vol. 310, 127503, 2021. DOI: 10.1016/j.jclepro.2021.127503.

22 L. J. Müller, A. Käthelön, M. Bachmann, A. Zimmermann, A. Sternberg, and A. Bardow, "A guideline for life cycle assessment of carbon capture and utilization," *Frontiers in Energy Research*, vol. 8, Feb., 15, 2020. DOI: 10.3389/fenrg.2020.000.

23 H. W. Jones, "High reliability at minimum cost," in *Proc. 2024 Annu. Reliability and Maintainability Symp.* (RAMS), Jan. 2024, pp. 1–5.

24 J. Kjärstad, R. Skagestad, N. H. Eldrup, and F. Johnsson, "Ship transport—a low cost and low risk CO_2 transport option in the Nordic countries," *International Journal of Greenhouse Gas Control*, vol. 54, no. 1, pp. 168–184, Nov., 2016.

25 A. Ilavya, K. Patel, and A. Bera, "Chapter two—underground carbon storage," in *Advances and Technology Development in Greenhouse Gases: Emission, Capture and Conversion, Greenhouse Gases Storage and Transportation*, 2024, pp. 25–44, DOI: 10.1016/B978-0-443-19067-4.00016-4.

26 A. N. Arnette, "Renewable energy and carbon capture and sequestration for a reduced carbon energy plan: An optimization model," *Renewable and Sustainable Energy Reviews*, vol. 70, pp. 254–265, Apr., 2017. DOI: 10.1016/j.rser.2016.11.218.

27 A. Heyes and B. Urban, "The economic evaluation of the benefits and costs of carbon capture and storage," *International Journal of Risk Assessment and Management*, vol. 22, no. 3–4, pp. 324–341, Oct., 2019. DOI: 10.04/IJRAM.2019.103337.

28 E. Hirsch and T. Foust, "Policies and programs available in the United States in support of carbon capture and utilization," *Energy Law Journal*, vol. 41, pp. 91–124, 2020.

29 C. Duncan, J. H. Primavera, N. A. Hill, D. C. Wodehouse, and H. J. Koldewey, "Potential for return on investment in rehabilitation-oriented blue carbon projects: Accounting methodologies and project strategies," *Frontiers in Forests and Global Change*, vol. 4, Art. no.775341, 2022.

30 G. De Ghetto, M. Restelli, and D. Rudellin, "A Simplified Method to Calculate the Economics of Field Development with Carbon Capture and Storage," in Proc. OMC Med Energy Conf. Exhib., Ravenna, Italy, Sep. 2021, pp. OMC-2021-236.

31 T. M. Andrews, A. W. Delton, and R. Kline, "High-risk high-reward investments to mitigate climate change," *Nature Climate Change*, vol. 8, pp. 890–894, 2018. DOI: 10.1038/s458-018-0266-y.

32 R. Nimmala, "Evaluating financial risks of carbon capture in the oil & gas industry using advanced machine learning techniques," *Journal of Artificial Intelligence & Cloud Computing*, vol. 3, no. 2, pp. 2–5, 2024. DOI: 10.47363/JAICC/2024(3)312.

33 G. Wood, "Policy risk, politics and low carbon energy," in *Handbook of Energy Politics*, J. I. Considine, Ed. Cheltenham, UK: Edward Elgar Publishing, 2018, pp. 405–434, DOI: 10.4337/9781784712303.00029.

34 F. Dong *et al.*, "Analysis of market risk volatility and warning in carbon trading market," *Journal of Cleaner Production*, vol. 452, p. 142014, May., 2024. DOI: 10.1016/j.jclepro.2024.142014.

35 G. A. Tabaaza and B. Lal, "Risk analysis in carbon capture and storage," in *Gas Hydrate in Carbon Capture, Transportation and Storage*, B. Lal, Ed. CRC Press, 2024, pp. 121–136.

36 A. M. Batkovskiy, A. V. Konovalova, E. G. Semenova, V. Y. Trofimets, and A. V. Fomina, "Risks of development and implementation of innovative projects," *Mediterranean Journal of Social Sciences*, vol. 6, no. 4(S4), pp. 243–253, 20.

37 K. Gumte, V. Akram, and B. N. Rath, "Handling bioenergy sector uncertainties with carbon credit revenue in developing nation's economy: An Indian case study," *Environment, Development and Sustainability*, 2024. DOI: 10.1007/s10668-024-05058-7.

38 P. Roefs, M. Moretti, K. Welkenhuysen, K. Piessens, and T. Compernolle, "CO_2-enhanced oil recovery and CO_2 capture and storage: An environmental economic trade-off analysis," *Journal of Environmental Management*, vol. 239, pp. 167–177, Jun., 2019. DOI: 10.1016/j.jenvman.2019.03.007.

39 D. Leeson, N. Mac Dowell, N. Shah, C. Petit, and P. S. Fennell, "A techno-economic analysis and systematic review of carbon capture and storage (CCS) applied to the iron and steel, cement, oil refining and pulp and paper industries, as well as other high purity sources," *International Journal of Greenhouse Gas Control*, vol. 61, pp. 71–84, Jun., 2017. DOI: 10.1016/j.ijggc.2017.03.020.

40 P. Lisbona, R. Gori, L. M. Romeo, and U. Desideri, "Techno-economic assessment of an industrial carbon capture hub sharing a cement rotary kiln as sorbent regenerator," *International Journal of Greenhouse Gas Control*, vol. 112, Dec., 103524, 2021. DOI: 10.1016/j.ijggc.2021.103524.

41 J. Ye, L. Yan, and X. Liu, "Economic feasibility and policy incentive analysis of carbon capture, utilization, and storage (CCUS) in coal-fired power plants based on system dynamics," *Environmental Science and Pollution Research*, vol. 30, pp. 37487–37375, 2023. DOI: 10.1007/s11356-022-24888-4.

42 J. C. Richstein and K. Neuhoff, "Carbon contracts-for-difference: How to de-risk innovative investments for a low-carbon industry?," *iScience*, vol. 25, no. 8, p. 104700, Aug., 2022. DOI: 10.1016/j.isci.2022.104700.

43 R. Pande, "Can the market in voluntary carbon credits help reduce global emissions in line with Paris agreement targets?," *Science*, vol. 384, no. 6696, pp. 690–691, May, 2024. DOI: 10.1126/science.adp5223.

44 C. Charoentanaworakun, K. Somprasong, A. Duongkaew, P. Wongchai, P. Katunyoo, and P. Thanaphanyakun, "Minimum carbon credit cost estimation for carbon geological storage in the Mae Moh Basin, Thailand," *Energies*, vol. 17, no. 9, p. 2231, 2024. DOI: 10.3390/en17092231.

45 H. M. Ahonen, J. Kessler, A. Michaelowa, A. Espelage, and S. Hoch, "Governance of fragmented compliance and voluntary carbon markets under the Paris agreement," *Politics and Governance*, vol. 10, no. 1, pp. 47–59, 2022.

46 J. Bai, W. Ma, Y. Wang, and J. Jiang, "Political incentives in market-based environmental regulation: Evidence from China's carbon emissions trading scheme," *Heliyon*, vol. 10, no. 4, p. e25730, Feb., 2024.

47 K. Zhou and Y. Li, "Carbon finance and carbon market in China: Progress and challenges," *Journal of Cleaner Production*, vol. 214, Mar. 20, pp. 536–549, 2019. DOI: 10.1016/j.jclepro.2018.12.298.

48 N. P. Espenan, "Improving voluntary carbon markets through standardization and blockchain technology," *Wyoming Law Review*, vol. 23, p. 141, 2023.

49 V. Battocletti, L. Enriques, and A. Romano, "The voluntary carbon market: Market failures and policy implications," *University of Colorado Law Review*, vol. 95, no. 3, p. 519, 2024.

50 A. Tripathi, A. Dixit, and Vipul, "Liquidity of financial markets: A review," *Studies in Economics and Finance*, vol. 37, no. 2, pp. 201–227, 2020. DOI: 10.1108/SEF-10-2018-0319.

51 L. Yang, X. Zhang, and K. J. McAlinden, "The effect of trust on people's acceptance of CCS (carbon capture and storage) technologies: Evidence from a survey in the People's Republic of China," *Energy*, vol. 96, pp. 69–79, Feb., 2016. DOI: 10.1016/j.energy.20.12.044.
52 Z. Zhang and D. Huisheng, "Carbon dioxide storage schemes: Technology, assessment and deployment," *Journal of Cleaner Production*, vol. 142, no. pt. 2, pp. 1055–1064, Jan., 2017. DOI: 10.1016/j.jclepro.2016.06.199.
53 Y. Zheng and B. Zhang, "Does China's national carbon market function well? A perspective on effective market design," *Journal of Chinese Governance*, vol. 8, no. 4, pp. 563–592, 2023. DOI: 10.1080/23812346.2023.2214023.
54 D. Kirikkaleli and T. S. Adebayo, "Do public-private partnerships in energy and renewable energy consumption matter for consumption-based carbon dioxide emissions in India?," *Environmental Science and Pollution Research*, vol. 28, pp. 30139–30302, 2021. DOI: 10.1007/s11356-021-12692-5.
55 M. Akhtar, N. A. Mufti, S. Mubin, M. Q. Saleem, S. Zahoor, and S. Ullah, "Identification of various execution modes and their respective risks for public–private partnership (PPP) infrastructure projects," *Buildings*, vol. 13, no. 8, p. 1889, 2023. DOI: 10.3390/buildings13081889.
56 E. Choi and A. Seiger, "Catalyzing capital for the transition toward decarbonization: Blended finance and its way forward," SSRN, 2020. Available: https://ssrn.com/abstract=3627858
57 C. Greig, K. Spokas, K. Waltzer, D. Ulama, and S. Hobart, "Systemic bankability is the key to unlocking energy transition speed and scale," SSRN, Feb. 2025. DOI: 10.2139/ssrn.5149329
58 K. Spokas, C. A. T. Force, D. Ulama, C. Greig, and K. Waltzer, "Systemic bankability is the key to unlocking energy transition speed and scale," unpublished manuscript/working paper.
59 U. S. Bhutta, A. Tariq, M. Farrukh, A. Raza, and M. K. Iqbal, "Green bonds for sustainable development: Review of literature on development and impact of green bonds," *Technological Forecasting and Social Change*, vol. 175, 2022 121378. DOI: 10.1016/j.techfore.2021.121378.
60 T. A. Meckel, A. P. Bump, S. D. Hovorka, and R. H. Trevino, "Carbon capture, utilization, and storage hub development on the Gulf Coast," *Greenhouse Gases: Science and Technology*, vol. 11, no. 4, pp. 619–632, Aug., 2021. DOI: 10.1002/ghg.2082.
61 A.-K. Furre, R. Meneghulo, P. Ringrose, and S. Kassold, "Building confidence in CCS: From Sleipner to the Northern lights project," *First Break*, vol. 37, no. 7, pp. 81–87, Jul., 2019. DOI: 10.3997/1365-2397.n0038.
62 A. C. Jones and D. J. Marples, *The Section 45Q Tax Credit for Carbon Sequestration*. Washington, DC: Congressional Research Service, 2023.
63 S. Bachu, "Identification of oil reservoirs suitable for CO_2-EOR and CO_2 storage (CCUS) using reserves databases, with application to Alberta, Canada," *International Journal of Greenhouse Gas Control*, vol. 44, pp. 2–165, Jan., 2016. DOI: 10.1016/j.ijggc.20.11.013.
64 Y. Zhao and K. Itakura, "A state-of-the-art review on technology for carbon utilization and storage," *Energies*, vol. 16, no. 10, p. 3992, May., 2023. DOI: 10.3390/en16103992.

15

Case Studies in Carbon Capture and Environmental Impact

15.1 Case Study 1: Large-Scale CCS in the Power Sector

Large-scale carbon capture and storage (CCS) projects in the power sector provide some of the most valuable lessons for understanding the technical, financial, and social realities of carbon capture deployment. Power plants, especially coal and natural gas units, are among the most significant point sources of CO_2 emissions worldwide, making them an early target for commercial-scale CCS demonstration. Coal-fired and natural-gas-fired power plants are extensive industrial facilities that generate electricity by burning coal or natural gas to produce steam or drive gas turbines. These plants are major point sources of CO_2 emissions. This case study examines the experiences of a crucial CCS-equipped power station [1]. It highlights engineering performance, environmental outcomes, financial challenges, and community responses.

One of the best-known examples is a commercial coal-fired power plant like Petra Nova that integrated post-combustion carbon capture technology, illustrated in Figure 15.1. The project aimed to capture over one million tonnes of CO_2 per year and store it in a deep geological formation [2]. From an engineering perspective, the project demonstrated that large-scale flue gas capture is technically achievable. The capture unit achieved high CO_2 removal efficiency during stable periods. In addition, the plant employed advanced solvent technology, heat integration strategies, and large-scale compressor operation [2, 3]. This setup provided a foundation for future CCS design and operation and contributed valuable data on energy consumption, solvent degradation, and long-term performance.

Environmentally, the project delivered significant reductions in CO_2 emissions during periods when the system operated at full capture capacity. In addition, monitoring data confirms safe, stable, and long-term CO_2 storage performance. However, the overall multi-year capture effectiveness fell short of initial expectations and estimates because the facility experienced frequent equipment outages and operational interruptions, many originating from technical challenges in integrating a complex CCS system with an aging power plant that was not originally designed for carbon capture [2, 3].

In addition, fluctuating electricity market conditions influenced the plant's ability to run the CCS system consistently. During periods of high electricity prices or market undersupply, the power plant reduced output because operating at full capacity was not economically justified; as a result, the CCS unit also operated below its intended capacity or was temporarily taken offline. High electricity prices, in turn, constrained the CCS unit's ability

Figure 15.1 Illustrative representation of a coal-fired power plant incorporating post-combustion CO_2 capture equipment, similar to the Petra Nova configuration.

to run continuously and achieve the intended and estimated capture rates [4]. These market-driven curtailments often forced the plant to ramp down or temporarily shut off the capture unit, reducing annual CO_2 capture volumes despite strong performance during fully operational periods. This behavior can be better understood by looking at the amine-based capture process shown in Figure 15.2. The system relies on a continuous flow of flue gas to the absorber and a steady steam supply to the regenerator. When the power plant reduces output, both gas flow and steam availability drop, forcing the CCS unit to operate below design capacity or temporarily shut down.

Financially, the project faced several difficulties common to early CCS projects. First, the cost of building the plant turned out to be much higher than initially expected. This extra cost occurred because the project involved first-of-a-kind engineering, requiring many components to be specially designed, tested, and adapted for a real power plant [5]. In addition, the power plant itself required significant modifications to allow the carbon capture unit to be connected and operated safely, further increasing the overall cost [6]. Once the system was running, the operating expenses were also higher than planned. The CCS process requires significant additional energy, increasing fuel use and operating costs. For instance, the chemical solvent used to capture CO_2 must be replaced regularly, leading to additional ongoing costs. Furthermore, the system requires highly trained engineers and technicians, which increases staffing expenses compared to a conventional power plant [7].

Even though the project received substantial government funding, it still faced economic pressure because the power plant did not operate continuously. When electricity prices were high or the plant reduced output, the CCS system also had to slow down or shut off, which meant the project earned less revenue than expected. All of these financial

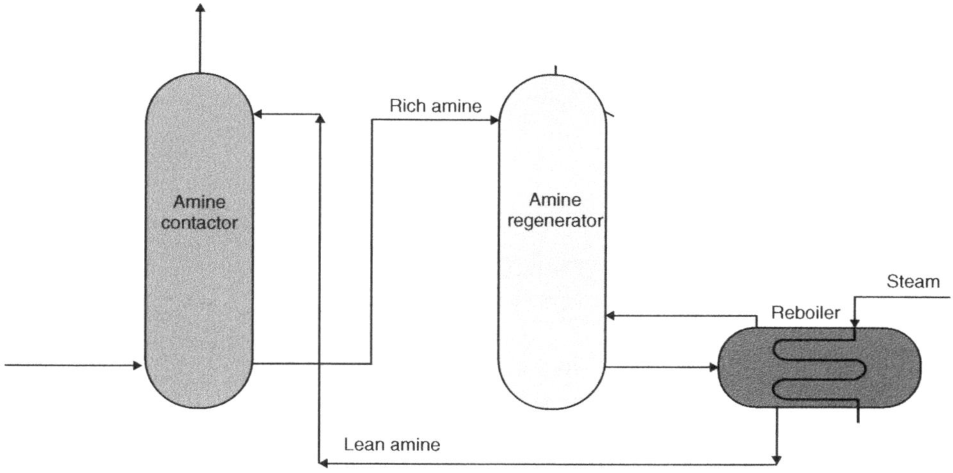

Figure 15.2 Simplified process flow of an amine-based CO_2 capture system, showing the absorber (contactor), regenerator (stripper), and reboiler. The process relies on steady flue-gas flow and continuous steam supply, making it sensitive to power-plant load changes.

challenges demonstrated that future CCS projects will need stable electricity markets, meaning predictable operating conditions, as well as long-term government policies that encourage investment. Precise and reliable CO_2 pricing, such as carbon taxes or emissions-trading systems, is also essential so companies can justify the high upfront costs of building and operating carbon capture facilities [8].

The project also highlighted the importance of community engagement for the success of CCS projects. Although there was no significant or organized opposition, some locals still had understandable concerns, especially about the safety of the CO_2 pipeline and whether the underground storage site would remain secure over many years [9]. To address these concerns, the project team focused on clear, open communication. They held public meetings where people could ask questions directly to engineers and environmental experts. The technical project team provided simple explanations of how the technology works and about the safety measures in place.

The project also shared results from online monitoring using sensors, including pressure data, leak-detection results, and groundwater-quality measurements. Making this information readily available supported that the storage site was behaving safely as expected. Over time, this transparency builds trust between project developers and the public, reduces uncertainty about carbon capture technology, and improves community acceptance of the project, highlighting the importance of early, honest engagement with local people for future CCS deployment [2]. This monitoring approach is illustrated in Figure 15.3, which shows how multiple systems work together to ensure storage-site safety. Pressure sensors, subsea monitoring tools, and reservoir-integrity measurements continuously track CO_2 behavior in different layers of the system, from the pipeline to the subsurface. By integrating these monitoring methods, operators can quickly identify irregularities and verify that the injected CO_2 remains securely stored.

The onshore monitoring station continuously checks pressure, temperature, and flow in the pipeline, injection well, and the reservoir so operators can spot early signs of problems,

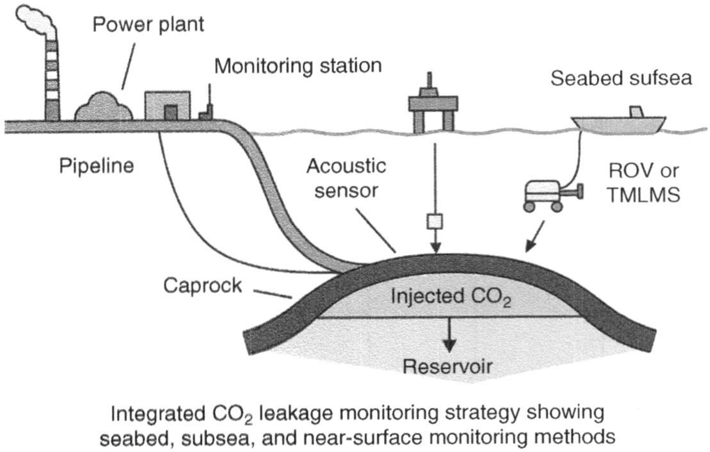

Figure 15.3 Overview of CO_2 storage monitoring systems, including onshore monitoring stations, subsea sensors, and inspection tools.

such as CO_2 leakage. Offshore, subsea tools such as underwater acoustic sensors and remotely operated vehicles (ROVs) monitor the water and seabed for any changes that might suggest CO_2 is escaping and leaking into the sea. Closer to the reservoir, sensors measure pressure and track how the injected CO_2 is moving under the caprock. Together, these surface and subsea monitoring methods give operators a full picture of the site and help confirm that the stored CO_2 remains safely contained.

Overall, this case study shows that CCS in the power sector can work well from a technical and environmental standpoint. The technology proved that it can successfully capture large amounts of CO_2 and store it safely underground. However, the project also makes clear that there are still major financial and operational challenges to be addressed before CCS can be widely used. For example, high construction and operating costs, unexpected equipment problems during operation, and variable electricity market conditions all made the project harder to run efficiently [2, 7, 8].

From these experiences, several important lessons have been learnt. Future CCS projects will need strong policy incentives, such as government funding or reliable carbon pricing, to make the technology affordable. They also require careful planning to integrate CCS systems smoothly with existing power plants, along with robust maintenance strategies to keep equipment reliable over time. Finally, continuous community engagement is essential because local acceptance increases when people understand the technology and see transparent safety and environmental information [2]. These lessons are now guiding the design and operation of a new generation of CCS projects worldwide, helping improve the performance of carbon capture plants and reducing risks.

15.2 Case Study 2: Industrial Capture and CO_2 Utilization

Industrial facilities, such as cement plants, fertilizer producers, steel mills, and chemical factories, are some of the most challenging sectors to decarbonize. This difficulty arises because they produce CO_2 through two main processes. Like power plants, these industries burn fuels for heat and energy, which releases CO_2. More significantly, many of them generate CO_2 directly from chemical reactions that are essential for producing their materials [10].

For example, cement production releases CO_2 when limestone is heated, and steelmaking produces CO_2 during the reduction of iron ore. In steelmaking, "reduction" is the process of removing oxygen from iron ore (which is mainly iron oxide) to produce metallic iron. Reduction is usually done in a blast furnace, where iron ore reacts with carbon monoxide. The carbon monoxide binds with the oxygen in the ore, forming CO_2 and leaving behind molten iron. This chemical reaction is a major source of CO_2 emissions in traditional steel production. Reduction process emissions cannot be eliminated simply by switching to renewable energy sources [11].

As a result, industrial CCS becomes a crucial tool in global climate strategies, as it can capture CO_2 from both fuel combustion and industrial chemical processes. This case study explores a real-world industrial carbon capture and utilization project, detailing its design, operational performance, and environmental benefits. It also examines the financial and practical challenges the project encountered, offering valuable lessons for future CCS implementation in hard-to-abate industrial sectors.

A clear example of industrial carbon capture is a large fertilizer plant that captures CO_2 during the production of ammonia and urea. Ammonia, a nitrogen-based fertilizer, is produced through the Haber-Bosch process, where nitrogen (N_2) is extracted from the air and reacted with hydrogen (H_2). This overall process involves combining nitrogen and hydrogen at high temperatures and pressures to create ammonia [12]. Urea, which is a nitrogen-rich fertilizer, is produced by reacting ammonia with CO_2. It is one of the most widely used fertilizers globally due to its ease of transport, storage, and application to crops [13]. A simplified schematic of the ammonia–urea production route, including CO_2 capture and utilization within the process, is shown in Figure 15.4.

In fertilizer plants, CO_2 is released as a byproduct of natural chemical reactions used to produce fertilizer and is emitted in a highly concentrated stream. This makes the capture process much easier and cheaper in a fertilizer plant than capturing CO_2 in a power plant [14]. In this project, an advanced capture unit was installed to separate high-purity CO_2, remove impurities, and compress CO_2 for safe transport through pipelines or direct

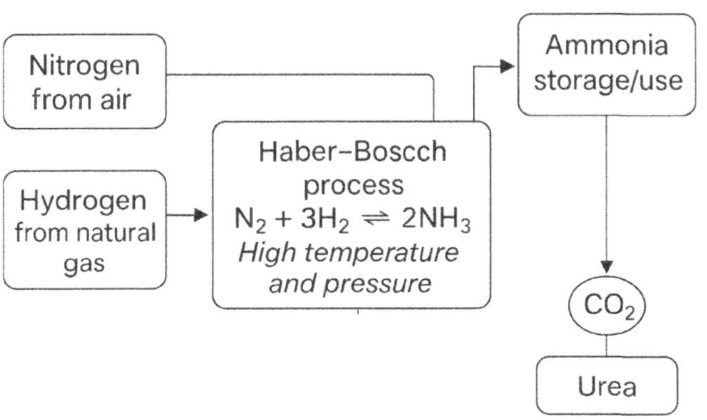

Figure 15.4 Process flow for ammonia and urea production.

use. A significant portion of the captured CO_2 is delivered to nearby industries for enhanced oil recovery (EOR), beverage processing, or as an ingredient in chemical manufacturing [15]. Carbon capture technology not only reduces emissions at the fertilizer plant but also supports CO_2 utilization and practical industrial applications that require CO_2.

From a technical perspective, the fertilizer-plant CCS project has performed extremely well. The system often captures even more CO_2 than it was originally designed for, mainly because the CO_2 coming from ammonia and urea production is already in a high-purity, concentrated form. This means the capture unit does not need to work as hard or use as much energy as a system that must separate CO_2 from dilute flue gas, like in a power plant [14]. Over several years of operation, the project has also generated valuable technical information. Engineers have been able to study how the capture solvent behaves over long periods, how to prevent corrosion in pipes and equipment, how to efficiently reuse heat from the chemical processes for the capture process, and how compressors perform when handling large volumes of CO_2 [15].

Environmentally, the project has made a major contribution by cutting CO_2 emissions from one of the region's most important heavy industrial facilities. Heavy industry refers to large, energy-intensive sectors, such as steel, cement, fertilizer, and chemicals, that involve heavy machinery, high temperatures, and large volumes of raw materials. These industries typically have high CO_2 emissions because their production processes rely on both fuel combustion and chemical reactions [16].

By capturing CO_2 instead of releasing it, the fertilizer plant turns what would normally be a waste stream into a useful product. This demonstrates how CO_2 utilization can create economic value while also lowering the climate impact of heavy industry [17]. However, the long-term environmental benefit depends greatly on how the captured CO_2 is ultimately used. For example, when CO_2 is sent to EOR, it is injected deep underground to help push out additional oil. While a portion of this CO_2 remains trapped in the rock formations, some of it may return to the surface mixed with the produced oil and gas. When these fuels are later burned, their carbon content, including the CO_2 originally injected, can eventually return to the atmosphere [18].

In contrast, when captured CO_2 is converted into chemicals, plastics, building materials, or fertilizers, it can remain stored inside these products for months, years, or even decades. However, once these products reach the end of their life cycle, the carbon they contain can still be released unless they are managed properly. This highlights the importance of recycling CO_2-based materials or capturing their emissions again to prevent the stored CO_2 from being released back into the atmosphere [19]. Overall, the project shows that careful tracking of CO_2 throughout its entire lifecycle is essential to ensure that utilization strategies deliver real and lasting climate benefits. Figure 15.5 illustrates how captured CO_2 can flow into various products such as plastics, polymers, and building materials, where it remains stored temporarily before potentially being released at end-of-life.

From an economic point of view, the picture is mixed. Industrial carbon capture refers to capturing CO_2 from factories and industrial plants such as cement, steel, fertilizer, or chemical production facilities. These plants often release CO_2 in high concentrations, which makes it easier and cheaper to capture compared to power plants [14, 17]. In this sense, industrial capture can be more cost-effective. However, even if the capture

Figure 15.5 Pathways of captured CO_2 utilized in industrial products, including polymers, plastics, and building materials, illustrating temporary carbon storage and the need for end-of-life management to avoid re-emissions.

process is cheaper, the overall business case is still challenging. Companies usually depend on government incentives such as tax credits, subsidies, or carbon pricing to reduce financial risks.

15.3 Case Study 3: BECCS Pilot Projects

Bioenergy with CCS, commonly called BECCS, is one of the few technologies in the world that can actually remove CO_2 from the atmosphere while producing useful energy. BECCS works by using biomass, such as wood chips, agricultural waste, or specially grown energy crops, as the fuel. Plants absorb CO_2 from the atmosphere as they grow. When the biomass is burned or processed, it releases the same CO_2. If we capture this CO_2 instead of letting it go back into the air, the entire cycle becomes net-negative, meaning the atmosphere ends up with less CO_2 than before. This concept makes BECCS very attractive in global climate strategies, especially in countries aiming for net-zero or net-negative targets [20].

This case study focuses on real pilot BECCS projects, particularly those in the United Kingdom and the United States. These early projects offer important lessons about engineering performance, biomass supply, environmental benefits, economic challenges, and community involvement.

One of the most widely discussed BECCS pilot projects is the small demonstration unit at the Drax Power Station in the United Kingdom. Drax, originally a large coal-fired power plant, had long been associated with carbon capture discussions because CCS technologies were initially developed for fossil-fuel power generation. After Drax converted most of its units to burn biomass instead of coal, the plant became a natural test site to explore whether conventional CCS systems could also work with biomass. In this context, the BECCS pilot at Drax tested a solvent-based CO_2-capture system on a slipstream of biomass flue gas. The purpose was to determine if CCS technologies, designed primarily for coal or natural-gas flue gases, could operate effectively with the different chemical composition and characteristics of biomass combustion exhaust. This pilot helped assess whether adding CCS to a biomass plant could deliver net-negative emissions [21]. Figure 15.6 presents an aerial view of the Drax Power Station, illustrating the industrial setting in which the UK's first BECCS pilot project was carried out. Although the CO_2-capture unit is attached to a conventional power plant, the system is called BECCS because the plant burns biomass, not fossil fuels.

Technically, the pilot succeeded in capturing CO_2 from biomass flue gas. This was an important step because biomass combustion produces flue gas with high moisture content, which can affect how well the solvent works [22]. Engineers observed that the system maintained stable capture performance and that solvent degradation was manageable. Over time, they collected valuable data on how temperature, moisture, and impurities influence capture efficiency. This information is now helping researchers improve future BECCS systems [21].

A second example comes from the United States, where the Illinois Industrial BECCS (IL-ICCS) project demonstrated negative-emission potential using CO_2 from ethanol production. Unlike power plants, ethanol plants produce a stream of almost pure CO_2 during fermentation. This high-purity CO_2 is much easier and cheaper to capture. The Illinois project captured over one million tonnes of CO_2 per year and injected it into a deep

Figure 15.6 Aerial view of the Drax Power Station in the UK, where a pilot BECCS unit tested solvent-based CO_2 capture on biomass flue gas to assess the feasibility of achieving net-negative emissions. (Courtesy of Shutterstock.com).

geological formation for permanent storage. Because the CO_2 came from renewable biomass, the result was a proven large-scale negative emission. This project demonstrated that BECCS can deliver climate benefits today, not just in the future [23].

Even though the technical results from BECCS pilot projects are encouraging, several challenges remain before the technology can be deployed at a meaningful scale. One of the most significant challenges is ensuring a sustainable biomass supply chain. Operating a large BECCS facility requires a massive and stable daily supply of biomass. This demand can jeopardize the environmental benefits of BECCS if it leads to increased deforestation, unsustainable land use, or excessive harvesting of natural resources [24].

A continuous supply chain also requires ongoing forestry operations, which must follow sustainable management practices to avoid degradation of ecosystems. Similarly, responsible agricultural harvesting is essential to prevent soil depletion, biodiversity loss, and other negative environmental impacts associated with overharvesting [25]. Transportation poses another challenge: if biomass must be moved long distances using fossil-fuel-based logistics, a substantial portion of the climate benefits can be offset by the associated emissions [26].

Environmentally, BECCS has the potential to deliver significant benefits, particularly when the biomass used is managed sustainably. During growth, as shown in Figure 15.7, plants absorb CO_2 from the atmosphere through photosynthesis. This absorbed CO_2 is called biogenic CO_2, meaning it originates from natural biological processes rather than from fossil fuels [27]. When this biomass is used for energy and the resulting biogenic CO_2 is captured and permanently stored underground, the carbon that was previously taken from the atmosphere is effectively removed from the carbon cycle. In this sense, BECCS is one of the very few technologies capable of actually reversing some of the impacts of past emissions, not just reducing future ones [20].

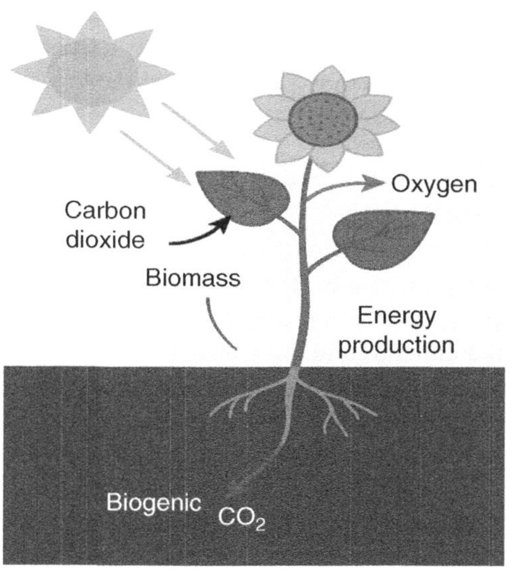

Figure 15.7 Photosynthesis and production of biogenic CO_2 in biomass. During growth, plants absorb atmospheric carbon dioxide (CO_2) using sunlight and convert it into biomass through photosynthesis, releasing oxygen (O_2) back into the atmosphere. When the biomass is later used for energy production, the resulting emissions are biogenic CO_2; carbon originally captured from the atmosphere rather than from fossil sources.

Economically, BECCS projects face many of the same challenges as other CCS technologies. Building the equipment needed to capture CO_2, upgrading power plants to handle biomass, and drilling wells to store the CO_2 underground all require very large investments. These projects are expensive to run as well. Because of the high costs, BECCS usually cannot make money on its own. Most projects need government support, carbon-removal credits, or strong carbon-pricing policies to be financially feasible. Without these incentives, energy companies are often hesitant to invest because the financial risks are too high. A good example is the Drax BECCS pilot project in the United Kingdom. Even though it has been shown that the technology works, it cannot operate economically at a large scale unless the government provides long-term financial support. This shows that strong policies are essential for turning BECCS from a small pilot into a real large-scale solution [28].

The Illinois project showed that BECCS can be cost-effective in some situations. Ethanol plants produce nearly pure CO_2, which makes capture cheap and energy-efficient. But this kind of opportunity is limited to specific industrial sectors. Expanding BECCS across the global power sector will require major infrastructure investment, stable biomass supply chain, and long-term policy mechanisms such as carbon taxes and emissions trading systems [29].

Social acceptance is another important factor. Community concerns about BECCS are usually related to land use, forest harvesting, transportation of biomass, and the safety of CO_2 storage sites. Public acceptance tends to improve when developers share monitoring data and engage openly with local residents [30]. For example, BECCS projects that publish information on groundwater safety, injection, and well performance tend to build stronger community trust. Early engagement with local communities is essential for avoiding misunderstandings and building support of the community [31].

In summary, BECCS pilot projects demonstrate both promise and complexity. The technology can capture CO_2 from biomass and deliver real negative emissions, but success depends on many interconnected factors: sustainable biomass supply, reliable capture systems, secure geological storage, supportive policy environments, and strong community involvement. BECCS is not simply a technical system; it is a complete environmental and social system that demands attention to many details outside engineering.

15.4 Case Study 4: Community-Led Environmental Monitoring

CCS projects are often large, complex systems that involve pipelines, injection wells, monitoring stations, and long-term geological storage [15]. Because these projects deal with CO_2 and underground reservoirs, it is natural for local communities to have questions and worries. People may wonder whether the CO_2 could leak, whether the pipelines are safe, or whether the underground rocks can hold the CO_2 for hundreds of years. These concerns are completely reasonable, especially when the technology is new to the region [30].

This case study is presented as a generalized example of community-led environmental monitoring and is not based on a single specific project, but rather integrates common practices and lessons observed across several CCS initiatives worldwide. It looks at a project where the local community was directly involved in environmental monitoring. Instead of simply being informed after decisions were made, the community became an active partner in the project's safety and reliability. This approach is known as community-led environmental monitoring, and it has recently become an important approach for increasing public trust in CCS technologies [32].

This case took place in a region where a CO_2 storage site had been operating for several years. The operators and government realized that, although the technical data showed the site was safe, some residents still did not feel fully confident. People were unsure about what was happening underground related to CO_2 storage and whether they could trust the reports issued by engineers. To address these concerns, a new approach was developed: opening the monitoring system to the community [15, 32].

The first step was to form a community advisory group. This group included local residents, teachers, farmers, youth representatives, environmental activists, and small business owners. They met regularly with project engineers, environmental scientists, and regulatory officials. The purpose of the advisory group was to learn about the project, ask questions, raise concerns, and help guide how information would be shared with the broader community [33].

To help the advisory group understand the technology, the project team organized training sessions. These sessions explained, in simple language, how CCS works, why monitoring is performed, and what the different measurements mean. For example the group learned how sensors measure pressure inside the injection well, how seismic monitoring detects small earth movements, and how groundwater samples are tested for any changes in quality. The group also visited the injection site and monitoring stations so they could see the equipment firsthand [32–34].

After the training, the community group was given access to an online monitoring dashboard. This dashboard, normally used only by engineers, displayed real-time data from the storage site. It showed graphs of injection pressure, temperature, well integrity, and groundwater chemistry. The data was color-coded to show normal conditions, unusual conditions, and potential risk levels. Community members could log in at any time from home and see exactly what the engineers were seeing. This kind of transparency helped reduce the feeling that information was being kept hidden [32–34].

In addition to the digital dashboard, the community group participated in groundwater sampling days. Together with environmental scientists, they collected water samples from nearby wells and helped test them for pH, dissolved minerals, and gas content. These joint activities helped build trust because residents could see for themselves that the groundwater remained stable and safe [32–34].

To support community understanding of the project, the team also held small, open community sessions where local residents could speak directly with engineers as shown in Figure 15.8. Instead of long technical presentations, the engineers used simple visuals, diagrams, and real monitoring data to explain how the storage site was performing. These smaller, discussion-based meetings gave people the chance to ask questions in a comfortable setting. Common questions included: "What happens if the pressure becomes too high?", "How do you know the CO_2 stays underground?", and "What is the emergency response plan?" By answering these questions clearly and openly, the project team demonstrated a genuine commitment to transparency and community engagement [34].

Figure 15.8 Small community briefing on CCS monitoring. Engineers meet with local residents in an open, discussion-based session, using simple visuals and real monitoring data to explain how the CCS project works and to answer community questions.

One of the most important aspects of this case study was the communication style. Instead of using complicated scientific terms, the project team explained things in everyday language. For example, instead of saying "reservoir integrity is stable," they said, "The underground rocks are holding the CO_2 safely, and the measurements show no movement outside the expected area." This helped everyone understand the information, not just technical experts [35].

Over time, the community advisory group became more confident in the CCS's success. This approach produced several important benefits. First, trust increased. When people can see the project reality and associated data themselves, they trust and accept the carbon capture project development more easily. Second, project transparency improved, which reduced rumors and the chance of public resistance. Third, local knowledge and concerns were incorporated into project decisions. Residents often know things and are concerned about the land and water flow patterns due to usage for the plant. Their input helped strengthen the environmental strength and benefits of the project [35].

However, the project also showed that community-led monitoring requires time, effort, and long-term commitment. Engineers must be patient, willing to communicate, and open to public questions. Building trust takes months or even years, not just one meeting. But the benefits are worth it, because community acceptance is essential for CCS projects to succeed. In conclusion, this case study demonstrates that community-led environmental monitoring can significantly improve the social acceptance of CCS projects. When residents are included as partners, not just observers, public understanding grows, and concerns are reduced [32, 33].

15.5 Comparative Analysis of Environmental Outcomes

The four case studies presented in this chapter, power-sector CCS, industrial CO_2 capture and utilization, BECCS pilot projects, and community-led environmental monitoring, show that carbon capture can play many different roles in reducing greenhouse gas emissions. However, these projects perform very differently in terms of environmental outcomes. This section compares the results across the four cases to identify what types of CCS projects deliver the strongest environmental benefits, what factors influence their performance, and what challenges must be addressed to ensure long-term sustainability.

A key finding from the comparison is that power-sector CCS projects deliver the highest total CO_2 reductions, simply because coal- and gas-fired power plants emit very large amounts of CO_2. When a power plant captures even a portion of its emissions and store it underground, the absolute amount of CO_2 stored underground each year can be substantial [36]. However, the overall environmental performance of these projects can vary considerably, since power plants must adjust their electricity generation up or down depending on real-time demand and electricity market conditions, which affects CO_2-capture efficiency [37]. If the power plant reduces output or shuts down temporarily, the CCS system also reduces its operation or stops. This means that even if the capture technology performs well during operation, the total annual CO_2 reduction may be lower than expected due to electricity price fluctuations. As a result, the environmental benefit of power-sector CCS depends heavily on but is limited to uninterrupted operation and stable electricity demand [2, 7, 8].

Industrial CCS projects, meaning systems that capture CO_2 directly at industrial facilities such as ammonia plants, fertilizer factories, or ethanol plants, often perform better and more consistently than CCS at power plants. The reason is that many industrial processes such as ammonia and fertilizer factories produce CO_2 in a very pure and concentrated form, which makes it much easier and cheaper to capture [38]. Power plants, by contrast, release CO_2 in dilute flue gas, where the CO_2 is mixed with large amounts of nitrogen, water vapor, and other combustion gases, making separation more difficult and energy-intensive [39]. Industrial CCS is also less affected by changing electricity demand: power plants must adjust their output depending on how much electricity the grid needs, but industrial plants often run at steady rates, so their CO_2 streams and capture systems are more stable. Because of these factors, industrial CCS typically has higher capture efficiency, lower energy use, and more reliable operation [39, 40].

The BECCS case study shows a third type of environmental outcome. BECCS projects offer the potential for negative emissions, meaning they can actually remove CO_2 from the atmosphere instead of simply reducing emissions at the source. This makes BECCS uniquely important in climate modeling, especially in scenarios where global emissions must reach net-negative levels. In pilot tests, BECCS systems have been able to capture CO_2 from biomass combustion or fermentation with high efficiency. However, the environmental success of BECCS depends heavily on the sustainability of the biomass supply chain. If forests are harvested without proper replanting, or if biomass is transported long distances with fossil-fuel trucks, the environmental benefits can decrease significantly. Similarly, if biomass production disrupts ecosystems, the overall sustainability of the project becomes questionable. Therefore, the environmental outcome of BECCS cannot be judged by capture performance alone [22, 23].

The fourth case study, community-led environmental monitoring, focuses not on capture efficiency but on the social and environmental integrity of CO_2 storage sites. Even if a CCS project performs effectively technically, the environmental outcome will not be meaningful if the project does not have public acceptance and community engagement. The long-term storage integrity of carbon dioxide is a major public concern, as it is essential to prevent leakage, protect groundwater resources, and ensure that the climate benefits of CCS remain permanent. Community-led actions help ensure that environmental outcomes are not only technically verified but also socially trusted. When residents have access to real-time pressure measurements, groundwater samples, and seismic data, they can verify that the storage site is safe. This transparency strengthens the environmental credibility of the project socially and brings public confidence in CCS as a long-term climate solution [32–35].

When comparing these four types of CCS efforts, several important patterns emerge. Power-sector CCS has the greatest potential for large-volume emission reductions but also faces the most operational challenges [36]. Industrial CCS is more stable and efficient [38]. BECCS offers powerful negative-emission potential but is highly dependent on sustainable biomass management [22, 23]. Community-led involvement and action do not directly influence CO_2 capture technologies, but they can significantly enhance safety, reduce risks through local participation, and build public trust; factors that are essential for the long-term environmental success of CCS projects [32, 33].

In the end, the comparison shows that every CCS project must be assessed based on its own local situation. Many factors, such as electricity costs, the types of industries in the area, public opinion, the availability of biomass, the geology of the region, and national or local government policies, can influence how well a CCS project performs environmentally. There is no single "best" CCS model that works everywhere. A solution that succeeds in one country or region might not be suitable in another. In summary, the comparative analysis of environmental outcomes shows that CCS is a diverse set of technologies with different strengths and limitations.

15.6 Key Takeaways and Lessons Learned

The four case studies presented in this chapter show that carbon capture, utilization, and storage (CCUS) can work in many different real-world situations, but each project comes with its own set of challenges, opportunities, and important lessons. By comparing the power sector, heavy industry, BECCS pilots, and community-led monitoring efforts, we can understand what helps CCS projects succeed and what still needs to improve. This section summarizes the most important lessons so future projects can benefit from past experiences [22, 23, 32, 33].

A major lesson from the power-sector case study is that technical success does not automatically mean overall success. The capture equipment at the power plant worked very well during stable operation and achieved high CO_2 removal rates. However, the plant struggled with unexpected interruptions, maintenance challenges, and changes in electricity market conditions [1, 4]. These issues reduced the total amount of CO_2 captured each year. This shows that CCS technology must be evaluated not only by how it performs on "good days," but by how reliably it works over long periods. To maintain a high level of safety and reliability, projects must be designed with strong online monitoring and real-time sensors as shown in Figure 15.3 to detect any abnormal process condition as early as possible before the problem becomes critical [41]. Otherwise, even the best capture technology will not deliver its promised results.

Another important lesson from the power sector is the role of financial stability. CCS is expensive to build and operate, and many of the costs come from upfront costs (CAPEX), added energy use, solvent handling and treatment, equipment wear, and maintenance [42]. If electricity prices decrease, the power plant may reduce output, and the CCS unit may not run at full capacity. This means CCS projects need long-term financial support, such as stable carbon pricing and government incentives.

The industrial capture case study shows a different picture. In heavy industry, such as fertilizer plants, steel mills, and cement factories, CO_2 streams are often more concentrated. This makes capture easier, cheaper, and more energy-efficient [43]. One of the lessons from this case study is that industrial CCS can sometimes achieve better performance at lower cost. When CO_2 is already present in a high-purity stream, the capture system does not need to work as hard as it does in a power plant. This reduces operating costs.

At the same time, industrial CO_2 utilization introduces a new challenge. When captured CO_2 is used in EOR, beverage production, or chemical manufacturing, it does not always stay stored forever. Some CO_2 eventually returns to the atmosphere when the product is used or burned. This demonstrates that the climate benefit of industrial CO_2 capture depends entirely on what ultimately happens to the captured CO_2. Lasting climate gains only occur when the CO_2 is stored permanently, recaptured after the product is used or burned, or recycled to prevent its later release [44].

The BECCS case study brings yet another perspective. BECCS is unique because it can create negative emissions, meaning it removes more CO_2 from the atmosphere than making emission. The technical results from BECCS pilot projects show that biomass-based CO_2 capture is possible and that high capture rates can be achieved. But large-scale BECCS introduces new challenges, especially related to the biomass supply chain. For continuous operation, BECCS facilities need very large, steady supplies of biomass. If biomass harvesting must remain high over long periods, the system can become unsustainable, and the environmental benefits of BECCS may be significantly reduced [22, 23].

The community-led monitoring case study highlights a completely different but equally important lesson: technology alone cannot guarantee public acceptance. Even if a CCS project is technically safe, people who live near pipelines, wells, or storage sites may still feel worried. Community concerns are normal and should not be ignored. This case study shows that when residents are given access to real monitoring data, included in decision-making, and encouraged to ask questions, trust begins to grow. The lesson here is that CCS developers must treat communities as partners, not as bystanders. Early engagement, clear communication, and transparency are essential for building confidence [32, 33].

When all four case studies are viewed together, several broad lessons become clear. First, the long-term reliability of CCS systems is far more important than short bursts of strong performance, because real climate benefits depend on consistent operation over many years. Second, high costs remain a major barrier, meaning stable policy support, such as carbon pricing, subsidies, or long-term incentives, is essential for financial viability. Third,

CCS is not a one-size-fits-all solution; industrial plants, power stations, and BECCS facilities each face different technical and environmental challenges and therefore require tailored strategies. Fourth, public acceptance plays a critical role, as even technically successful projects can struggle if communities feel excluded or poorly informed. Fifth, long-term storage safety must be demonstrated clearly and communicated transparently, with open access to monitoring data to build trust and reduce fears of leakage. In conclusion, the case studies in this chapter show that CCS is not a single technology but a set of interconnected systems that require good engineering, strong policy support, and meaningful community involvement.

15.7 Key Highlights (Chapter Summary)

◇ **Chapter 15 – Key Highlights:**

1) Power plants can technically capture large amounts of CO_2, but long-term performance is affected by operational interruptions and changing electricity markets.
2) Industrial CO_2 streams are easier and cheaper to capture because they are often high-purity, making CCS more efficient and reliable in heavy industry.
3) The climate benefit depends on the final fate of the CO_2; permanent storage gives real reductions, but some utilization pathways rerelease CO_2 later.
4) BECCS can deliver negative emissions, making it unique among CCS approaches, but only when biomass is sourced sustainably.
5) Biomass supply chains strongly influence environmental impact, requiring sustainable forestry, responsible agriculture, and full life-cycle assessment.
6) Including communities in monitoring builds trust, reduces misinformation, and helps the success of the project.
7) Public open access to real-time monitoring data increases transparency and strengthens social acceptance of CCS projects.
8) Long-term monitoring and verification are critical because the environmental success of CCS depends on proving that stored CO_2 remains safely underground for decades to centuries.
9) Strong policy frameworks and predictable regulations are essential for scaling CCS, since financial stability, permitting, and industry investment all depend on clear long-term government support.
10) CCS applied in different sectors, such as power generation, heavy industry, BECCS facilities, and community-integrated projects, requires distinct approaches, because each setting presents its own technical, economic, and environmental challenges.

Keywords: Carbon Capture, CCS Case Studies, Industrial Emissions, Power Sector CCS, BECCS, CO_2 Utilization, Geological Storage, Environmental Outcomes, Negative Emissions, Community Monitoring, Public Acceptance, Sustainability Assessment, Lifecycle Analysis, Operational Performance, Climate Mitigation

References

1 C. Shearer, R. Fofrich, and S. J. Davis, "Future CO_2 emissions and electricity generation from proposed coal-fired power plants in India," *Earth's Future*, vol. 5, no. 4, pp. 408–416, Apr., 2017. DOI: 10.1002/2017EF000542.

2 K. Fujiwara, "Petra nova carbon capture, utilization, and storage project: Capturing CO_2 from the flue gas stream of a coal-fired power plant and increasing oil production from a legacy oil field," *Journal of the Japanese Association for Petroleum Technology*, vol. 84, no. 2, pp. 85–96, 2019. DOI: 10.3720/japt.84.85.

3 J. Jenkins, *A Case Study of the Petra Nova Carbon Capture Project*. Chicago, IL, USA: The Paulson Institute, 2015.

4 A. MacDonald, C. Clack, A. Alexander, *et al.*, "Future cost-competitive electricity systems and their impact on U.S. CO_2 emissions," *Nature Climate Change*, vol. 6, pp. 526–531, 2016. DOI: 10.1038/nclimate2921.

5 O. O. Dosumu, O. Adediran, E. O. Nwulu, and U. B. Chibunna, "Carbon capture and storage (CCS) in the U.S.: A review of market challenges and policy recommendations," *International Journal of Multidisciplinary Research and Growth Evaluation*, vol. 5, no. 1, pp. 1–15, 2021.

6 R. S. Elias, M. I. M. Wahab, and L. Fang, "Retrofitting carbon capture and storage to natural gas-fired power plants: A real-options approach," *Journal of Cleaner Production*, vol. 192, pp. 722–734, Aug., 2018. DOI: 10.1016/j.jclepro.2018.05.019.

7 A. Allahyarzadeh Bidgoli, N. Hamidishad, and J. I. Yanagihara, "The impact of carbon capture storage and utilization on energy efficiency, sustainability, and production of an offshore platform: Thermodynamic and sensitivity analyses," *Journal of Energy Resources Technology*, vol. 144, no. 11, p. 112102, Nov., 2022. DOI: 10.1115/1.4053980.

8 A. Torvanger and J. Meadowcroft, "The political economy of technology support: Making decisions about carbon capture and storage and low carbon energy technologies," *Global Environmental Change*, vol. 21, no. 2, pp. 303–312, May, 2011. DOI: 10.1016/j.gloenvcha.2011.01.017.

9 H. C. Mantripragada, H. Zhai, and E. S. Rubin, "Boundary dam or Petra Nova—which is a better model for CCS energy supply?," *International Journal of Greenhouse Gas Control*, vol. 82, pp. 59–68, Mar., 2019. DOI: 10.1016/j.ijggc.2019.01.004.

10 S. C. Nwokolo, R. Singh, S. Khan, A. Kumar, and S. Luthra, "Decarbonizing hard-to-abate sectors in Africa," in *Africa's Path to Net-Zero, CSR, Sustainability, Ethics & Governance*, S. Luthra, S. Kumar, and A. Kumar, Eds. Cham: Springer, 2023, pp. 95–114, DOI: 10.1007/978-3-031-44514-9_6.

11 S. Jahanshahi, J. G. Mathieson, M. A. Somerville, *et al.*, "Development of low-emission integrated steelmaking process," *Journal of Sustainable Metallurgy*, vol. 1, pp. 94–114, 2015. DOI: 10.1007/s40831-015-0008-6.

12 J. Humphreys, R. Lan, and S. Tao, "Development and recent progress on ammonia synthesis catalysts for Haber–Bosch process," *Advanced Energy and Sustainability Research*, vol. 2, no. 2, 2000043, 2021. DOI: 10.1002/aesr.202000043.

13 L. E. Scherger, V. Zanello, and C. Lexow, "Impact of urea and ammoniacal nitrogen wastewaters on soil: Field study in a fertilizer industry (Bahía Blanca, Argentina)," *Bulletin*

of Environmental Contamination and Toxicology, vol. 107, pp. 565–573, 2021. DOI: 10.1007/s00128-021-03280-x.

14 B. Stout, R. Lal, and C. Monger, "Carbon capture and sequestration: The roles of agriculture and soils," *International Journal of Agricultural and Biological Engineering*, vol. 9, no. 1, pp. 1–8, 2016. DOI: 10.3965/j.ijabe.20160901.2280.

15 K. Sotoodeh, *Piping, Valves, and Process Engineering in Carbon Capture Systems: Applications, Sustainability, and Environmental Protection*, 1st ed.. Elsevier, 2026, ISBN: 978–0443510786

16 E. Abe, Y. Suzuki, H. Abromeit, J. P. Addis, P. W. S. Andrews, and E. Brunner, "Traditional and heavy industries," *International Bibliography of Business History*, vol. 63, no. 4, p. 75, 2013.

17 Q. Zhuang, B. Clements, and Y. Li, "From ammonium bicarbonate fertilizer production process to power plant CO_2 capture," *International Journal of Greenhouse Gas Control*, vol. 10, pp. 56–63, Sep., 2012. DOI: 10.1016/j.ijggc.2012.05.019.

18 J. R. Stewart and R. S. Haszeldine, Carbon accounting for carbon dioxide enhanced oil recovery. *Scottish Carbon Capture and Storage (SCCS)*, 2014.

19 E. Van Roijen, S. A. Miller, and S. J. Davis, "Building materials could store more than 16 billion tonnes of CO_2 annually," *Science*, vol. 387, no. 6730, pp. 176–182, Jan., 2025. DOI: 10.1126/science.adq8594.

20 B. H. M. Silveira, H. K. M. Costa, and E. M. Santos, "Bioenergy with carbon capture and storage (beccs) in Brazil: A review," *Energies*, vol. 16, no. 4, 2021, 2023. DOI: 10.3390/en16042021.

21 F. B. Affan and A. Maaroof, "A critical evaluation of the UK Drax power station," *Journal of Ecosystem & Ecography*, vol. 6, no. 3, 2016. DOI: 10.4172/2157-7625.1000206.

22 A. Almena-Ruiz, J. Sparks, P. Thornley, and M. Röder, "Opportunities and challenges for bioenergy with carbon capture and storage (BECCS) systems supporting net-zero emission targets," *Renewable and Sustainable Energy Reviews*, vol. 150, p. 111453, 2021.

23 J. Walker, J. Aurandt, and B. Pieper, "Marquis industrial complex: Employing beccs to decarbonize the economy," in *Proceedings of the 16th International Conference on Greenhouse Gas Control Technologies (GHGT-16)*, vol. 23, Oct. 2022.

24 B. Sharma, R. G. Ingalls, C. L. Jones, and A. Khanchi, "Biomass supply chain design and analysis: Basis, overview, modeling, challenges, and future," *Renewable and Sustainable Energy Reviews*, vol. 24, pp. 608–627, Aug., 2013. DOI: 10.1016/j.rser.2013.03.049.

25 P. J. Verkerk, R. Mavsar, M. Giergiczny, M. Lindner, D. Edwards, and M. J. Schelhaas, "Assessing impacts of intensified biomass production and biodiversity protection on ecosystem services provided by European forests," *Ecosystem Services*, vol. 9, pp. 155–165, Sep., 2014. DOI: 10.1016/j.ecoser.2014.06.004.

26 P. T. Lautala, M. R. Hilliard, E. Webb, *et al.*, "Opportunities and challenges in the design and analysis of biomass supply chains," *Environmental Management*, vol. 56, pp. 1397–1415, 2015. DOI: 10.1007/s00267-015-0565-2.

27 Z. M. Harris, S. Milner, and G. Taylor, "Chapter 5—biogenic carbon—capture and sequestration," in *Greenhouse Gas Balances of Bioenergy Systems*, H. H. E. M. Verhagen, Ed. Academic Press, 2018, pp. 55–76, DOI: 10.1016/B978-0-08-101036-5.00005-7.

28 S. E. Tanzer, K. Blok, and A. Ramírez, "Decarbonising industry via BECCS: Promising sectors, challenges, and techno-economic limits of negative emissions," *Current Sustainable/Renewable Energy Reports*, vol. 8, pp. 253–262, 2021.

29 C. A. Barcelos *et al.*, "High-efficiency conversion of ionic liquid-pretreated woody biomass to ethanol at the pilot scale," *ACS Sustainable Chemistry & Engineering*, vol. 9, no. 11, pp. 4042–4053, 2021.

30 J. A. E. Nielsen, K. Stavrianakis, and Z. Morrison, "Community acceptance and social impacts of carbon capture, utilization and storage projects: A systematic meta-narrative literature review," *PLoS One*, vol. 17, no. 8, Aug., e0272409, 2022. DOI: 10.1371/journal.pone.0272409.

31 B. W. Terwel, F. Harinck, N. Ellemers, and D. L. Daamen, "Going beyond the properties of CO_2 capture and storage (CCS) technology: How trust in stakeholders affects public acceptance of CCS," *International Journal of Greenhouse Gas Control*, vol. 5, no. 2, pp. 181–188, Mar., 2011. DOI: 10.1016/j.ijggc.2010.10.001.

32 A. Kipp, A. Cunsolo, D. Gillis, A. Sawatzky, and S. L. Harper, "The need for community-led, integrated and innovative monitoring programmes when responding to the health impacts of climate change," *International Journal of Circumpolar Health*, vol. 78, no. 2, 1517581, 2019. DOI: 10.1080/22423982.2018.1517581.

33 C. C. Conrad and K. G. Hilchey, "A review of citizen science and community-based environmental monitoring: Issues and opportunities," *Environmental Monitoring and Assessment*, vol. 176, pp. 273–291, 2011. DOI: 10.1007/s10661-010-1582-5.

34 K. Hobson, R. Mayne, and J. Hamilton, "Monitoring and evaluating eco-localisation: Lessons from UK low carbon community groups," *Environment and Planning A: Economy and Space*, vol. 48, no. 7, pp. 1393–1410, 2016. DOI: 10.1177/0308518X16640531.

35 S. L.'. O. Seigo, S. Dohle, and M. Siegrist, "Public perception of carbon capture and storage (CCS): A review," *Renewable and Sustainable Energy Reviews*, vol. 38, pp. 848–863, Oct., 2014. DOI: 10.1016/j.rser.2014.07.017.

36 P. Freund, "Making deep reductions in CO_2 emissions from coal-fired power plant using capture and storage of CO_2," *Proceedings of the Institution of Mechanical Engineers, Part A: Journal of Power and Energy*, vol. 217, no. 1, pp. 1–7, Feb., 2003. DOI: 10.1243/095765003321148628.

37 B. Zohuri and P. McDaniel, "The electricity: An essential necessity in our life," in *Advanced Smaller Modular Reactors*. Cham: Springer, 2019, pp. 1–18, DOI: 10.1007/978-3-030-23682-3_1.

38 K. H. M. Al-Hamed and I. Dincer, "A comparative review of potential ammonia-based carbon capture systems," *Journal of Environmental Management*, vol. 287, Jun., 112357, 2021. DOI: 10.1016/j.jenvman.2021.112357.

39 C. Giavarini, F. Maccioni, and M. L. Santarelli, "CO_2 sequestration from coal fired power plants," *Fuel*, vol. 89, no. 3, pp. 623–628, Mar., 2010. DOI: 10.1016/j.fuel.2009.09.035.

40 M. Mirhosseini, "Sensitivity analysis, adaptability improvement and control of grid-connected photovoltaic power plants under grid frequency variations," *Solar Energy*, vol. 184, pp. 260–272, May, 2019. DOI: 10.1016/j.solener.2019.03.072.

41 E. Wilson and D. Gerard, Carbon capture and sequestration: Integrating technology, monitoring, regulation. 2007.

42 P. E. Hardisty, M. Sivapalan, and P. Brooks, "The environmental and economic sustainability of carbon capture and storage," *International Journal of Environmental Research and Public Health*, vol. 8, no. 5, pp. 1460–1477, May, 2011. DOI: 10.3390/ijerph8051460.

43 P. C. Psarras, S. Comello, P. Bains, P. Charoensawadpong, S. Reichelstein, and J. Wilcox, "Carbon capture and utilization in the industrial sector," *Environmental Science & Technology*, vol. 51, no. 19, pp. 11440–11449, Aug., 2017. DOI: 10.1021/acs.est.7b01723.

44 P. Styring, E. A. Quadrelli, and K. Armstrong, *Carbon Dioxide Utilisation: Closing the Carbon Cycle*. Elsevier, 2014.

16

Future Perspectives: Justice, Sustainability, and System Redesign

16.1 Summary of Crosscutting Themes

This section brings together the major ideas discussed throughout the book and shows how they connect to the future of carbon capture. One of the central ideas is that carbon capture is not only a technical solution but also part of a much larger environmental and social system. Effective carbon management requires understanding how technology interacts with policy, economics, communities, and ecosystems [1].

Another important point is the need for fairness and shared global responsibility when dealing with climate change [2]. All countries are affected by climate change, but not all have contributed to the problem in the same way [3]. Some nations and industries have produced large amounts of greenhouse gas emissions for more than a century, mainly through heavy use of coal, oil, and gas. At the same time, they have gained major economic benefits from these activities, building wealth and strong industrial systems. Because of this history, these high-emission countries and industries have a greater responsibility to take the lead in reducing emissions and supporting climate solutions [4].

Leading the energy transition means investing more in clean technologies, including renewable energy, carbon capture, and energy-efficient systems [5]. It also means helping lower-income countries, which often have fewer resources to deal with climate challenges, even though they have contributed far less to the climate change problem [6]. Support for developing countries can include financial assistance, access to technology or technology transfer, training programs, and partnerships that help local communities develop low-carbon energy systems without slowing their economic development [7].

Sharing knowledge and expertise is another key part of fairness. New technologies should not remain in the hands of only a few wealthy companies or nations. Instead, research, data, and practical experience should be shared openly so that countries around the world can benefit and build their own climate solutions [8]. This helps speed up global action and prevents inequalities from becoming worse.

Fairness also means ensuring that the benefits of carbon capture and other climate innovations reach many people, not just a small group of investors or corporations. Clean air, stable jobs, safer communities, and climate protection should be shared benefits across society. Climate solutions must be designed so that workers, local communities, and developing regions also gain new opportunities rather than being left behind. In simple terms, fighting climate change must be a team effort. Those who have contributed the most to the problem should do the most to fix it and help others join the transition. When responsibility, technology, and benefits are shared fairly, the global response becomes stronger, more effective, and more just for everyone [9].

A third important theme is integration; the idea that effective climate action requires multiple solutions working together rather than reliance on a single approach. Carbon capture is a valuable tool, but it becomes far more impactful when integrated with other emission-reduction strategies, such as renewable energy generation and low-carbon industrial processes. On its own, carbon capture cannot solve climate change [10]. While it can remove CO_2 from industrial sources or even directly from the air, continued high emissions across other sectors would limit its overall effect. For meaningful impact, carbon capture must be part of a broader, coordinated climate strategy that combines multiple technologies and policy actions.

One of the most important technologies to integrate with carbon capture is renewable energy, particularly wind and solar power. These energy sources generate electricity without releasing CO_2, therefore reducing emissions and decreasing the volume of carbon that needs to be captured. In addition, renewable power can supply the clean electricity and heat required to operate carbon capture facilities, enabling carbon capture and storage (CCS) systems to function without dependence on fossil-fuel energy [11]. Another key area of integration is nature-based solutions, including afforestation (planting new forests), ecosystem restoration, and wetland protection. Natural systems absorb CO_2 directly from the atmosphere, and when properly managed, they enhance and complement technological carbon capture efforts [12].

Figure 16.1 illustrates the integrated approach to carbon management, where renewable energy and nature-based solutions work alongside advanced direct air capture (DAC) technologies. Wind turbines and solar panels provide clean power to operate a DAC unit at the

Figure 16.1 Integration of renewable energy, direct air capture, and nature-based solutions for comprehensive carbon management and climate mitigation.

center, while surrounding forests and waterways symbolize natural carbon sinks that absorb and store CO_2. Together, these elements represent a holistic pathway toward sustainable climate mitigation.

Energy efficiency is another important modern theme, which means using less energy to perform the same tasks. When factories, transportation systems, and buildings become more efficient, they emit fewer greenhouse gases, which reduces the overall burden on carbon capture systems and allows for smaller and more manageable capture solutions [13]. In addition, circular economy strategies help reduce emissions by lowering waste, reusing materials, and recycling products. When materials stay in use for longer, industries produce fewer emissions from manufacturing and waste disposal [14].

No single solution is sufficient on its own. Climate change is a complex challenge that requires coordinated action across multiple strategies. Carbon capture must therefore operate alongside renewable energy, energy efficiency, nature-based solutions, and circular economy practices. . Carbon capture must work side by side with renewable energy, efficient systems, natural solutions, and circular economy practices [15].In the future, our energy and industrial systems should be designed so that all these approaches support one another. When carbon capture is integrated with clean energy, efficient technologies, and nature-based actions, the result is a stronger, more reliable, and more effective system for reducing emissions. Integration improves system performance and provides a stronger pathway toward meeting global climate goals.

The book also emphasizes the importance of innovation. New developments in materials, digital tools such as monitoring and data analysis, and improved system designs are helping to make carbon capture technologies more efficient and safer [16, 17]. However, technical innovation alone is not enough. These advances must be supported by strong governance, clear regulations, and active engagement with local communities to ensure projects are developed responsibly and gain public trust. Together, these themes show that the future of carbon capture depends on both scientific progress and a commitment to justice, cooperation, and thoughtful system redesign.

16.2 Rethinking Carbon Capture Beyond Technology

Carbon capture is often described as a technical solution designed to remove carbon dioxide (CO_2) from power plants, factories, or even directly from the air. These technologies include absorption systems, membranes, and advanced materials that separate CO_2 from other gases so it can be transported and stored safely underground or reused in industrial processes [18]. While these tools are extremely important, focusing only on the technology gives an incomplete picture of carbon capture's true role in addressing climate change. Climate change is not a problem caused by technology alone, nor can it be solved by technology alone. It is the result of deeply interconnected systems involving energy production, transportation, food supply, industrial processes, land use, consumption habits, and political decision-making. For this reason, carbon capture must be understood as part of a much larger system rather than as a stand-alone technical fix.

Figure 16.2 shows how carbon capture must be understood as part of a broad, interconnected system rather than as an isolated technological solution. The central position of

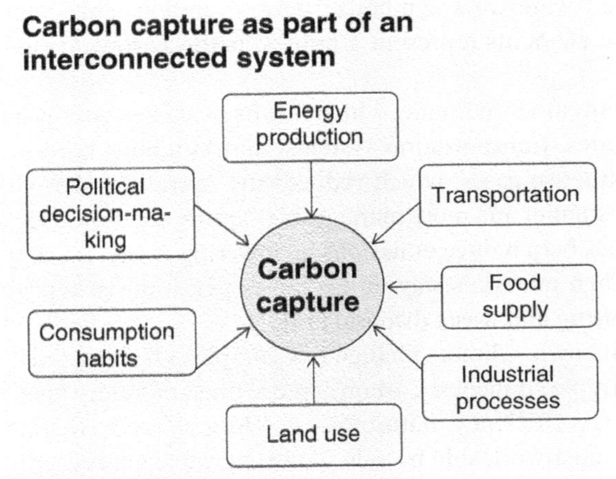

Figure 16.2 Carbon capture as a central element within an interconnected climate-mitigation system linking energy, industry, food production, land use, transportation, consumption patterns, and political decision-making.

carbon capture in the diagram highlights its role as one component within a network of interacting sectors. The chart emphasizes that meaningful climate action depends on coordinated progress across all these areas, demonstrating that carbon capture technologies can only achieve their full potential when integrated with wider societal, economic, and policy-driven transformations.

A major risk in relying too heavily on carbon capture is treating it as an excuse to delay more fundamental changes. Fossil-fuel industries or governments might argue that they can continue current practices because future CCS deployment will remove emissions. This mindset can weaken the urgency to expand renewable energy, improve energy efficiency, redesign cities for lower emissions, or reduce overall consumption. Carbon capture should therefore be viewed as a supporting tool, not a replacement for broader climate action. Real progress depends on combining CCS with deep reductions in emissions across all sectors of society [19].

Rethinking carbon capture beyond technology also means recognizing the human, social, economical, and political dimensions of CCS projects. Capture facilities, pipelines, and storage sites are not built in empty spaces; they are located near communities, workplaces, farmland, coastlines, or natural habitats [20]. These projects can raise concerns over safety, land use, property rights, and environmental protection. Without proper communication and public involvement, communities may distrust CCS developments, especially if they feel they carry the risks while receiving few benefits. A successful carbon capture strategy must therefore include transparent decision-making and meaningful public participation. Communities should be informed early, consulted openly, and given opportunities to influence design choices. Building trust requires listening to concerns, sharing scientific information in an accessible language, and ensuring that safety and monitoring systems are clearly communicated [21].

Equity and fairness are also central to a truly sustainable carbon capture system. Some regions, particularly low-income or marginalized communities, already experience higher environmental burdens such as air pollution or industrial waste. CCS projects located in or near these areas must avoid repeating patterns of environmental injustice. Benefits such as job creation, infrastructure investment, education programs, and improved local services should be shared with host communities along with carbon capture plant development in the area. Internationally, fairness matters as well. Wealthier nations that have produced most historical emissions generally possess greater technical and financial capacity to develop CCS technologies. These countries have an ethical responsibility to share knowledge, training, and investment with lower-income nations so that carbon mitigation efforts do not widen global inequalities [22, 23].

Looking beyond technology also includes understanding carbon as part of a larger environmental cycle rather than simply as waste to be buried underground. Captured CO_2 can sometimes be used as a resource in what is known as carbon utilization. Examples include producing building materials, synthetic fuels, chemicals, or mineral products as illustrated in Figure 16.3. Although CO_2 reuse does not usually result in permanent storage, it represents an alternative pathway where captured CO_2 is used in products, temporarily retaining carbon and helping to reduce emissions in other sectors [24]. Connecting carbon capture to the circular economy, where materials are reused and waste is minimized, helps integrate CCS into a more sustainable industrial system. However, carbon capture, storage, and utilization should not become an excuse to increase fossil-fuel use or create products that quickly release CO_2 back into the atmosphere [25].

Another important perspective involves the connection between carbon capture and land and water resources. CCS facilities require physical space, water for cooling or processing, energy inputs, and infrastructure corridors for equipment, piping, and pipelines.

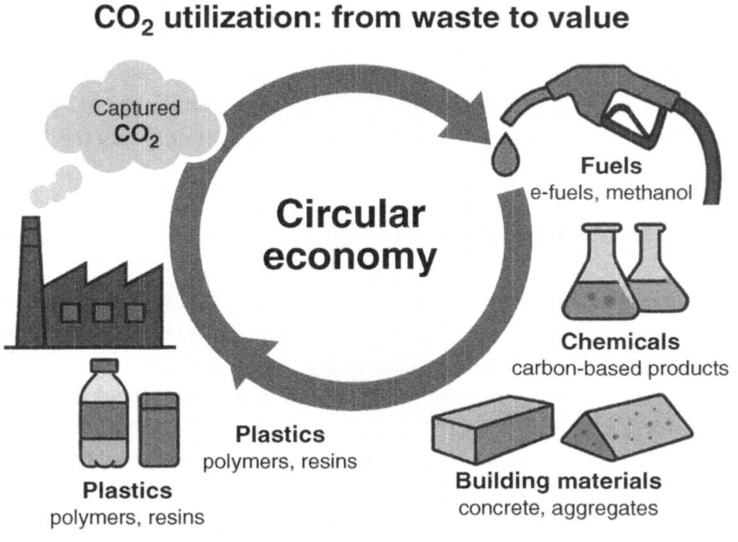

Figure 16.3 Schematic representation of CO_2 utilization pathways within the circular carbon economy, illustrating conversion routes from captured CO_2 to fuels, chemicals, polymer products, and construction materials.

These demands can create competition with agriculture, conservation areas, or community development [26]. Planning CCS expansion requires careful regional and environmental assessments to prevent unintended ecological harm. Integrating CCS with renewable energy systems and using low-water or low-energy capture technologies can help reduce these conflicts by reducing energy consumption and increasing overall sustainability [19].

Education plays a major role in rethinking carbon capture. Students, engineers, policymakers, and community leaders need access to accurate and reliable information about CCS, its capabilities, limitations, risks, and benefits. Public understanding is essential to avoid exaggerated expectations or unnecessary fears. Educational programs that combine engineering with environmental studies, economics, ethics, and public policy prepare a new generation of professionals to approach CCS as a multidimensional challenge rather than a purely technical task [27].

From a governance standpoint, rethinking carbon capture means designing policies that guide CCS deployment responsibly. Governments must ensure strong safety regulations, long-term monitoring requirements, and legal accountability for storage operations [28]. They must also align CCS policies with overall climate goals so that capture supports net-zero transitions rather than prolonging dependence on fossil fuels [29]. Transparent reporting systems and international cooperation help maintain credibility and strengthen public trust.

In a broader sense, moving beyond technology means embracing systemic redesign. Instead of inserting carbon capture into existing fossil-based systems, societies can use CCS as part of a wider transformation toward low-carbon industrial clusters, renewable energy integration, sustainable agriculture, and nature-based climate solutions. This approach treats carbon mitigation as a network of connected actions, technical, social, environmental, and institutional, working toward common climate and equity goals. Ultimately, carbon capture's greatest contribution is not simply removing CO_2 molecules from exhaust gases or air streams. Its real value lies in supporting a transition toward a cleaner, fairer, and more sustainable future. When combined with renewable energy expansion, efficiency improvements, community participation, responsible governance, and ethical international cooperation, CCS becomes one tool within a much broader sustainability toolbox.

16.3 Long-Term Justice and Global Equity

Addressing climate change is not only a technical challenge but also a matter of justice and fairness. CCS must therefore be evaluated not just by how effectively it reduces emissions, but also by how it affects people and nations differently. Long-term justice and global equity require recognizing that the causes of climate change and the ability to address it are unevenly distributed across the world [9].

High-income countries have produced the majority of historical greenhouse gas emissions during their industrial development. These emissions created wealth and economic growth while contributing significantly to today's climate crisis. In contrast, many low-income countries have contributed far less to global emissions yet face some of the most severe impacts of climate change, including droughts, floods, sea-level rise, and food insecurity. Justice, therefore, demands that those who contributed most to the problem

Figure 16.4 Visual comparison of industrial carbon capture deployment in high-income regions and climate vulnerability in low-income communities, highlighting the global equity challenges of climate mitigation.

play a leading role in financing solutions, transferring technology, and supporting global climate adaptation and mitigation efforts [6, 9].

Figure 16.4 illustrates the unequal global dimensions of climate change and carbon capture deployment. The left side shows an industrialized setting with a large power plant, renewable energy installations, and a modular DAC system, representing regions with strong financial capacity, advanced technology, and infrastructure to implement carbon removal solutions. The right side depicts a rural community in a climate-vulnerable region suffering from degraded land and water stress, symbolizing populations that have contributed least to historical emissions yet face the most severe climate impacts. The split composition highlights the disparity between responsibility for emissions and exposure to climate consequences, emphasizing the need for equitable international cooperation so that carbon capture technologies benefit all regions rather than reinforcing existing inequalities.

Carbon capture technology is expensive and complex to develop at large scale. Wealthier nations and multinational industries currently dominate research, development, and deployment of CCS systems [30]. Without international cooperation, this could create new forms of inequality in which advanced technologies remain accessible only to those who can pay for them. A globally just CCS strategy should ensure that developing nations have access to training, funding, and infrastructure support so that carbon management efforts benefit the entire world rather than reinforcing existing divisions [31].

Equity also applies to the local level. Communities located near carbon capture facilities, transport pipelines, and storage sites may experience risks related to land use, safety concerns, or environmental disturbances. These communities must be treated fairly. Long-term justice requires meaningful community consultation, transparent communication of risks, and direct sharing of benefits. Examples include job creation, infrastructure investment for the community, and community education initiatives. When local residents feel respected and included, trust in CCS projects increases, and social acceptance becomes stronger [32].

A just approach to CCS also considers impacts across generations. Carbon storage sites are designed to remain stable for hundreds to thousands of years, creating obligations that extend far beyond the lifetime of current projects or political administrations. Future generations will rely on today's engineers, regulators, and companies to design secure monitoring systems and long-term stewardship plans. Ethical responsibility, therefore, includes ensuring that liabilities do not disappear when projects end or companies dissolve, leaving future societies to manage unresolved risks [33].

Global equity further depends on how CCS fits within overall climate strategies. Carbon capture can help reduce emissions from hard-to-decarbonize industries, but it must not slow the adoption of renewable energy or energy efficiency improvements, especially in developing regions [7]. A fair climate transition ensures that fast access to clean energy technologies is available globally rather than locking lower-income countries into prolonged dependence on fossil fuels combined with expensive CCS solutions. Figure 16.5 shows hard-to-abate industrial sectors and carbon capture infrastructure: the cement production facility is shown on the left, the steel manufacturing complex in the center, and the carbon capture installation on the right, illustrating how CCS can be applied to reduce emissions from these processes.

Equity refers to fairness in how benefits and burdens are shared across communities, regions, and social groups. In the context of climate solutions, equity means ensuring

Figure 16.5 Representative industrial facilities from hard-to-abate sectors (cement and steel) alongside carbon capture installations used to mitigate process emissions.

that the costs of climate mitigation do not fall disproportionately on people or ecosystems that already face social, economic, or environmental disadvantages. While CCS can help reduce global emissions, these technologies require substantial resources, including energy, water, specialized materials, and land. If not planned carefully, the demand for these resources can create local pressures, particularly in regions already experiencing water scarcity or contested land use. Therefore, sustainable CCS development must integrate equity considerations into project planning. This includes assessing local water availability, protecting sensitive ecosystems, engaging affected communities, and ensuring fair compensation for land use [23, 32].

Addressing justice and equity in carbon capture ultimately requires recognition that climate change is a global problem demanding shared solutions and shared responsibility. This means that all countries should have a fair opportunity to benefit from climate technologies, not just those with the greatest financial or technical resources. Because carbon capture systems are expensive and complex, poorer or developing countries may struggle to access them without international support. This is why strong global cooperation is essential. International climate agreements, global funding programs, and global research partnerships help ensure that knowledge, technology, and financial resources are shared globally more equitably. Wealthier countries can contribute by providing funding, technical assistance, training programs, and open-access research platforms that allow all regions to participate in climate-mitigation efforts [32, 34].

In conclusion, achieving long-term justice and global equity in carbon capture requires recognizing the shared responsibility of all nations for addressing climate change. Effective action must go beyond technical performance to consider social impacts, protect local communities, support developing countries, and safeguard the needs of future generations. Technical success alone is not sufficient. Carbon capture projects must also deliver fair social outcomes through transparency, public engagement, meaningful community participation, and equal access to benefits across society [32, 34]. When principles of justice and equity guide carbon capture strategies, CCS becomes more than a tool for reducing emissions. It also contributes to building public trust, strengthening international cooperation, and fostering a truly collective global response to climate change.

16.4 Regenerative Environmental Governance

Regenerative environmental governance means managing environmental policies and decisions in a way that does more than just prevent damage. It actively restores ecosystems, supports communities, and strengthens long-term sustainability. Environmental governance refers to the laws, policies, and decision-making processes that guide how societies manage natural resources and respond to environmental challenges. Traditional governance models often focus on limiting damage and reducing pollution, enforcing safety standards, or managing waste. While these approaches are necessary, they are no longer sufficient for addressing complex, long-term challenges such as climate change [35]. Regenerative environmental governance moves beyond simply minimizing harm and instead aims to actively restore ecosystems, strengthen communities, and create long-lasting social and environmental benefits.

In the context of CCS, regenerative governance means designing rules and policies that ensure carbon management contributes to environmental restoration and social well-being rather than merely controlling emissions [36]. CCS projects should be evaluated not only on technical performance but also on their ability to improve local environmental conditions, support biodiversity, create sustainable employment, and strengthen community participation.

A core principle of regenerative governance is inclusive and participatory decision-making. Communities affected by CCS facilities, pipelines, or storage sites must be involved from the earliest planning stages through project implementation and operation. Public consultations, transparent environmental assessments, and ongoing community dialogue ensure that local knowledge, concerns, and priorities are not only heard but actively considered in project design and decision-making. This participation can influence site selection, safety measures, environmental protections, and benefit-sharing arrangements. When communities are treated as partners rather than passive stakeholders, projects are more likely to build public trust and operate in socially responsible, locally accepted ways [37].

Another key element of regenerative governance involves strong regulatory oversight combined with long-term stewardship obligations. Governments must establish clear safety standards for capture, transport, and storage operations, as well as strict requirements for environmental protection and long-term monitoring of storage sites. Regulatory frameworks should require that companies remain responsible for their projects even after injection activities end [28]. This includes establishing long-term financial provisions for site monitoring, risk management, and corrective actions to protect future generations from potential environmental issues, such as carbon dioxide leakage from subsurface storage sites.

Regenerative governance also emphasizes integrating CCS planning with broader environmental and climate policies. Carbon capture should not be deployed in isolation but as part of national and international sustainability strategies that prioritize renewable energy development and integration with carbon capture, energy efficiency improvements, and nature-based solutions [38]. Coordinated policies ensure that CCS strengthens and does not replace efforts to reduce emissions at the source, supports the expansion of clean energy systems, and contributes directly to long-term decarbonization and ecosystem protection goals [39].

Strong governance must also encourage knowledge sharing and international cooperation. Climate change is a global challenge requiring collective solutions. Through international agreements, technology partnerships, and shared research platforms, countries can exchange best practices, improve safety standards, and support equitable deployment of CCS [40]. Developing countries in particular benefit from programs that provide technical training, regulatory support, and access to technology and finance.

Transparency is a key pillar of regenerative governance. People must have open access to information about CCS projects, including safety studies, environmental monitoring results, and financial arrangements. This openness strengthens accountability and builds public trust. Community-based programs organized by the government and project developers such as carbon dioxide storage monitoring activities can further support transparency and credibility of the carbon capture projects. When decisions are made openly and based on clear evidence, governance and project development become stronger, more trustworthy, and more responsive to public concerns [41].

Economic incentives are another important part of regenerative governance. Well-designed financial policies can encourage companies to adopt carbon capture systems that deliver real environmental and social benefits. These incentives may include subsidies for low-impact capture technologies from the government, carbon pricing systems that reward verified permanent carbon storage, and public funding programs that connect CCS projects with ecosystem restoration or local job creation [42]. When economic mechanisms are aligned with environmental goals, they help guide business decisions toward long-term sustainability and social responsibility.

Ultimately, regenerative environmental governance views carbon capture as a tool integrated into the broader aspects of social and ecological considerations. It recognizes that technology alone cannot solve climate challenges [35, 36]. Instead, positive outcomes depend on governance systems that encourage and promote collaboration, protect communities, and restore ecosystems. When governance frameworks are regenerative, carbon capture projects can contribute to healthier environments and more resilient societies.

16.5 Innovations on the Horizon

Rapid innovation is shaping the future of CCS, making the technology more efficient and environmentally responsible. While many capture systems already operate at commercial scale, ongoing research and development are opening new paths that can significantly increase performance while lowering resource use and costs [43]. These innovations span materials science, process engineering, digital technologies, system integration with renewable energies, and nature-based solutions.

One major area of progress is the development of advanced capture materials [44]. Traditional solvent-based capture systems are highly energy-intensive because large amounts of heat are required to separate CO_2 from exhaust gases. This heat is typically supplied by generating steam from boiling water, which results in high energy use and significant water consumption. In contrast, newer materials, such as solid sorbents, membranes, and hybrid materials including metal–organic frameworks (MOFs) and polymer membranes, offer much higher CO_2 selectivity and require far less energy for regeneration [45]. These technologies can reduce overall energy demand, allow for smaller and more compact equipment, and make carbon capture feasible for smaller industrial facilities that were previously unable to adopt CCS. Figure 16.6 shows the basic structure of a MOF, where metal ions (nodes) are connected by organic linkers to form a rigid porous network with internal cavities used for gas adsorption, including carbon dioxide capture.

Electrochemical and low-temperature capture technologies are another promising development. Instead of relying mainly on heat from steam to release captured CO_2, these systems use electrical signals or changes in temperature and pressure to regenerate the capture materials. Although changing pressure and temperature still requires energy, for example, compressing gases or providing cooling and moderate heating, but the energy demand is typically much lower than that of traditional steam-based systems. Because these processes can be driven directly by renewable electricity from sources such as wind or solar, they offer a cleaner and more flexible way to operate carbon capture. By linking CCS to low-carbon

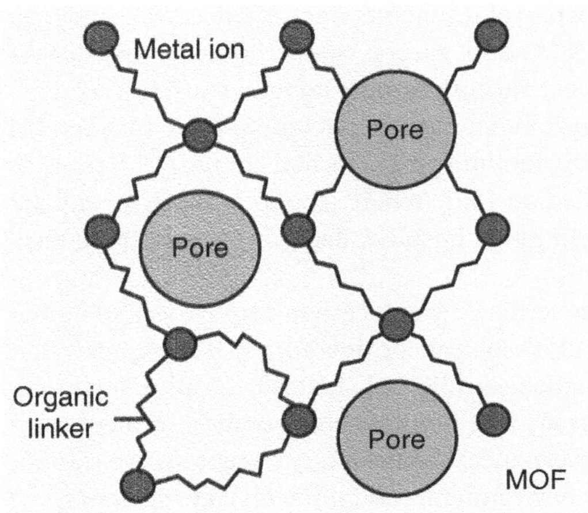

Figure 16.6 Conceptual representation of a metal–organic framework (MOF) structure, illustrating metal nodes, organic linkers, and pore spaces responsible for CO_2 or other gases adsorption and separation.

power systems, electrochemical approaches can reduce overall energy use and lower the lifecycle emissions associated with capture operations [46].

Advances in CO_2 transport are also making CCS more practical and easier to deploy. Pipelines remain the main method for moving large volumes of CO_2, but new transport options now provide greater flexibility [47]. For example, cryogenic shipping cools CO_2 into liquid form so it can be transported by ship, making it suitable for coastal facilities and offshore storage locations that lack pipeline connections. Another key innovation is the development of modular pipeline systems. These involve shorter, standardized pipeline sections that can be installed step-by-step instead of constructing long, permanent pipelines all at once. For instance, in an industrial cluster with several smaller factories, modular pipelines can initially connect only the first few CO_2 capture sites to a nearby compression station. As more facilities adopt CCS, additional pipeline segments can be added to expand the network. This approach reduces upfront investment costs, speeds up deployment, and allows CO_2 transport infrastructure to grow gradually as CCS projects scale up. Improved compression technologies further reduce the energy required for transport, improving overall efficiency [48]. Figure 16.7 illustrates the main CO_2 conditioning and transport pathways discussed in this section, highlighting the use of liquid shipping and supercritical pipeline transport.

On the storage side, major progress has been made in how underground CO_2 storage sites are studied and monitored. Advanced seismic imaging is now used to create detailed underground maps, while satellites, fiber-optic sensors, and real-time pressure measurements continuously track reservoir conditions for safe and reliable CO_2 storage. These tools help detect small changes or potential leaks at an early stage, improving safety and reliability. In addition, artificial intelligence and machine-learning techniques support predictive models that estimate how CO_2 plumes will move underground over time. This improves site selection, safety planning, and long-term risk management [48].

Carbon utilization paths are also rapidly expanding. Researchers are developing methods to convert CO_2 into building materials such as carbonated concrete and mineral aggregates that permanently lock carbon into stable forms. Other technologies enable conversion into

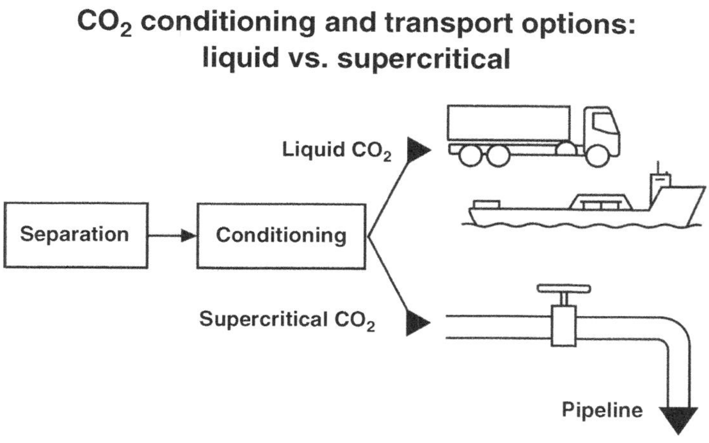

Figure 16.7 Overview of CO_2 conditioning and transport options, illustrating liquid CO_2 shipment by truck or ship and supercritical CO_2 transport through pipelines.

fuels and chemicals such as methanol or synthetic hydrocarbons. While many utilization pathways do not result in permanent storage, they can displace fossil-derived feedstocks and support the development of circular carbon systems [49].

Another important frontier is integrated industrial ecosystems, often called carbon management hubs or clusters. These systems combine capture facilities, transport pipelines, shared storage infrastructure, renewable energy inputs, hydrogen production, and carbon utilization plants into regional networks. By sharing infrastructure and resources, hubs reduce costs, improve efficiency, and minimize environmental footprints. For example, using a single pipeline to transport both CO_2 and hydrogen instead of constructing two separate pipelines greatly reduces the environmental impact associated with materials, construction, and land disturbance. They also support workforce development by creating concentrated centers of technical expertise and employment [47, 48].

Figure 16.8 illustrates the future repurposing of large-scale transport system (LTS) pipelines for combined hydrogen (H_2) and carbon dioxide (CO_2) transport after technical revalidation. The surrounding symbols represent key considerations in pipeline repurposing, including economic viability to ensure cost-effective operation, safety and cybersecurity protection to prevent physical or digital failures, storage integration to enable efficient connection between transport and injection facilities, and capacity assessment to verify that pipelines can safely handle the required volumes while meeting regulatory standards. Repurposing means using existing infrastructure for a new purpose instead of building everything from scratch. Repurposing is applicable because existing natural-gas pipeline networks already provide extensive, long-distance transport infrastructure that can potentially be reused for low-carbon energy carriers such as hydrogen (H_2) and carbon dioxide (CO_2).

Digitalization is playing an increasingly important role across all stages of CCS. Technologies such as digital twins, sensor networks, and automated control systems help improve capture efficiency, enhance safety, and detect maintenance issues before equipment failures occur. At the same time, artificial intelligence tools analyze environmental monitoring data, such as leakage measurements or air-quality readings, to automatically

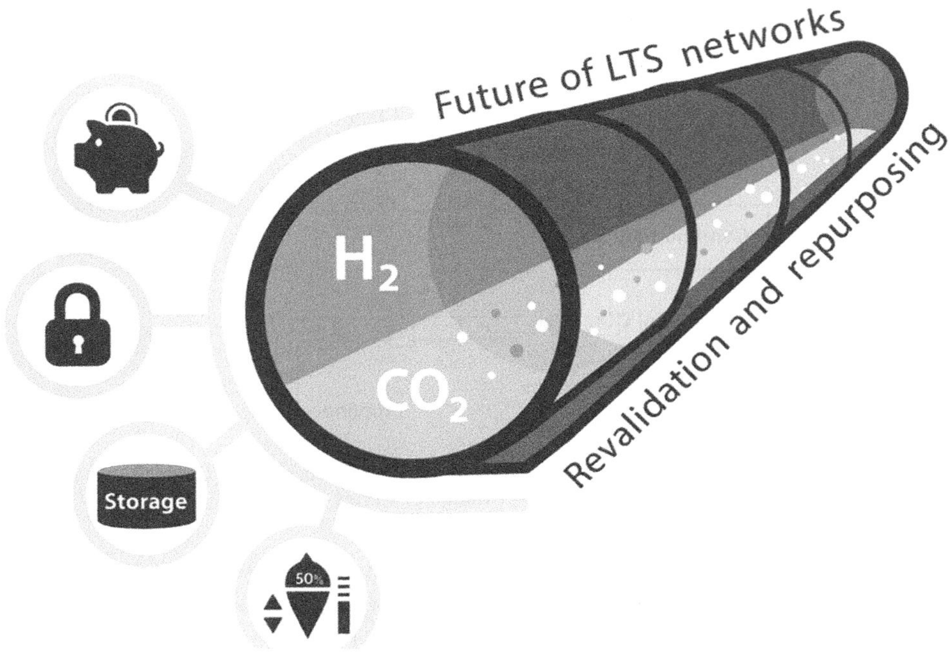

Figure 16.8 Conceptual illustration of pipeline repurposing and revalidation for future hydrogen (H_2) and carbon dioxide (CO_2) transport in large-scale transport system (LTS) networks.

prepare regulatory reports, increase transparency about project performance, and build public confidence that CCS is operating safely [48, 50].

Nature-based innovations are also being developed to complement engineered carbon capture systems. These approaches use natural biological and chemical processes to remove CO_2 from the atmosphere and store it in soils, plants, or oceans. Examples include soil-enhancement techniques such as biochar addition and regenerative farming practices that increase the ability of agricultural soils to absorb and retain carbon. Drone-assisted reforestation uses automated aerial planting systems to rapidly distribute seeds over large or hard-to-access areas, speeding up forest restoration compared with manual planting. Restoration of coastal ecosystems allows vegetation and sediments to naturally capture carbon through plant growth and long-term soil burial. In marine environments, ocean monitoring technologies support methods such as alkalinity enhancement, where mineral additions increase seawater's capacity to absorb and store CO_2 more effectively [51, 52].

Looking ahead, modular CCS systems represent an important innovation pathway. Modular CCS systems work like LEGO building blocks. Each capture unit is a standardized, container-sized module that can be easily added, removed, or combined with others, depending on how much CO_2 needs to be captured. Because the modules are pre-built and only connected on site, they are fast to install and require much less construction than traditional CCS plants. This flexibility means facilities can start with a small system and expand it later as needed, or relocate modules if operations change. Modular CCS makes

Figure 16.9 Conceptual illustration of a modular direct air capture facility composed of standardized, containerized capture units that can be deployed, relocated, or expanded to match site-specific CO_2 capture requirements.

carbon capture easy to deploy and accessible for smaller or distributed industries [48]. Figure 16.9 illustrates a representative modular DAC layout in which standardized capture units are installed in scalable arrays to support flexible deployment and capacity expansion.

Despite these advances, important challenges remain. New technologies must still be scaled from pilot projects to large commercial systems, their full environmental impacts over the entire lifecycle must be carefully evaluated, and public trust must be maintained through transparency and community engagement [48, 53]. In addition, major efforts are needed, especially from both engineering and regulatory perspectives, to coordinate the development of shared CO_2 transport and storage infrastructure. Continued investment and strong governance support are essential. Innovation must be guided by long-term sustainability goals, not just short-term cost reductions [28].

In summary, new innovations point toward a more flexible, efficient, and integrated future for carbon management. Advances in capture materials are lowering energy use and reducing equipment size, while digital tools improve system monitoring, safety, and maintenance planning. Developments in transport networks are making it easier to connect small and large CO_2 sources to storage hubs, and innovations in storage monitoring are strengthening long-term safety and public confidence. At the same time, circular economy approaches are expanding opportunities to reuse CO_2 in useful products. Together, these advances are evolving CCS from a single-purpose technology into a comprehensive climate solution.

16.6 Strategic Recommendations and Calls to Action

The future success of CCS depends not only on technological advances but also on coordinated actions by governments, industries, academic institutions, communities, and international organizations. This section presents key strategic recommendations and calls to action to ensure that CCS contributes effectively to climate mitigation while supporting justice, sustainability, and long-term system transformation.

First, governments must align CCS development with clear climate goals, meaning that carbon capture policies should support national and international targets for net-zero emissions rather than simply reduce emissions in the short term [54]. In practice, this means using CCS mainly for "hard-to-abate" sectors such as cement, steel, chemicals, and waste treatment, where emissions cannot easily be eliminated through electrification or renewable energy alone because carbon is released directly from industrial processes [55]. Public funding and regulations should therefore direct CCS toward these sectors instead of extending the lifetime of fossil-fuel power generation. Strong climate targets, transparent monitoring and reporting systems, and clear regulatory frameworks are essential to ensure CCS is deployed where it creates the greatest long-term climate benefit and supports a genuine transition to net-zero emissions.

Second, strong industry leadership and investment across all aspects of carbon capture are essential. Companies must invest not only in capture equipment but also in workforce training, environmental protection measures, and active community engagement. Environmental protection measures should include emissions monitoring, leak detection, safe waste handling, water protection, and habitat conservation [56]. Corporate strategies should place CCS within broader sustainability plans, linking capture projects to energy-efficiency improvements, increased use of renewable energy, and circular-economy practices. Open and transparent reporting of emissions performance and long-term storage monitoring results helps build public trust and strengthens corporate accountability [19].

Third, research and education must be expanded. Universities and research centers play a critical role in advancing capture materials, storage monitoring systems, and lifecycle assessment tools. Interdisciplinary programs that combine engineering, environmental science, economics, policy, and ethics prepare future professionals to manage complex sustainability challenges. Educational outreach to students and the general public helps people understand both the capabilities and limitations of CCS. By providing clear, factual information, these efforts reduce misunderstandings and limit the spread of misinformation [27, 48].

Fourth, communities must be actively involved in CCS projects. Project developers should engage local residents at early stages through public consultations, open-access information platforms, and participatory decision processes. Host communities should benefit directly from projects through employment opportunities, infrastructure improvements, educational programs, and environmental restoration initiatives. Respect for community rights and local concerns strengthens social acceptance and long-term project stability [27].

Fifth, international cooperation is essential. Climate change goes beyond national boundaries, and CCS research, financing, and governance must therefore be organized through global partnerships. High-emitting countries that are normally wealthier should provide financial resources, technical support, and training to developing nations [1, 4].

Therefore, developing countries can have low-carbon technologies directly, instead of following the traditional development path that relied heavily on coal, oil, and gas, an approach that led to today's carbon-intensive growth and high emissions [57]. Shared research programs, open technology platforms for developing and testing carbon removal solutions such as DAC, and multilateral funding mechanisms, such as international climate funds supported by multiple governments and development banks, help ensure that mitigation tools are accessible worldwide in a fair and equitable way.

Sixth, strong governance and stewardship frameworks must be put in place and enforced. Because CO_2 is stored underground for decades or even centuries, companies and governments must ensure that storage sites are carefully monitored long after projects end [58]. This includes long-term monitoring programs, routine inspections, and financial guarantees or trust funds to cover maintenance and any corrective actions if problems occur. Independent verification organizations and transparent public reporting systems are also necessary to confirm that storage sites remain safe and that responsibilities are not abandoned. These measures protect future generations and ensure that carbon storage remains secure over the long term.

Finally, a systems-thinking mindset must guide all CCS actions. This means carbon capture should not be used as a stand-alone solution but should be integrated with other climate measures such as renewable energy development, energy-efficiency improvements, and nature-based solutions [10, 51]. Governments and investors should prioritize projects that deliver multiple benefits at the same time, reducing emissions, supporting ecosystem restoration, creating jobs, and promoting social development, rather than focusing only on the amount of CO_2 captured.

The call to action is clear: CCS must be developed in a responsible, inclusive, and strategic way. Progress in technology must go hand in hand with strong governance, public participation, education, and international cooperation. Communities must be involved, regulations must be transparent, and global partnerships must ensure fair access to climate solutions and technologies. When CCS is guided by principles of justice and sustainability, it can become more than just a tool for reducing emissions; instead, it can also support environmental recovery, strengthen communities, and contribute to long-term social and economic resilience.

16.7 Key Highlights (Chapter Summary)

◇ Chapter 16 – Key Highlights:

1) Carbon capture must be viewed as part of an integrated sustainability system, not as an isolated technological solution, working alongside renewable energy, energy efficiency, nature-based solutions, and circular economy practices.
2) Systemic redesign is essential, meaning that energy, industry, governance, and community systems must be restructured together to deliver long-term climate stability, social equity, and ecological resilience rather than short-term emissions reductions alone.

3) Climate justice requires leadership from high-emitting countries and industries. This leadership includes providing greater financial support, sharing clean technologies, and offering training and education to build local skills in lower-income countries that are often the most vulnerable to climate change.
4) Community participation is essential for successful CCS deployment. Involving local residents in open discussions, clearly communicating potential risks, and ensuring that communities share in project benefits help build trust and social acceptance.
5) Carbon utilization and circular economy approaches use captured CO_2 as a raw material for making products such as fuels, chemicals, and building materials. This reduces waste and supports more sustainable production by keeping carbon in use rather than releasing it back into the atmosphere.
6) Regenerative environmental governance moves beyond pollution control, promoting ecological restoration and community resilience.
7) Technological innovation continues to improve CCS viability, including advanced capture materials, digital tools, transport optimization, geological monitoring systems, industrial carbon hubs, and modular capture units.
8) Education and public engagement are critical, enabling workforce development, informed decision-making, and public trust.
9) Strong long-term governance frameworks are necessary, including strict monitoring requirements and international cooperation to ensure safe and responsible storage across generations.
10) Strategic action requires cooperation among governments, industry, universities, and civil society. By aligning CCS development with principles of justice, sustainability goals, and global climate commitments, these groups can ensure that carbon capture really supports a fair, effective, and regenerative transition to a low-carbon future.

Keywords: Justice, Sustainability, Systemic Redesign, Global Equity, Regenerative Governance, Community Participation, Climate Responsibility, Carbon Capture, Carbon Utilization, Circular Economy, Environmental Stewardship, International Cooperation, Capacity Building, Technology Transfer, Ethical Accountability, Social Inclusion, Public Engagement, Integrated Climate Strategies, Sustainable Development, Long-Term Stewardship, Intergenerational Equity, Environmental Justice, Policy Integration, Climate Resilience, Systems Thinking

References

1 Å. Tørvanger and J. Meadowcroft, "The political economy of technology support: Making decisions about carbon capture and storage and low carbon energy technologies," *Global Environmental Change*, vol. 21, no. 2, pp. 303–312, May, 2011. DOI: 10.1016/j.gloenvcha.2011.01.017.

2 E. Persson, K. Eriksson, and Å. Knaggård, "A fair distribution of responsibility for climate adaptation—translating principles of distribution from an international to a local context," *Philosophies*, vol. 6, no. 3, 68, 2021. DOI: 10.3390/philosophies6030068.

3 W. Leal Filho *et al.*, "Influences of climate change and variability on estuarine ecosystems: An impact study in selected European, South American and Asian countries," *International Journal of Environmental Research and Public Health*, vol. 19, no. 1, 585, 2022. DOI: 10.3390/ijerph19010585.

4 H. Li, Y. Ye, H. Li, and Q. Ye, "Emergence, distribution dynamics and drivers of global high-emission countries since the industrial revolution," *International Journal of Global Warming*, vol. 29, no. 4, pp. 383–394, 2023. DOI: 10.1504/IJGW.2023.130137.

5 O. B. Adekoya, O. T. Kenku, J. A. Olivide, M. A. S. Al-Faryan, and O. D. Ogunjemilua, "Does economic complexity drive energy efficiency and renewable energy transition?," *Energy*, vol. 278, 127712, Sep., 2023. DOI: 10.1016/j.energy.2023.127712.

6 R. Wu, J. Wang, S. Wang, and K. Feng, "The drivers of declining CO_2 emissions trends in developed nations using an extended STIRPAT model: A historical and prospective analysis," *Renewable and Sustainable Energy Reviews*, vol. 149, 111328, Oct., 2021. DOI: 10.1016/j.rser.2021.111328.

7 P. Fragkos, "Assessing the role of carbon capture and storage in mitigation pathways of developing economies," *Energies*, vol. 14, no. 7, 1879, 2021. DOI: 10.3390/en14071879.

8 J. Yao, W. Liu, and K. Bi, "The global knowledge evolution of carbon capture, utilization, and storage technology: A network analysis of patent mining," *IEEE Transactions on Engineering Management*, vol. 72, pp. 192–204, 2024. DOI: 10.1109/TEM.2023.3324490.

9 K. Itooka, T. Hashimoto, M. Akai, S. Aoyama, C. Shinohara, and H. Horikoshi, "Towards an implementation of carbon dioxide capture and storage: Aligning climate change mitigation with principles of justice and fairness," in Proc. 17th Greenhouse Gas Control Technologies Conf. (GHGT-17), Oct. 2024, pp. 1–10.

10 M. S. Zantye, A. Arora, and M. F. Hasan, "Renewable-integrated flexible carbon capture: A synergistic path forward to clean energy future," *Energy & Environmental Science*, vol. 14, no. 7, pp. 3986–4008, 2021. DOI: 10.1039/D1EE00925H.

11 H. Ishaq, O. Siddiqui, G. Chehade, and I. Dincer, "A solar and wind driven energy system for hydrogen and urea production with CO_2 capturing," *International Journal of Hydrogen Energy*, vol. 46, no. 6, pp. 4749–4760, Jan., 2021. DOI: 10.1016/j.ijhydene.2020.01.208.

12 P. R. Oraon, V. Sagar, and K. Beauty, "Ecological restoration of degraded land through afforestation activities," in *Land and Environmental Management Through Forestry*, A. K. Pandey, P. K. Joshi, and S. K. Sharma, Eds, 1st ed., 2023, pp. 201–216.

13 R. Li, L. Li, and Q. Wang, "The impact of energy efficiency on carbon emissions: Evidence from the transportation sector in Chinese 30 provinces," *Sustainable Cities and Society*, vol. 82, p. 103880, Jul., 2022. DOI: 10.1016/j.scs.2022.103880.

14 A. Hailemariam and M. O. Erdiaw-Kwasie, "Towards a circular economy: Implications for emission reduction and environmental sustainability," *Business Strategy and the Environment*, vol. 32, no. 4, pp. 1951–1965, 2023. DOI: 10.1002/bse.3266.

15 F. Wang *et al.*, "Climate change: Strategies for mitigation and adaptation," *The Innovation Geoscience*, vol. 1, no. 1, p. 100015, 2023. DOI: 10.1016/j.xinngeo.2023.100015.

16 P. Bielka, "The future of carbon capture and storage technology: Innovative approach with digital twin," *Journal of Geotechnology and Energy*, vol. 40, no. 2, pp. 5–10, 2023, ISSN: 2720-3581

17 A. K. Priya, B. Devarajan, A. Alagumalar, and H. Song, "Artificial intelligence enabled carbon capture: A review," *Science of the Total Environment*, vol. 886, p. 163913, Aug., 2023. DOI: 10.1016/j.scitotenv.2023.163913.

18 T. Wilberforce, A. G. Olabi, E. T. Sayed, K. Elsaid, and M. A. Abdelkareem, “Progress in carbon capture technologies,” *Science of the Total Environment*, vol. 761, p. 143203, Mar., 2021. DOI: 10.1016/j.scitotenv.2020.143203.

19 M. G. Fikru and J. W. A. Azure, “Renewable energy technologies and carbon capture retrofits are strategic complements,” *Technological Forecasting and Social Change*, vol. 196, p. 122850, Nov., 2023. DOI: 10.1016/j.techfore.2023.122850.

20 H. J. Buck, “Social science for the next decade of carbon capture and storage,” *The Electricity Journal*, vol. 34, no. 7, Aug.–Sep., 107003, 2021. DOI: 10.1016/j.tej.2021.107003.

21 J. A. Nielsen, K. Stavrianakis, and Z. Morrison, “Community acceptance and social impacts of carbon capture, utilization and storage projects: A systematic meta-narrative literature review,” *PLoS One*, vol. 17, no. 8 e0272409, 2022, 10.1371/journal.pone.0272409.

22 M. E. Belfiori and M. Macera, “Climate inequality: Carbon capture for redistribution,” CESifo Working Paper Series, Working Paper no. 11239, 2024. [Online]. Available: SSRN, paper ID 4922412.

23 D. Rojas-Rueda, K. McAuliffe, and E. Morales-Zamora, “Addressing health equity in the context of carbon capture, utilization, and sequestration technologies,” *Current Environmental Health Reports*, vol. 11, pp. 225–237, 2024. DOI: 10.1007/s40572-024-00447-6.

24 N. Norouzi and S. Choubanshezafar, “An overview on the carbon utilization technologies with an approach to the negative emission construction material,” *International Journal of New Chemistry*, vol. 8, no. 3, pp. 298–328, Summer, 2021. DOI: 10.22034/ijnc.2020.123020.1102.

25 A. Bellini, B. Andersen, N. J. Klungseth, and A. Tadayon, “Achieving a circular economy through the effective reuse of construction products: A case study of a residential building,” *Journal of Cleaner Production*, vol. 450, p. 141753, Apr., 2024. DOI: 10.1016/j.jclepro.2024.141753.

26 A. Briones-Hidrovo, J. R. Copa Rey, A. C. Dias, L. A. C. Tarelho, and S. Beauchet, “Assessing a bio-energy system with carbon capture and storage (BECCS) through dynamic life cycle assessment and land—water—energy nexus,” *Energy Conversion and Management*, vol. 268, p. 116014, Sep., 2022. DOI: 10.1016/j.enconman.2022.116014.

27 Y. Vasilev, A. Cherepovitsyn, A. Tsvetkova, and N. Komendantova, “Promoting public awareness of carbon capture and storage technologies in the Russian federation: A system of educational activities,” *Energies*, vol. 14, no. 5, 1408, Mar., 2021. DOI: 10.3390/en14051408.

28 H. Zhang, “Regulations for carbon capture, utilization and storage: Comparative analysis of development in Europe, China and the Middle East,” *Resources, Conservation and Recycling*, vol. 173, Oct. 2021, art.no. 105722, 10.1016/j.resconrec.2021.105722.

29 D. Y. Shu, S. Deutz, B. A. Winter, N. Baumgärtner, L. Leenders, and A. Bardow, “The role of carbon capture and storage to achieve net-zero energy systems: Trade-offs between economics and the environment,” *Renewable and Sustainable Energy Reviews*, vol. 178, May 2023, art.no. 113246, 10.1016/j.rser.2023.113246.

30 K. Mintz-Woo and J. Lane, “Why and where to fund carbon capture and storage,” *Science and Engineering Ethics*, vol. 27, 70, 2021. DOI: 10.1007/s11948-021-00344-3.

31 B. Maher and J. Symons, “The international politics of carbon dioxide removal: Pathways to cooperative global governance,” *Global Environmental Politics*, vol. 22, no. 1, pp. 44–68, Feb., 2022. DOI: 10.1162/glep_a_00643.

32 S. Kime, V. Jacome, D. Pellow, and R. Deshmukh, "Evaluating equity and justice in low—carbon energy transitions," *Environmental Research Letters*, vol. 18, no. 12, 123003, Nov., 2023. DOI: 10.1088/1748-9326/ad08f8.

33 D. Scott, "Ethics of climate engineering: Chemical capture of carbon dioxide from air," in *Ethics of Chemistry*, M. Schummer and T. Børsen, Eds, Vol. 14. World Scientific, 2021, pp. 375–401, DOI: 10.1142/9789811233548_0014.

34 B. K. Sovacool, "Expanding carbon removal to the Global South: Thematic concerns on systems, justice, and climate governance," *Energy and Climate Change*, vol. 4, 100103, Dec., 2023. DOI: 10.1016/j.egyc.2023.100103.

35 C. Hilson and A. Savaresi, "Regenerative approaches and environmental law: Beyond sustainability?," in *A Research Agenda for Environmental Law*, J. van Zeben and C. Hilson, Eds. Cheltenham, UK: Edward Elgar Publishing, 2025, ch. 6, pp. 59–72, DOI: 10.4337/9781035324408.0011.

36 S. Nicholas and P. M. DeMarco, "Absorb carbon with natural landscapes and regenerative agriculture," in *AESS Interdisciplinary Environmental Studies and Sciences Series*, R. I. Appalachia and P. M. DeMarco, Eds. Cham: Springer, 2024, pp. 165–182, DOI: 10.1007/978-3-031-61921-2_9.

37 R. Kotagodahatti, K. Hewage, H. Karunathilake, and R. Sadiq, "Long-term feasibility of carbon capturing in community energy systems: A system dynamics-based evaluation," *Journal of Cleaner Production*, vol. 377, 134460, Dec., 2022. DOI: 10.1016/j.jclepro.2022.134460.

38 M. S. Zantye, A. Arora, and M. M. F. Hasan, "Renewable-integrated flexible carbon capture: A synergistic path forward to clean energy future," *Energy & Environmental Science*, vol. 14, no. 7, pp. 3985–4007, 2021. DOI: 10.1039/D0EE03807E.

39 A. M. Tedesco, S. López-Cubillos, R. Chazdon, M. Mills, F. C. Santos, and A. J. Dean, "Beyond ecology: Ecosystem restoration as a process for social-ecological transformation," *Trends in Ecology & Evolution*, vol. 38, no. 7, pp. 643–653, Jul., 2023. DOI: 10.1016/S0169-5347(23)00036-8.

40 R. Leal-Arcas and A. Filis, "International cooperation on climate change mitigation: The role of climate clubs," *European Energy and Environmental Law Review*, vol. 30, no. 5, pp. 195–218, 2021. DOI: 10.54648/eelr2021020.

41 G. Aristiati and K. Salehin, "Transparent carbon capture and storage using blockchain technology," *E3S Web of Conferences*, vol. 475, 01003, 2024. DOI: 10.1051/e3sconf/202447501003.

42 D. Havyatt, R. Nepal, and D. Johnstone, "Chapter 11: The political economy of carbon capture, use, and storage," in *Energy Transition and Carbon Neutrality in ASEAN, Series in Energy Transition, Carbon Neutrality, and Sustainability*, 2024, pp. 271–303, DOI: 10.1142/9789811288050_0011.

43 C. B. Peres, P. M. R. Resende, L. J. R. Nunes, and L. C. de Morais, "Advances in carbon capture and use (CCU) technologies: A comprehensive review and CO_2 mitigation potential analysis," *Clean Technologies*, vol. 4, no. 4, pp. 1193–1207, 2022. DOI: 10.3390/cleantechnol4040073.

44 M. Ozkan and R. Custelcean, Eds., "The status and prospects of materials for carbon capture technologies," *MRS Bulletin*, vol. 47, pp. 390–394, 2022. DOI: 10.1557/s43577-022-00364-9.Guest Eds

45 F. Sher *et al.*, "Advanced metal–organic frameworks for superior carbon capture, high-performance energy storage and environmental photocatalysis—a critical review," *Journal of Materials Chemistry A*, vol. 12, pp. 27932–27973, 2024. DOI: 10.1039/D4TA03877K.

46 A. Fly, I. Kirkpatrick, and R. Chen, "Low temperature performance evaluation of electrochemical energy storage technologies," *Applied Thermal Engineering*, vol. 189, 116750, May, 2021. DOI: 10.1016/j.applthermaleng.2021.116750.

47 K. Sotoodeh, "Material failure and corrosion behaviour of CO_2–H_2 hybrid pipelines in high-pressure transport: Mechanisms, risks, and mitigation," *Journal of Failure Analysis and Prevention*, 2025. DOI: 10.1007/s11668-025-02344-y.

48 K. Sotoodeh, *Piping, Valves, and Process Engineering in Carbon Capture Systems: Applications, Sustainability, and Environmental Protection*, 1st ed.. Amsterdam, Netherlands: Elsevier, 2026. ISBN: 978–0443510786.

49 Y. Zhao and K.-I. Itakura, "A state-of-the-art review on technology for carbon utilization and storage," *Energies*, vol. 16, no. 10, 3992, 2023. DOI: 10.3390/en16103992.

50 S. Wang,et al.. "Digitalization solution for carbon capture, transportation, utilization, and storage," in Proc. Greenhouse Gas Control Technologies Conf. (GHGT-17), Oct. 2024, posted Jan. 2025, SSRN Paper ID 5069953.

51 H. Jain, "Synergizing carbon reduction technologies and nature-based solutions for a climate-neutral future," *Mitigation and Adaptation Strategies for Global Change*, vol. 30, 42, 2025. DOI: 10.1007/s11027-025-10237-6.

52 C. M. Fortuna *et al.*, "Top predator status and trends: Ecological implications, monitoring and mitigation strategies to promote ecosystem-based management," *Frontiers in Marine Science*, vol. 11, 1282091, 2024. DOI: 10.3389/fmars.2024.1282091.

53 F. Mulyasari, A. K. Harahap, A. O. Rio, R. Sule, and W. G. A. Kadir, "Potentials of the public engagement strategy for public acceptance and social license to operate: Case study of carbon capture, utilisation, and storage Gundih Pilot project in Indonesia," *International Journal of Greenhouse Gas Control*, vol. 108, 103312, June, 2021. DOI: 10.1016/j.ijggc.2021.103312.

54 S. A. Usas and L. Ricardez-Sandoval, "An optimal sustainable planning strategy for national carbon capture deployment: A review on the state of CO_2 capture in Canada," *The Canadian Journal of Chemical Engineering*, vol. 102, no. 7, pp. 2332–2351, July, 2024. DOI: 10.1002/cjce.25249.

55 T. Zhang, M. Zhang, L. Jin, M. Xu, and J. Li, "Advancing carbon capture in hard-to-abate industries: Technology, cost, and policy insights," *Clean Technologies and Environmental Policy*, vol. 26, pp. 2077–2094, 2024. DOI: 10.1007/s10098-024-02810-5.

56 J.-L. Fan, Z. Li, Z. Ding, K. Li, and X. Zhang, "Investment decisions on carbon capture utilization and storage retrofit of Chinese coal-fired power plants based on real option and source–sink matching models," *Energy Economics*, vol. 126, 106972, Oct., 2023. DOI: 10.1016/j.eneco.2023.106972.

57 M. Farooq *et al.*, "Carbon capture for sustainable environment in developing countries," in *Energy and Environmental Security in Developing Countries*, M. Asif, Ed. Cham, Switzerland: Springer, 2021, pp. 481–504 ch. 21, DOI: 10.1007/978-3-030-63654-8_21.

58 L. Huang and X. Yang, "Evaluating different geophysical monitoring techniques for geological carbon storage," in *Geophysical Monitoring for Geologic Carbon Storage*, L. Huang, Ed, Geophysical Monograph Series. Hoboken, NJ, USA: Wiley/AGU, 2022, pp. 1–24 , ch. 1, DOI: 10.1002/9781119156871.ch1.

17

Artificial Intelligence, Digitalization, and Smart Carbon Capture Systems

17.1 Role of Digitalization in Carbon Capture

17.1.1 Digitalization in CCS: Concept and Data Infrastructure

Digitalization refers to the use of advanced data technologies, including sensors, software platforms, cloud-based systems, and artificial intelligence (AI), to monitor, analyze, and optimize industrial processes [1]. In carbon capture and storage (CCS), digitalization is rapidly becoming a central technology for enhancing operational efficiency, safety, reliability, and environmental performance. Rather than relying solely on manual operation and fixed time-based inspection programs, modern CCS facilities increasingly depend on smart sensors generating real-time data combined with automated control and decision-support systems that continuously adapt to changing process and mechanical conditions [2, 3].

At its most basic level, digitalization begins with data collection. Thousands of sensors are installed throughout capture plants, piping and pipelines, mechanical equipment, and CO_2 storage sites to measure temperature, pressure, flow rates, chemical composition, vibration, and many other parameters. These sensors transmit data to centralized control systems in real time. This creates a continuous digital picture of the CCS process and system performance.[4]. This constant stream of data allows operators to detect small changes that might otherwise go unnoticed, such as early signs of equipment malfunction, or minor pipeline pressure drops that could signal a leak [5].

In digitally enabled CCS facilities, data collected from field instruments, including pressure, temperature, flow, vibration, and corrosion sensors, are transmitted through Industrial Internet of Things (IIoT) networks to centralized control and monitoring platforms. These systems integrate field data into distributed control systems (DCS), supervisory control and data acquisition (SCADA) platforms, and cloud-based databases, where information is continuously stored, filtered, and validated [6].

Cloud-based platforms enable secure data storage and management as well as real-time access for operational teams and analytics applications, including machine learning (ML) tools [7]. Decision-support algorithms embedded within control systems continuously evaluate incoming data streams and initiate either automated corrective actions. These actions stabilize process conditions and prevent failures [8]. This integrated digital architecture transforms traditional process monitoring into a dynamic, data-driven operational model capable of continuous process monitoring, rapid response, and proactive failure prevention.

17.1.2 Artificial Intelligence, Optimization, and Predictive Maintenance

Once collected, data are processed using advanced analytics and AI [8]. ML algorithms observe historical operating patterns and learn how the system behaves under different conditions [9]. Over time, the software becomes increasingly accurate at predicting optimal operating parameters. For example, digital platforms can adjust solvent circulation rates or temperatures to maximize CO_2 capture efficiency while minimizing energy use. This real-time optimization reduces operational costs and lowers the carbon footprint of the capture process itself.

Digitalization also significantly improves plant reliability and maintenance planning [10]. Traditionally, industrial equipment has been inspected or serviced at fixed time intervals, regardless of whether any actual problems exist. This approach can result in unnecessary inspections when no degradation is present or, conversely, missed early warning signs of developing failures. In the worst cases, maintenance becomes purely reactive, with equipment or components being repaired only after failures occur [11].

Digital systems enable a transition toward predictive maintenance, where continuous process and mechanical data monitoring identifies abnormal trends such as changes in vibration signatures, temperature increases, or pressure fluctuations that indicate wear, fatigue, or damage. Advanced ML models can estimate the remaining useful life (RUL) of components and predict probable failure timelines, allowing maintenance teams to schedule inspection and repair at the most effective time before serious breakdowns occur. This proactive approach prevents costly unplanned shutdowns while significantly reducing risks to safety and the environment [12].

17.1.3 Digital Safety, Compliance, and Sustainability in CCS

In CO_2 transport systems, digitalization plays an essential safety role. Long-distance pipelines are equipped with pressure sensors, acoustic sensors, and satellite monitoring tools that track operating conditions continuously. AI platforms analyze this information to detect unusual behavior patterns that could indicate corrosion damage, material fatigue, or leaks. If irregularities appear, automated alert systems notify operators immediately, allowing quick response before minor issues turn into major environmental hazards [13]. Carbon storage sites also benefit from digital technologies. Geological storage operations rely on subsurface monitoring systems, including seismic imaging, well pressure tracking, and satellite surface observation. These tools generate enormous quantities of data describing how injected CO_2 moves underground over time [14].

Figure 17.1 illustrates the transition from traditional manual valve operation on the left to a fully digitalized CCS monitoring and control system on the right, where real-time sensor data are transmitted to centralized digital platforms for continuous supervision and decision support. The symbols on the left represent key process measurements, including pressure, temperature, flow or level, thermal conditions, vibration, and operational/electrical signals, that are captured by field sensors and transmitted to control rooms. Instead of requiring on-site manual adjustments, system performance is dynamically observed and optimized through automated commands, enabling safer, more efficient, and proactive operation of CCS facilities.

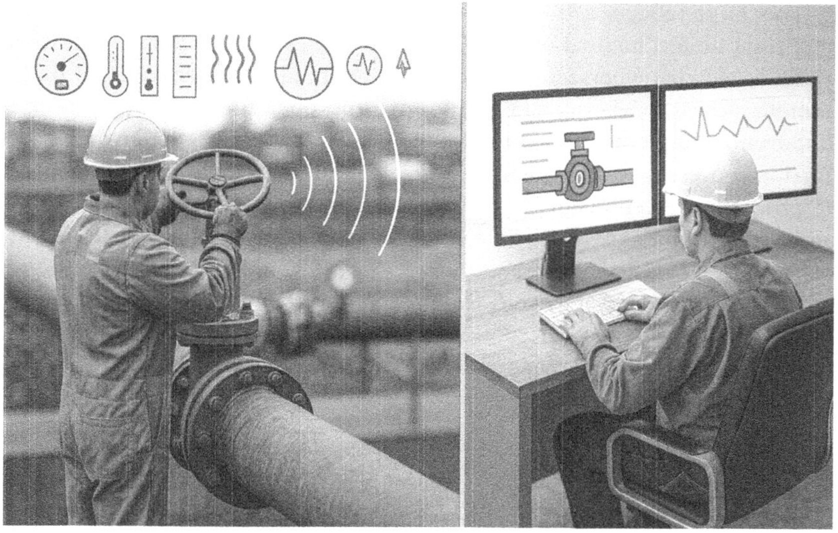

Figure 17.1 Transition from manual field operation to fully digitalized CCS monitoring and control.

Another important application of digitalization is in environmental compliance and transparency. In modern CCS facilities, digital monitoring systems automatically measure and record emissions, equipment status, and safety indicators on a continuous basis [15]. Safety indicators are measurable signals, such as pressure deviations, temperature rises, vibration trends, valve leakage rate, and corrosion rate measurement, that show how safely a facility is operating. For example, if pressure suddenly drops while vibration increases, the digital system recognizes these as safety indicators of a potential leak. Regulatory authorities can access these verified datasets remotely and at any time. At the public level, simplified digital dashboards can be used to communicate environmental performance to nontechnical audiences [16]. For example, a CCS facility may publish a public online dashboard that shows daily CO_2 capture rates or leak detection status using simple graphs or color-coded indicators. Members of the public and local communities can easily check whether the plant is operating safely and within environmental limits through these digital graphs.

Digitalization also enhances workplace safety. Sensors combined with AI-enabled monitoring systems can track hazardous gas concentrations, detect facility corrosion, or identify areas of mechanical stress [17]. Autonomous drones and robotic inspection tools further reduce human exposure to high-risk environments such as confined pipeline corridors, offshore platforms, or underground storage facilities. These machines can perform routine visual inspections, ultrasonic testing, and leak detection more quickly and safely than manual teams [18].

Beyond technical improvements, digitalization transforms how CCS workforces operate the plants. Modern carbon management systems require engineers and operators trained in data analysis, software platforms, automation technologies, and cybersecurity. This shift reflects a transition under Industry 4.0, where digital skills become as important as mechanical or chemical expertise [19]. Digitalization in CCS brings new risks, especially cybersecurity threats due to reliance on centralized databases and remote access. Cyberattacks could disrupt operations or compromise safety [20]. Importantly, digitalization supports more sustainable CCS deployment not only by optimizing processes such as capture efficiency but by reducing resource consumption and process failure prevention. Smarter process control lowers unnecessary energy demand and failure risks, while preventive maintenance minimizes equipment replacement and material waste [21].

In summary, digitalization reshapes CCS from a traditional industrial process into a dynamic, intelligent system capable of continuous monitoring, learning, and improvement. Sensors, AI integration, digital twins, and ML together create a more efficient, reliable, and safe carbon management plant.

17.2 Foundations of AI and Machine Learning in Carbon Capture and Storage

17.2.1 Artificial Intelligence and Data Foundations for CCS

AI and ML are now central tools for improving how CCS systems are designed, operated, and maintained. AI refers to computer systems that can perform tasks normally requiring human intelligence, such as recognizing patterns, making predictions, and learning from experience. ML is a subset of AI that focuses specifically on algorithms that improve automatically through exposure to data, rather than being programmed with fixed rules. In CCS, these technologies allow facilities to move from static operation toward dynamic, adaptive systems that continuously optimize performance [22].

Figure 17.2 illustrates how AI and ML are applied throughout the carbon capture, utilization, and storage (CCUS) value chain. Sensor data from capture facilities is transmitted to digital-twin platforms, where ML algorithms analyze operating conditions, including temperature, pressure, flow rates, and solvent performance. This analysis helps optimize CO_2 capture efficiency and energy usage. AI systems also enhance utilization processes by adjusting relevant process parameters to improve the performance of CO2 conversion to valuable products. During geological storage, AI evaluates CO_2 injection data in the piping and tubing, along with reservoir pressure and CO_2 migration behavior, to ensure safe long-term containment and compliance with regulations.

The foundation of AI-enabled CCS lies in data generation. Capture plants, pipelines, and storage sites produce vast amounts of process, mechanical, and material-related information through sensors that measure temperature, pressure, flow rates, solvent composition, energy usage, vibrations, and underground reservoir conditions. Collecting and organizing reliable data is the first step in applying ML. Poor data quality directly results in inaccurate predictions, making data validation, filtering, and storage systems essential components of a digital CCS infrastructure [22].

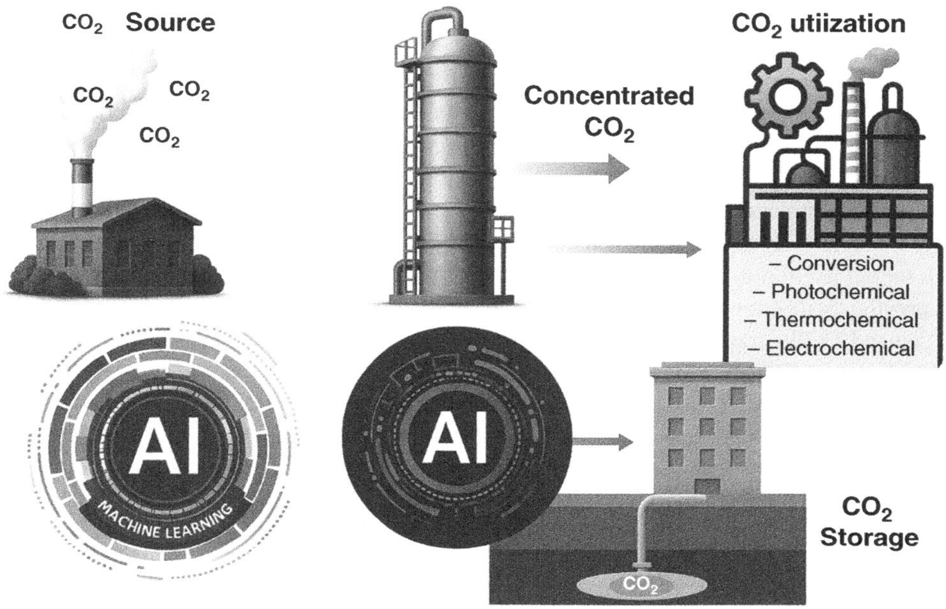

Figure 17.2 Conceptual illustration of artificial intelligence (AI) and machine-learning integration within carbon capture, utilization, and storage (CCUS) systems, showing data-driven optimization of capture performance, utilization processes, and safe geological storage through digital-twin modeling and real-time analytics.

17.2.2 Machine Learning Methodologies and Training

Once reliable data are collected, ML models are trained to recognize patterns between operating conditions and system performance [23]. For example the software learns how variations in solvent concentration or flue-gas temperature affect the amount of CO_2 captured by comparing thousands of past measurements. The software learns that when solvent concentration increases, the CO_2 capture rate also increases, because more active material is available to absorb CO_2. It also observes that when the flue-gas temperature rises above the optimal range, capture efficiency decreases, since higher temperatures reduce solvent absorption capacity. Each time new operating data are collected, the ML model updates its understanding of these cause-and-effect relationships. Over time, the system can predict which combination of solvent concentration and temperature will produce the highest capture efficiency, and it uses these predictions to automatically adjust operating conditions during real-time operation.

Three main types of ML methods are commonly used in CCS applications: The first type is "***Supervised Learning,***" which uses labeled data, where the correct outcomes are already known. Labeled data means operators provide historical examples that connect operating conditions such as pressure and temperature to results, such as high CO_2 capture efficiency, solvent degradation, or equipment failures. The ML model studies these examples and learns how to predict the expected outcome when new operating data are received [24].

The second type is ***"Unsupervised Learning,"*** which works without labeled results. The second type is unsupervised learning, which works without labels or predefined outcomes provided by operators. Instead, the ML system learns directly from ongoing operational data by observing normal process behavior over time. It identifies typical operating patterns and groups similar conditions together based on past experience. By continuously comparing new data to these learned patterns, the system can recognize when something unusual occurs, such as unexpected changes in vibration, pressure, or temperature, that may indicate early equipment faults or leaks. This judgment is made automatically by the ML model itself, not from predefined operator-input failure examples, allowing detection of previously unknown or emerging problems [24].

Figure 17.3 illustrates the difference between supervised and unsupervised ML. In supervised learning (left), the system already knows the correct outcome associated with each set of input parameters because it is trained using labeled historical data. These data are shown with different shapes and colors to represent known operating categories, such as low operating pressure, normal operating conditions, and high operating pressure. In unsupervised learning (right), no labels are provided, so all data points appear identical; the algorithm instead groups them into clusters based solely on similarities in their measured features, allowing it to discover natural operating patterns and identify abnormal behavior without prior examples.

The third type is ***"reinforcement learning,"*** which is based on learning through trial and error. The AI interacts with a digital-twin simulation of the CCS system, where it tests different control actions such as adjusting temperatures, pressures, or solvent flow rates. Each action is evaluated using "rewards" based on positive performance outcomes, for example achieving higher CO_2 capture efficiency, lower energy use, or improved safety conditions. Actions that lead to better results receive higher rewards, encouraging the

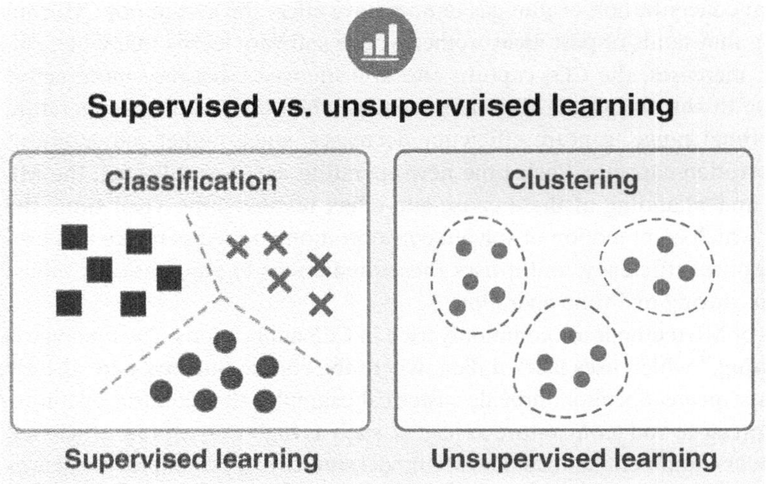

Figure 17.3 Comparison of supervised learning (labeled data used for classification) and unsupervised learning (unlabeled data grouped into clusters based on inherent similarity patterns).

system to repeat similar strategies. Over time, the AI learns which operating decisions consistently produce the best overall performance and then applies these optimized strategies to guide real-time plant control [24].

17.2.3 AI for Process Optimization and Predictive Maintenance

AI also plays an important role in process modeling and simulation [25]. Traditional process modeling refers to the use of commercial software such as Aspen Plus and Aspen HYSYS, which simulate plant behavior using detailed physical equations based on thermodynamics, fluid mechanics, heat transfer, and mass transfer. These models require extensive input data and simplifying assumptions and may not fully capture the complexity of real operating conditions or equipment aging. ML models complement these tools by learning directly from plant operating data and identifying patterns that are difficult for equation-based models to represent. Process engineers apply ML using platforms and software such as Python-based libraries (TensorFlow, Keras, PyTorch, Scikit-Learn), MATLAB ML Toolbox, Aspen PIMS AI modules, or AVEVA advanced analytics tools [26].

In capture operations, AI supports process optimization by analyzing large volumes of real-time data and calculating optimal operating requirements. These optimum process requirements are sent to the plant control system, which automatically applies the adjustments to equipment such as valves and pumps. This kind of multiple coordination of many variables at the same time is difficult for human operators to perform manually. On the other hand, AI can determine the optimal solvent circulation rate, regeneration heat input, and absorber pressure to maximize CO_2 capture while using the least possible energy. The control system continuously applies these optimized settings, resulting in smoother operation, lower energy consumption, and improved overall sustainability performance [25].

ML also improves equipment diagnostics and reliability management. ML algorithms analyze sensor data such as vibration patterns, temperature changes, and unusual energy consumption to detect early signs of equipment wear or damage. These signals are difficult for humans to notice in real time. This enables a maintenance approach known as predictive maintenance, where repairs are scheduled based on actual equipment condition rather than fixed time intervals. Instead of replacing parts too early or waiting for failures to occur, maintenance is performed only when the data indicate a real problem, reducing unnecessary downtime, material waste, and safety risks [5, 9].

Although AI brings powerful advantages, it must be applied carefully. AI systems can reflect limitations or biases if training data are incomplete or unrepresentative. Models also require continuous updating as operating conditions change. Outdated algorithms may make inaccurate recommendations. The ethical use of AI is an important consideration when applying digital technologies in CCS operations. AI systems often rely on extensive data collection, including information about workforce activities such as work location tracking, camera-based monitoring, biometric access controls, and performance logging systems, often referred to as workforce monitoring technologies. While these tools can improve safety and operational efficiency, they may also raise privacy concerns if workers feel they are being constantly observed or if personal data are collected without a clear purpose or consent. Cybersecurity protections, including strong authentication controls

Figure 17.4 Conceptual illustration of cybersecurity measures for protecting artificial intelligence and digital-twin systems, including authentication controls and data encryption to prevent unauthorized access and data misuse.

and data encryption, as illustrated in Figure 17.4, are essential to prevent unauthorized system access, data manipulation, or malicious interference with AI-driven digital platforms [27]. Training future CCS professionals in AI literacy is equally critical [28].

In summary, the application of AI and ML in CCS is built on the availability of reliable data, accurate process modeling, and intelligent decision-support tools. High-quality sensor data allows AI to understand how CCS processes behave, while predictive models transform this data into practical information that helps optimize operations and maintenance planning [29]. These information is then translated into actions through advanced control systems that safely adjust plant operation in real time. At the same time, ethical governance, such as protecting worker privacy and ensuring strong cybersecurity, ensures that digital technologies are used responsibly and maintain public trust [30]. Importantly, AI systems do not replace engineering judgment; rather, they serve as powerful tools that enhance human decision-making by providing faster analysis and more reliable support for complex operational challenges.

17.3 Digital Twins for Process Optimization

A digital twin is a virtual model of a physical system that continuously updates itself using real-time data from sensors installed on actual equipment. In CCS systems, digital twins provide a powerful tool for understanding how complex processes behave, testing improvement strategies, and optimizing operations safely without disturbing real facilities. They represent one of the most important advancements linking digital technologies with real-world industrial performance [31].

17.3.1 Digital-Twin Architecture and Data Integration

At its core, a digital twin is a virtual (computer-based) copy of a real engineering system. It combines three main elements: the physical system (piping, valves, compressors), data collection, and computer modeling. First, sensors installed on equipment such as compressors, pipelines, valves, or storage reservoirs continuously measure operating conditions like pressure, temperature, flow rate, vibration, and corrosion indicators [31]. This real-time data is sent to specialized software platforms. Specialized software platforms form the core of a digital twin, providing the digital environment where sensor data is processed, models are executed, and the virtual replica of the physical system is continuously updated [32]. Figure 17.5 illustrates the interaction between physical assets, that is piping system in this case, with sensor data streams, and the digital-twin environment described above.

17.3.2 Simulation, Anomaly Detection, and Process Optimization

Next, the software uses mathematical and engineering models based on well-known physical laws such as fluid flow, heat transfer, and mechanical stress to calculate how the system should behave under normal conditions. These are called physics-based equations because they come directly from fundamental engineering principles. The physics-based equations calculate normal operating behavior, and continuous comparison with real sensor measurements to find possible leakage or faults [33]. In other words, the software applies statistical methods to analyze patterns and trends in the data, helping to identify unusual behavior or performance changes over time. Finally, AI and ML algorithms learn from historical and real-time data to recognize early warning signs of problems, predict future equipment failures, and optimize maintenance planning [22, 23].

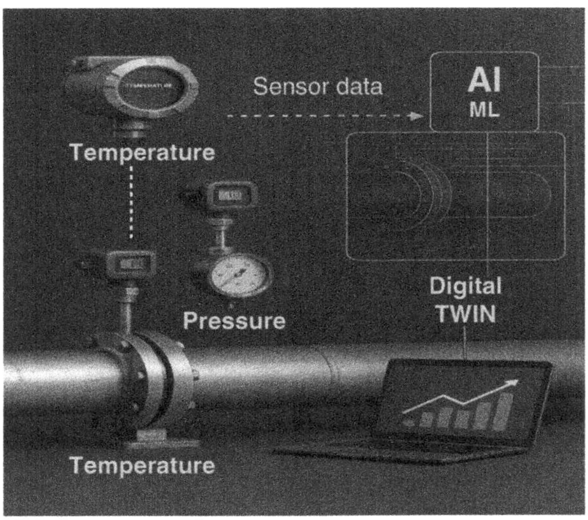

Figure 17.5 Digital twin architecture showing integration of physical assets, sensors, and analytics platform.

As new sensor data continuously arrive, the digital twin automatically updates, keeping the virtual model synchronized with the real system. This enables engineers to simulate system behavior, meaning they can safely test "what-if" scenarios without disturbing actual operations. For example, they can examine how a piping network would respond to changes in pressure, temperature, or flow rate by adjusting these parameters within the digital model instead of the physical plant [34]. In this way, potential risks can be explored and solutions optimized virtually, leading to improved safety, reliability, and overall operational efficiency.

For instance, in a pipeline, the relationship between flow rate (Q), velocity (V), and pipe cross-sectional area (A) is defined by:

$$Q = V \times A \ [35]$$

If sensor readings show a flow rate and velocity that are inconsistent with the known pipe diameter (area), the digital twin can detect anomalies. Such deviations may indicate problems such as internal blockage, buildup or fouling, partial valve closure, unexpected leakage, or measurement errors [36, 37].

In capture plants, digital twins allow operators to test operating conditions virtually. For example, engineers can simulate changes in solvent flow rates, absorber temperatures, gas compositions, or regeneration energy inputs and immediately observe predicted outcomes such as capture efficiency and energy use. This allows identification of optimal operating windows without interrupting actual plant operations or risking equipment safety.

17.3.3 Operational Applications, Safety, and Sustainability

Digital twins support startup and commissioning by allowing engineers to practice operating procedures, checking and adjusting how the plant's automatic control systems respond to changes in operating conditions, and identifying problems safely before real systems go online. They are also used as training simulators, helping operators learn how the plant responds during normal operation and emergency situations without any safety risks. In addition, digitalization (including digital twins) cannot fully replace real testing, but it can significantly reduce the amount of testing needed and make testing safer and more efficient [38].

Equipment reliability is another area where digital twins offer major advantages. Virtual models can simulate wear and degradation processes by comparing expected performance curves with real operational data. As soon as deviations appear, such as leakage from a facility, the digital twin flags potential problems and estimates remaining equipment life. Maintenance interventions can then be scheduled more efficiently, helping to prevent unexpected failures and slow the progression of damage or failure mechanisms before they become critical [39].

Figure 17.6 presents a conceptual visualization of digital-twin implementation for CO_2 pipeline systems. In CO_2 transport systems, digital twins simulate pipeline flow, pressure dynamics, thermal changes, and corrosion development [40]. For geological storage sites, digital twins become especially critical, especially in the simulation of the CO_2 storage site. Digital twins function as advanced decision-support systems. Instead of responding reactively to operational issues, plant managers can use digital twin forecasts to anticipate

Figure 17.6 Conceptual illustration of a digital twin applied to CO_2 pipeline transport systems, linking physical infrastructure with virtual simulation environments for flow, pressure, thermal behavior, corrosion monitoring, and operational decision support.

challenges and evaluate alternative strategies. Rather than exposing real systems to risky or extreme conditions, scenarios can be simulated virtually using the digital twin, allowing optimal decisions to be identified through "what-if" analyses.

Integration with ML, as a branch of AI, further enhances digital-twin functionality. ML algorithms process digital-twin data to identify complex relationships and trends among system variables that are not fully captured by classical physics-based models alone [41]. In addition, reinforcement learning ML techniques can utilize the digital-twin environment to train, test, and validate control strategies, enabling continuous improvement of operational decision-making while avoiding risks to real physical assets [24].

From an environmental sustainability perspective, digital twins enhance plant safety and reliability by reducing failures associated with leakage, corrosion, and other operational parameters that can adversely affect the environment. By enabling early fault detection and predictive maintenance, they minimize unplanned shutdowns, material losses, and waste generation. In addition, process optimization supported by digital-twin models improves energy efficiency, which is a key driver of environmental performance and sustainability [42].

Digital twins also strengthen transparency and regulatory reporting. Regulators increasingly require evidence of operational stability and CO_2 storage security. Digital-twin models enable visualization of CO_2 plume movement, pressure evolution within storage formations, and safety performance metrics, providing critical inputs for environmental reporting, monitoring programs, and regulatory compliance documentation [43].

However, challenges remain. Digital twins require vast volumes of high-quality data and consistent sensor reliability. Poor instrumentation or missing data can weaken model accuracy. There is also the risk of overreliance on automation, meaning that digital twins should assist, not replace, professional decision-making. Cybersecurity protection is another critical requirement, since real-time synchronization systems are vulnerable to digital intrusion if not properly secured. Training is essential to maximize digital-twin benefits [44]. Engineers, technicians, and regulators must learn how to interpret twin analyses, verify AI recommendations, and maintain the systems when digital-twin model behavior shows deviation from assumptions.

In summary, digital twins are powerful tools that help improve CCS systems. They allow safe testing of different operating scenarios, predict equipment problems, support training and emergency planning, reduce costs, and improve environmental performance. By combining engineering models, real-time sensor data, and AI, digital twins turn CCS facilities into smart, adaptive systems that continuously learn and operate more safely and efficiently.

17.4 Predictive Maintenance and Leak Detection Systems

17.4.1 Predictive Maintenance Fundamentals and Equipment Health Monitoring

Predictive maintenance and leak detection systems are among the most critical digital applications in CCS. These technologies use sensor data, AI, and advanced monitoring tools to detect early signs of equipment deterioration or pipeline leakage before serious failures occur. Instead of waiting for breakdowns or performing routine maintenance based solely on fixed time schedules, predictive systems assess equipment condition to guide efficient, timely interventions (online maintenance). This approach improves safety, reliability, and cost reduction [11, 45].

Figure 17.7 illustrates the principle of distributed acoustic sensing (DAS) applied to predictive maintenance and pipeline leak detection in CCS transport systems. Fiber-optic sensors installed along the pipeline continuously on cables record acoustic and vibration signals generated by fluid flow and mechanical activity. These data are transmitted to an integrator and analytical platform where signal-processing and AI algorithms distinguish normal operational noise from abnormal acoustic patterns associated with leaks or structural faults, enabling early detection and rapid response.

In traditional industrial operations, maintenance approaches are typically either reactive or preventive. Reactive maintenance involves repairing equipment only after a failure occurs, which can be costly and risky, as it may interrupt operations or pose environmental hazards. Preventive maintenance, on the other hand, involves planning activities such as inspections, lubrication, or calibration before failures occur, which can prevent unnecessary component replacements and downtime. Predictive maintenance provides a more effective solution by continuously monitoring equipment to detect early signs of potential issues. This allows for timely intervention before any failures occur, helping to minimize risks and reduce unnecessary maintenance costs [11].

At CCS capture facilities, hundreds of sensors gather data on critical process and mechanical parameters, including vibration, temperature, pressure, chemical performance, and energy

Figure 17.7 Conceptual illustration of distributed acoustic sensing (DAS) for predictive maintenance and leak detection in CCS pipelines. Acoustic signals collected by fiber-optic sensors are processed to differentiate normal operating conditions from leak-related signatures.

consumption. ML algorithms analyze this digital-twin sensor data to identify patterns that may indicate equipment wear or malfunction. For example small increases in vibration readings from compressors may suggest bearing degradation. Similarly, abnormal solvent temperature trends in the regeneration unit can signal potential heat-exchanger performance problems during solvent separation from carbon dioxide. AI systems compare real-time operating data with historical performance records and provide early warnings to maintenance teams when deviations indicate declining equipment reliability [46].

Digital twins, together with AI and ML, allow engineers to predict how long specific components can operate safely. This predicted timeframe is known as RUL. By estimating RUL, maintenance activities can be precisely scheduled, reducing the risk of unexpected shutdowns and avoiding unnecessary costs associated with maintenance performed either too early or too late [47]. Additionally, planning for spare parts becomes more accurate, as replacements are carried out only when really necessary. Consequently, predictive maintenance enhances operational stability by ensuring that facilities are maintained on time, which minimizes operational disruptions. Furthermore, it lowers the costs associated with facility operations and maintenance, as well as reducing material waste that may occur from replacing components only after a failure has happened.

17.4.2 Leak Detection, Robotic Inspection, and Safety Compliance

Leak detection is a key function of digital monitoring systems, especially in CO_2 pipeline networks, where undetected leaks can pose serious safety and environmental risks. Carbon dioxide can be dangerous to humans at high concentrations. Pipelines are equipped with fiber-optic sensing systems that continuously monitor operating conditions along the pipeline route. Satellite monitoring and drone inspections add another layer of safety by detecting surface changes or unusual vegetation patterns that may indicate underground CO_2 leaks. Vegetation anomalies refer to abnormal plant growth patterns, such as discoloration or die-off, that can indicate underground CO_2 leakage [45].

Figure 17.8 Schematic representation of integrated satellite monitoring over a CCS facility used to assess land-use changes, vegetation health, and potential environmental impacts beyond site boundaries.

This integrated monitoring approach is illustrated in Figure 17.8, which depicts satellite surveillance of a CCS facility, where remote sensing data are used to detect land-use changes, assess vegetation health, and identify surface anomalies indicative of potential CO_2 environmental impacts.

AI systems analyze monitoring data to separate normal operating changes from real fault conditions. Temporary pressure variations caused by routine temperature or flow changes are ignored, while persistent abnormalities trigger alarms for inspection. Early detection allows operators to isolate pipeline sections and correct problems before leaks become hazardous [48].

Robotic inspection systems enhance predictive and detection capabilities even further. Autonomous drones inspect difficult-to-access pipeline corridors and offshore facilities, collecting high-resolution images and thermal data [49]. Mobile robotic devices known as pipeline inspection gauges (pigs) travel inside pipelines to map wall thickness, identify corrosion pits, mechanical damage, or weld imperfections [50].

Storage sites also benefit from similar technologies. Long-term underground storage requires continuous pressure monitoring. AI analyzes these data sources together to confirm that injected CO_2 remains trapped within geological formations. If previously stable readings begin to change abnormally, digital monitoring systems flag the issue for investigation. Safety improvements by using digitalization are significant. Personnel exposure to hazardous working conditions is reduced because robotic inspectors and remote sensing tools perform dangerous tasks. In addition, process and mechanical abnormalities are found before they become critical and lead to catastrophic failure [1, 10, 17]. Economic benefits accompany technical and safety improvements. Predictive maintenance reduces emergency repair costs, extends equipment service life, and minimizes downtime [11].

Transparency and compliance also improve through digital tracking systems. Maintenance records, inspection results, and emergency response activities are digitally archived and can be audited by regulatory authorities [16]. However, challenges remain. Predictive systems depend heavily on sensor quality and algorithm accuracy. Incorrect calibration or incomplete datasets can lead to missed warnings or false alarms [4]. Cybersecurity to protect digital networks from intrusion, sabotage, or data manipulation is essential. Secure encryption and access restrictions are necessary to protect critical CCS infrastructure [23, 27]. Education and training for engineers and operators on working with digital tools remain vital [50]. In summary, predictive maintenance and leak detection systems represent a significant evolution in CCS safety and reliability management.

17.5 Automation and Smart Safety Systems

Automation is a key outcome of digitalization in CCS operations. Sensor data are transmitted through wireless and wired networks to centralized control and data platforms, where automated systems analyze process conditions and make control decisions without continuous human intervention. Control commands are then sent directly to final elements such as valves, compressors, pumps, and other equipment to adjust plant operation in real time. Automation and smart safety systems, therefore, play a vital role in modern CCS facilities [51].

Automation uses computerized control systems, robotics for smart inspection tasks instead of direct human involvement, and AI to efficiently manage and analyze industrial process data. Smart safety systems integrate long-distance monitoring by operators from centralized control rooms, early-warning detection using ML algorithms, and emergency-response measures such as automated alarms to prevent incidents and minimize risks to workers, surrounding communities, and the environment. Together, these technologies enhance operational safety and reliability while improving overall system efficiency across CCS infrastructure [51, 52].

17.5.1 Operational Automation and Robotic Inspection

Robotics is becoming increasingly important for routine inspection and maintenance activities in CCS operations. Autonomous drones survey facilities and plant sites using high-resolution cameras and thermal imaging to detect corrosion, mechanical damage, or abnormal heating patterns [49]. An example of drone-based inspection and gas-plume monitoring used in CCS safety surveillance is shown in Figure 17.9. The image illustrates the principle of drone-based infrared (IR) monitoring for gas leak detection in CCS facilities. Drones equipped with optical gas-imaging or multispectral cameras can detect gas leaks based on the fact that different gas species absorb IR radiation at specific characteristic wavelengths. The camera records these absorption patterns and converts them into visual contrasts that allow operators to locate potential leakage zones. The overlaid compound labels shown in the figure (e.g. CO_2, CH_4, NO_x, CO, and

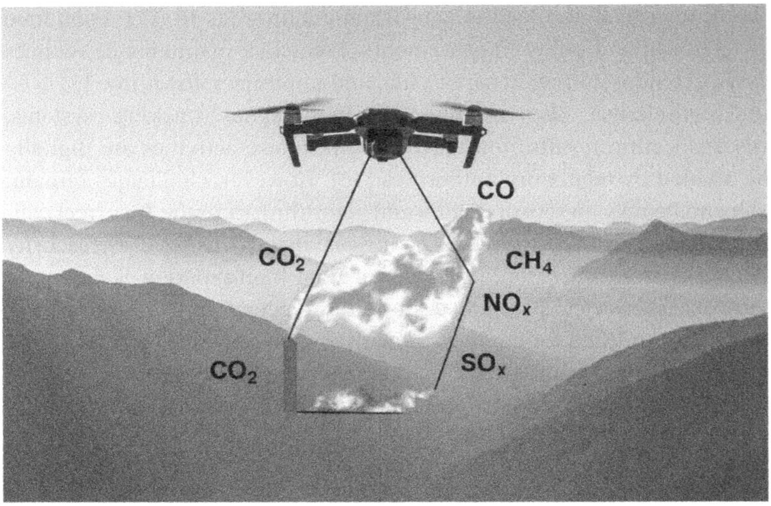

Figure 17.9 Conceptual illustration of drone-based infrared inspection used for gas plume visualization and leak detection in CCS operations. Optical gas-imaging cameras identify gases by their characteristic infrared absorption signatures. Gas labels shown are illustrative only and do not represent actual measurements.

SO_x) are illustrative only and represent possible detectable gases rather than real measured concentrations at the depicted site [45].

Ground-based inspection robots are deployed in confined or hazardous environments where manual access is unsafe, such as inside storage tanks or along difficult-to-access pipeline corridors. In addition, internal inspection devices known as "pigs" travel through pipelines to measure wall thickness and identify corrosion, weld defects, dents, or deformations that could compromise pipeline structural integrity [50].

17.5.2 Smart Safety Systems, Environmental Monitoring, and Cybersecurity

One important aspect of automation in digitalization is the use of smart sensors and safety systems that focus on real-time hazard detection and rapid emergency response. For example gas leak detection sensors monitor the air for elevated carbon dioxide (CO_2) levels that may indicate leaks or pose health risks to personnel. Pressure sensors, acoustic sensors that are sensitive to the sounds of cracks and leaks, and fiber-optic cables capable of detecting temperature changes, strain, or vibrations along pipelines continuously scan for early signs of leaks or ground movement. When abnormal conditions are detected, automated alarms are triggered instantly, alerting control rooms and personnel at the operational site for an emergency response [45, 46].

Many CCS facilities now incorporate automated safety shutdown systems. When sensors detect hazardous conditions such as abnormal pressure increases or equipment overheating, automated control systems immediately respond by isolating affected equipment sections through final control elements such as emergency shutdown valves and by activating emergency venting or pressure-relief procedures without waiting for human intervention.

These rapid automated responses significantly reduce the risk of incidents escalating into major safety events or uncontrolled releases of hazardous substances, including carbon dioxide [46, 52].

Worker safety is one of the greatest benefits of automation. Historically, industrial accidents often occurred during equipment inspections or maintenance in confined or high-risk environments such as storage tanks. Robotics and remote monitoring minimize the need for human presence in hazardous zones [52].

Smart sensors also play a direct role in environmental protection by monitoring parameters that can affect surrounding ecosystems. Water-quality sensors installed near CO_2 injection sites, along pipelines, and around surface facilities monitor groundwater and surface-water chemistry to detect any potential contamination. Seismic monitoring systems track microseismic activity to verify safe CO_2 injection and storage behavior within the reservoir. Soil-gas sensors measure CO_2 concentrations at the ground surface to ensure containment integrity and provide early warning of any leakage into the soil. In addition, satellite imaging systems observe land use and vegetation health to confirm that CCS operations do not cause environmental damage beyond facility boundaries [53].

Figure 17.10 presents an integrated, multi-scale environmental monitoring system applied to a CCS site, combining near-surface, surface, airborne, and satellite sensing technologies. Soil-gas probes measure CO_2 concentrations directly at the subsurface to provide early detection of leakage, while microbial sampling systems assess changes in soil geochemistry and biological activity in both surface and subsurface levels. Water-monitoring sensors installed within shallow aquifers and surface-water bodies continuously track groundwater and stream chemistry to identify any potential contamination associated with

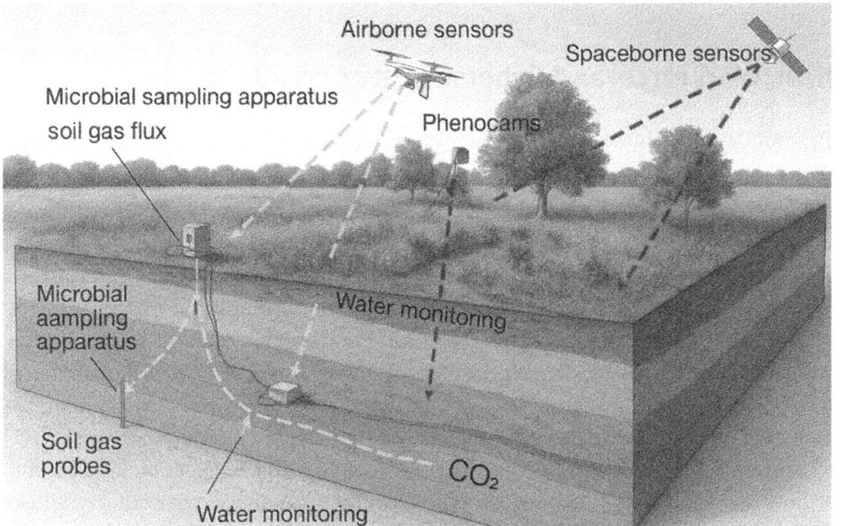

Figure 17.10 Integrated environmental monitoring of a CCS site using subsurface soil-gas probes, microbial and water-quality sensors, surface phenocams, airborne platforms, and satellite remote sensing to verify CO_2 storage integrity and protect surrounding ecosystems.

injection operations. At the surface, fixed phenocams observe vegetation dynamics and ecosystem health indicators. Broader spatial surveillance is provided by airborne and spaceborne (satellite) sensors, which capture landscape-scale data on land use, vegetation patterns, and surface anomalies. These sensors enable long-term verification that CCS activities remain environmentally secure beyond facility boundaries.

Automated safety platforms also support public transparency and regulatory compliance by providing continuous real-time data that can be used for regulatory reporting and public disclosure [16]. However, automation introduces new challenges that must be carefully managed. Because faults or software errors can spread quickly in automated systems, human supervision remains essential. Trained operators constantly monitor AI platforms and intervene whenever abnormal behavior is detected or when automated decisions conflict with engineering judgment [50].

Cybersecurity is another critical concern, as unauthorized access to digital networks or cyberattacks could compromise the functionality of automated safety systems and interfere with plant operations. Robust cybersecurity measures are therefore essential parts of smart safety architectures, including encrypted communication systems, secure access controls, isolated control networks, and continuous cybersecurity monitoring to prevent unauthorized entry and data manipulation [23, 27].

Workforce training must adapt alongside automation adoption. Operators and safety engineers need skills in digital platform management, robotics operation, and cybersecurity awareness [50]. In summary, automation and smart safety systems are transforming CCS operations into safer, more reliable, and environmentally secure establishments. By utilizing smart controls, robotics, AI-driven monitoring, and rapid-response to emergency situations, these systems minimize the risk of accidents, improve emergency preparedness, and protect nearby communities and ecosystems.

17.6 Ethical, Workforce, and Cybersecurity Considerations

As digitalization becomes deeply integrated into CCS operations, new ethical, social, and security challenges emerge alongside technological benefits. Deployment of digital technologies requires careful attention to workforce transitions, data protection, system transparency, and cybersecurity [54].

Digitalization significantly reshapes the CCS workforce. Traditional manual and routine inspection roles are increasingly replaced or supported by robotics and automated control systems [49]. While this shift improves safety by reducing human exposure to hazardous environments, it also transforms job descriptions for the workforce. As a result of digitalization, demand is growing for data analysts, automation engineers, cybersecurity specialists, drone operators, AI system supervisors, and digital maintenance technicians [55].

This transformation presents both opportunities and challenges. Workers lacking digital competencies may face job displacement or employment insecurity, particularly those whose careers have focused on manual inspection and traditional plant operations. Consequently, a central ethical responsibility of both industry and government is to invest in workforce retraining and advanced skills development that support employees in transitioning to emerging digital roles [56]. Collaborative partnerships among industry, universities, and professional organizations, such as the American Society of Mechanical Engineers (ASME), are essential for establishing flexible training pathways that integrate core process engineering expertise with data analytics, automation technologies, and cybersecurity awareness. Digital workforce development must also emphasize interdisciplinary competence. Future CCS professionals need not only technical expertise but also an understanding of environmental governance, ethical AI practices, and community and social engagement [57].

Digital CCS systems collect large amounts of data about plant operation, the environment, and workers. While this data is useful for safety and efficiency, it can create privacy concerns for employees. Many workers may not feel comfortable sharing personal information or being continuously monitored through tools such as biometric access systems, tracking software, or wearable safety devices. If these systems are not managed carefully, employees may feel watched rather than protected. For this reason, clear data rules are needed to explain what information is collected, why it is collected, who can see it, how long it is stored, and how it can be used. Strong data governance helps protect worker privacy while still allowing digital systems to improve safety and performance [58].

AI systems can also create transparency problems. AI programs may be used to help make decisions about plant safety, work schedules, or how tasks are assigned to employees. However, many advanced AI tools work like "black boxes," meaning people cannot easily understand how the system makes its decisions. Engineers and regulators may see the final results, but not the reasoning behind them [59]. This lack of understanding can reduce trust and make it difficult to check for mistakes or unfair decisions. Therefore, ethical management of AI requires explainable AI. In fact, AI systems can be designed to be understandable by humans through the use of explainable AI (XAI) methods. Instead of giving only a final answer or prediction, these systems also provide a clear explanation of why a decision was made [60]. Bias risk is another ethical issue. ML models learn from historical datasets that may reflect outdated operating practices or social inequities [61].

As CCS becomes more digitally interconnected, cybersecurity becomes a central safety challenge. Digital networks involved in automated capture plants and pipeline monitoring systems rely on continuous internet or network connectivity. Cyberattacks targeting these systems could disrupt operations, compromise safety systems, falsify monitoring data, and damage public confidence. Cybersecurity threats to digital CCS systems include cyberattacks on control rooms, manipulation of sensor data, stoppage of emergency shutdown systems, and theft of sensitive industrial information. Even small cyber intrusions could disrupt CO_2 pipeline monitoring or prevent detection of equipment failures. To counter these risks, CCS operators must implement robust cybersecurity architectures. Core components include encrypted communication networks, multifactor authentication systems, and isolated control networks that prevent external intrusions [23, 27]. Cyber resilience also requires trained personnel. Engineers and safety officers should be educated on secure system use and tricking people into giving access or information.

In conclusion, digital technologies bring major benefits to CCS performance and safety, but they also create important ethical responsibilities. These include supporting workers as job roles change, protecting personal and operational data, and maintaining strong cybersecurity. Responsible digitalization requires ongoing training programs, AI systems that people can understand and trust, clear privacy protections, and reliable cyber defenses.

17.7 Key Highlights (Chapter Summary)

◇ **Chapter 17–Key Highlights:**

1) Digitalization enables real-time monitoring and optimization of CCS facilities through extensive sensor networks, data storage and analysis platforms, and automated control systems that improve safety, reliability, and CCS infrastructure performance.
2) Machine learning transforms CCS operations by identifying complex patterns in real-time data and comparing them with historical records to determine whether the system is operating normally or deviating from safe conditions that could pose safety risks.
3) Digital twins create virtual replicas of CCS systems that allow operators to safely test and simulate different operating scenarios to understand possible outcomes without physically affecting real facilities.
4) Predictive maintenance is an outcome of machine learning that enables maintenance activities to be scheduled based on actual equipment condition and failure prediction rather than fixed inspection intervals.
5) Advanced leak detection technologies improve safety and environmental protection by combining fiber-optic sensors, satellite monitoring, and drones with AI analytics to rapidly detect faults in pipelines and carbon dioxide storage systems.
6) Drones and robotic systems minimize the need for human inspections, improving worker safety by reducing exposure to hazardous environments, particularly during high-risk and confined-space maintenance tasks.
7) Smart safety systems enable real-time hazard detection, rapid emergency response, and automated shutdown protocols to prevent escalation of unsafe conditions.
8) Workforce transformation requires extensive reskilling and education, shifting employee roles toward digital operations, data analysis, robotics supervision, and cybersecurity.
9) Ethical governance of AI and digital platforms is essential to ensure secure employment and workforce development through education, prevent misuse of workers' private data, and promote transparency in AI-based algorithms and their decision-making processes.
10) Cybersecurity protocols are fundamental to CCS safety, protecting digitally interconnected infrastructure from data manipulation and cyberattacks.

Keywords: Digitalization, Artificial Intelligence, Machine Learning, Digital Twins, Industry 4.0, Predictive Analytics, Process Optimization, Sensor Networks, Big Data, Automation, Robotics, Smart Safety Systems, Leak Detection, Predictive Maintenance, Autonomous Inspection, Pipeline Monitoring, Geological Monitoring, Real-Time Control, Energy Efficiency, Data Privacy, Ethical AI, Workforce Reskilling, Cybersecurity, Human–AI Collaboration, Emergency Response Systems, Systems Integration, Data Governance, Environmental Monitoring, Operational Reliability, Sustainable CCS

References

1 R. Singh *et al.*, "Highway 4.0: Digitalization of highways for vulnerable road safety development with intelligent IoT sensors and machine learning," *Safety Science*, vol. 143, 105407, Nov., 2021. DOI: 10.1016/j.ssci.2021.105407.

2 Q. Xu, M. Zhong, and X. Li, "How does digitalization affect energy? International evidence," *Energy Economics*, vol. 107, 105879, Mar., 2022. DOI: 10.1016/j.eneco.2022.105879.

3 T. T. Thanh, L. T. Ha, H. P. Dung, and T. T. Huong, "Impacts of digitalization on energy security: Evidence from European countries," *Environment, Development and Sustainability*, vol. 25, pp. 11599–11644, 2023. DOI: 10.1007/s10668-022-02545-7.

4 U. Ahrend, M. Aleksy, M. Berning, J. Gebhardt, F. Mendoza, and D. Schulz, "Sensors as the basis for digitalization: New approaches in instrumentation, IoT-concepts, and 5G," *Internet of Things*, vol. 15, 100406, Sep., 2021. DOI: 10.1016/j.iot.2021.100406.

5 A. Jaramillo-Alcazar, J. Govea, and W. Villegas-Ch, "Anomaly detection in a smart industrial machinery plant using IoT and machine learning," *Sensors*, vol. 23, no. 19, 8286, 2023. DOI: 10.3390/s23198286.

6 A. Nechibvute and H. D. Mafukidze, "Integration of SCADA and industrial IoT: Opportunities and challenges," *IETE Technical Review*, vol. 41, no. 3, pp. 312–325, 2024. DOI: 10.1080/02564602.2023.2246426.

7 Y. Peng and I.-C. Wu, "A cloud-based monitoring system for performance analysis in IoT industry," *The Journal of Supercomputing*, vol. 77, pp. 9266–9289, 2021. DOI: 10.1007/s11227-021-03640-8.

8 G. Lazaroiu, A. Androniceanu, I. Grecu, G. Grecu, and O. Neguriţă, "Artificial intelligence-based decision-making algorithms, internet of things sensing networks, and sustainable cyber-physical management systems in big data-driven cognitive manufacturing," *Oeconomia Copernicana*, vol. 13, no. 4, pp. 1047–1080, 2022.

9 B. Brumen, A. Černezel, and L. Bošnjak, "Overview of machine learning process modeling," *Entropy*, vol. 23, no. 9 , 2021. DOI: 10.3390/e23091123.1123

10 M. Jabłoński and A. Jabłoński, "Shaping the safety culture of high reliability organizations through digital transformation," *Energies*, vol. 14, no. 16, 4721, 2021. DOI: 10.3390/en14164721.

11 K. Sotoodeh, *The Comprehensive Valve Dictionary: Terminology, Applications, and Insights for Oil, Gas, and Renewable Energy*, 1st ed.. Boca Raton, FL, USA: CRC Press (Taylor & Francis), 2025, ISBN: 9781394294916.

12 R. van Dinter, B. Tekinerdogan, and Ç. Catal, "Predictive maintenance using digital twins: A systematic literature review," *Information and Software Technology*, vol. 151, 107008, Nov., 2022. DOI: 10.1016/j.infsof.2022.107008.

13 K. Sotoodeh, "Material failure and corrosion behavior of CO_2–H_2 hybrid pipelines in high-pressure transport: Mechanisms, risks, and mitigation," *Journal of Failure Analysis and Prevention, Springer*, 2025. DOI: 10.1007/s11668-025-02344-y.

14 A. Chawla *et al.*, "IoT-based monitoring in carbon capture and storage systems," *IEEE Internet of Things Magazine*, vol. 5, no. 4, pp. 106–111, 2023.

15 V. N. Vlasenko and A. S. Shirobokov, "Digitalization of state environmental management: Legal aspects," *RUDN Journal of Law*, vol. 25, no. 2, pp. 601–619, 2021.

16 V. Heimburg and M. Wiesche, “Digital platform regulation: Opportunities for information systems research,” *Internet Research*, vol. 33, no. 7, pp. 72–85, 2023. DOI: 10.1108/INTR-05-2022-0321.

17 H. Pasman, H. Sun, M. Yang, and F. Khan, “Chapter one—opportunities and threats to process safety in digitalized process systems—an overview,” *Methods in Chemical Process Safety*, vol. 6, pp. 1–23, 2022. DOI: 10.1016/bs.mcps.2022.05.007.

18 D. Moreno-Jacobo, G. Toledo-Nin, A. Ochoa-Zezzatti, V. Torres, and F. Estrada-Otero, “Evaluation of drones for inspection and control in Industry 4.0,” in *Technological and Industrial Applications Associated with Intelligent Logistics*, A. Ochoa-Zezzatti, D. Oliva, and A. J. Perez, Eds, Lecture Notes in Intelligent Transportation and Infrastructure. Cham, Switzerland: Springer, 2021, pp. 337–347, DOI: 10.1007/978-3-030-68655-0_29.

19 A. Chandratreya, “Digital transformation and the future workforce,” in *Transforming the Service Sector With New Technology*, A. Nayyar, M. Naved, and R. Rameshwar, Eds. IGI Global Scientific Publishing, 2025, pp. 19–40.

20 R. Mustofa, M. Rafiquzzaman, and N. U. I. Hossain, “Analyzing the impact of cyber-attacks on the performance of digital twin-based industrial organizations,” *Journal of Industrial Information Integration*, vol. 41, 100633, 2024. DOI: 10.1016/j.jii.2024.100633.

21 H. Zhou and J. Liu, “Digitalization of the economy and resource efficiency for meeting the ESG goals,” *Resources Policy*, vol. 86, Part A, 104199, 2023. DOI: 10.1016/j.resourpol.2023.104199.

22 J. K. Sahith and B. Lal, “Leveraging machine learning and artificial intelligence for enhanced carbon capture and storage (CCS),” in *Gas Hydrate in Carbon Capture, Transportation and Storage*, A. K. H. Thomas and N. Kalarikkal, Eds, 2024, pp. 159–196.

23 I. H. Sarker, “Machine learning for intelligent data analysis and automation in cybersecurity: Current and future prospects,” *Annals of Data Science*, vol. 10, pp. 1473–1498, 2023. DOI: 10.1007/s40745-022-00444-2.

24 E. F. Morales and H. Jair Escalante, “A brief introduction to supervised, unsupervised, and reinforcement learning,” in *Biosignal Processing and Classification Using Computational Learning and Intelligence: Principles, Algorithms, and Applications*, R. S. Alvarado and J. I. Sierra, Eds. Academic Press, 2022, pp. 111–129, DOI: 10.1016/B978-0-12-820125-1.00017-8.

25 S. Morales, R. Clarisó, and J. Cabot, “Towards a DSL for AI engineering process modeling,” in *Product-Focused Software Process Improvement (PROFES 2022)*, D. Taibi, M. Kuhrmann, T. Mikkonen, J. Klünder, and P. Abrahamsson, Eds, Lecture Notes in Computer Science, Vol. 13709. Cham, Switzerland: Springer, 2022, DOI: 10.1007/978-3-031-21388-5_4.

26 F. H. Alshammari, “Trends in intelligent and AI-based software engineering processes: A deep learning-based software process model recommendation method,” *Computational Intelligence and Neuroscience*, 1960684, 2022. DOI: 10.1155/2022/1960684.

27 R. Ilieva and G. Stoilova, “Challenges of AI-driven cybersecurity,” in Proc. 2024 XXXIII International Scientific Conference Electronics (ET), Sep. 2024, pp. 1–4.

28 M. Bortz *et al.*, “AI in process industries—current status and future prospects,” *Chemie Ingenieur Technik*, vol. 95, no. 7, pp. 975–988, 2023. DOI: 10.1002/cite.2022020247.

29 P. Mathiyalagan and S. Pokhriyal, “Improvements in industrial sensors: Increasing reliability and efficiency,” in Proc. 3rd Int. Conf. on Optimization Techniques in the Field of Engineering (ICOFE 2024), 2025, pp. 1–18.

30 D. Mishra and S. Kaushal, "Building public trust in AI-enhanced energy systems through strategic communication," in *Leveraging AI for Innovative Sustainable Energy: Solar, Wind and Green Hydrogen*, A. Nayyar, M. Naved, and R. Rameshwar, Eds. IGI Global Scientific Publishing, 2025. ch007, DOI: 10.4018/979-8-3373-0045-0.ch007.

31 A. Rubio-Rico, F. Mengod-Bautista, A. Lluna-Arriaga, B. Arroyo-Torres, and V. Fuster-Roig, "The industrial digital energy twin as a tool for the comprehensive optimization of industrial processes," *Processes*, vol. 11, no. 8, 2353, 2023. DOI: 10.3390/pr11082353.

32 D. D. Enyewu, M. A. M. Capretz, and G. T. Bitsumlák, "Toward smart-building digital twins: BIM and IoT data integration," *IEEE Access*, vol. 10, pp. 97–110, 2022.

33 A. Kantaros, T. Ganetsos, E. Pallis, and M. Papoutsidakis, "From mathematical modeling and simulation to digital twins: Bridging theory and digital realities in industry and emerging technologies," *Applied Sciences*, vol. 15, no. 16, 9213, 2025. DOI: 10.3390/app15169213.

34 F. Pires, B. Ahmad, A. P. Moreira, and P. Leitão, "Digital twin based what-if simulation for energy management," in Proc. 2021 4th IEEE International Conference on Industrial Cyber-Physical Systems (ICPS), 2021, pp. 1–6.

35 K. Sotoodeh, "Comparing dual plate and swing check valves and the importance of minimum flow for dual plate check valves," *American Journal of Industrial Engineering*, vol. 5, no. 1, pp. 31–45, 2018. DOI: 10.12691/ajie-5-1-5.

36 K. Sotoodeh, "Erosion prevention for piping and valves in offshore oil and gas industry: A case study," *Journal of Failure Analysis and Prevention, Springer*, 2023. DOI: 10.1007/s11668-023-01620-z.

37 K. Sotoodeh, *Equipment and Components in the Oil and Gas Industry, Volume 2: Components*, 1st ed.. Boca Raton, FL, USA: CRC Press (Taylor & Francis), 2024. ISBN: 978-1032731476

38 M. Singh *et al.*, "Applications of digital twin across industries: A review," *Applied Sciences*, vol. 12, no. 11, 5727, 2022. DOI: 10.3390/app12115727.

39 A. I. Khalyasmaa, A. I. Stepanova, S. A. Eroshenko, and P. V. Matrenin, "Review of the digital twin technology applications for electrical equipment lifecycle management," *Mathematics*, vol. 11, no. 6, 1315, Mar., 2023. DOI: 10.3390/math11061315.

40 H. Chen *et al.*, "Digital twin based CO_2 pipeline transportation: A review," in *Proc. 14th Int. Conf. on Computer Engineering and Networks (CENet 2024)*, G. Yin, X. Liu, J. Su, and Y. Yang, Eds, Lecture Notes in Electrical Engineering, Vol. 1388. Singapore: Springer, 2025, pp. 93–104, DOI: 10.1007/978-981-96-4229-8_10.

41 M. Fahim, V. Sharma, T.-V. Cao, B. Canberk, and T. Q. Duong, "Machine learning-based digital twin for predictive modeling in wind turbines," *IEEE Access*, vol. 10, pp. 151–162, 2022.

42 R. Hashmi, H. Liu, and A. Yavari, "Digital twins for enhancing efficiency and assuring safety in renewable energy systems: A systematic literature review," *Energies*, vol. 17, no. 11, 2456, 2024. DOI: 10.3390/en17112456.

43 P. Bielka, "The future of carbon capture and storage technology: Innovative approach with digital twin," *Journal of Geotechnology and Energy*, vol. 40, no. 2, 2023. DOI: 10.7494/jge.2023.40.2.5413.

44 D. Holmes, M. Papathanasaki, L. Maglaras, M. A. Ferrag, S. Nepal, and H. Janicke, "Digital twins and cyber security—solution or challenge?," in Proc. 2021 6th South-East Europe Design Automation, Computer Engineering, Computer Networks and Social Media Conference (SEEDA-CECNSM), 2021, pp. 1–6.

45 K. Sotoodeh, *Corrosion Challenges and Material Selection in Low-Carbon Energy Systems: Hydrogen and CO2 Environment*, 1st ed.. Amsterdam, Netherlands: Elsevier, 2026. ISBN: 978-0443485350.

46 K. Sotoodeh, *Piping, Valves, and Process Engineering in Carbon Capture Systems: Applications, Sustainability, and Environmental Protection*, 1st ed.. Amsterdam, Netherlands: Elsevier, 2026. ISBN: 978-0443510786

47 D. He, J. Zhao, Z. Jin, C. Huang, F. Zhang, and J. Wu, "Prediction of bearing remaining useful life based on a two-stage updated digital twin," *Advanced Engineering Informatics*, vol. 65, Part A, 103123, 2025. DOI: 10.1016/j.aei.2025.103123.

48 D. B. Steffelbauer, J. Deuerlein, D. Gilbert, E. Abraham, and O. Piller, "Pressure-leak duality for leak detection and localization in water distribution systems," *Journal of Water Resources Planning and Management*, vol. 148, no. 3, 2022. DOI: 10.1061/(ASCE) WR.1943-5452.0001515.

49 A. Verma, A. Kidawat, N. D. Dubey, F. Naseer, and S. Pradhan, "A review on various types of in-pipe inspection robot," *Materials Today: Proceedings*, vol. 50, pt. 5, pp. 1425–1434, 2022, 10.1016/j.matpr.2021.08.335.

50 L. S. Treiman, C.-J. Ho, and W. Kool, "The consequences of AI training on human decision-making," *Proceedings of the National Academy of Sciences*, vol. 121, no. 33, e2408731121, Aug., 2024. DOI: 10.1073/pnas.2408731121.

51 D. Mathew, N. C. Brintha, and J. T. W. Jappes, "Artificial intelligence powered automation for industry 4.0," in *New Horizons for Industry 4.0 in Modern Business*, A. Nayyar, M. Naved, and R. Rameshwar, Eds, Contributions to Environmental Sciences & Innovative Business Technology. Cham, Switzerland: Springer, 2023, pp. 1–14, DOI: 10.1007/978-3-031-20443-2_1.

52 K. Sotoodeh, *Safety Engineering in the Oil and Gas Industry*, 1st ed.. Boca Raton, FL, USA: CRC Press (Taylor & Francis), 2023. ISBN: 978-1032479736

53 V. Garrido-Momparler and M. Peris, "Smart sensors in environmental/water quality monitoring using IoT and cloud services," *Trends in Environmental Analytical Chemistry*, vol. 35, e00173, 2022. DOI: 10.1016/j.teac.2022.e00173.

54 P. P. Lang, "Digitalization of the society: Legal and ethical features," in *Digital Technologies in the New Socio-Economic Reality*, S. I. Ashmarina and V. V. Mantulenko, Eds, Lecture Notes in Networks and Systems, Vol. 304. Cham, Switzerland: Springer, 2022, pp. 681–690, DOI: 10.1007/978-3-030-83175-2_83.

55 S. Wang *et al.*, "Digitalization solution for carbon capture, transportation, utilization, and storage," *Transportation, Utilization, and Storage*, Dec. 24, 2024.

56 O. Krutova, P. Koistinen, T. Turja, H. Melin, and T. Särkikoski, "Two sides, but not of the same coin: Digitalization, productivity and unemployment," *International Journal of Productivity and Performance Management*, vol. 71, no. 8, pp. 3507–3533, Nov., 2022. DOI: 10.1108/IJPPM-05-2020-0233.

57 A. Zauskova, A. Kusa, M. Kubovics, S. Seckova, and M. Urminova, "Current state and prediction of the future of digitization as a part of Industry 4.0," *Serbian Journal of Management*, vol. 17, no. 1, Apr., 2022. DOI: 10.5937/sjm17-36468.

58 A. V. Serova and O. V. Shcherbakova, "The employee's right to privacy transformation: Digitalization challenges," *Kutafin Law Review*, vol. 9, no. 3, pp. 437–465, 2022.

59 G. Reischauer, T. Hess, T. Sellhorn, *et al*., "Transparency in an age of digitalization and responsibility," *Schmalenbach Journal of Business Research*, vol. 76, pp. 483–494, 2024. DOI: 10.1007/s41471-024-00203-4.

60 R. Mosleh *et al*., "Explainable artificial intelligence (XAI) techniques for energy and power systems: Review, challenges and opportunities," *Energy and AI*, vol. 9, p. 100169, Aug., 2022. DOI: 10.1016/j.egyai.2022.100169.

61 M. G. Hanna *et al*., "Ethical and bias considerations in artificial intelligence/machine learning," *Modern Pathology*, vol. 38, no. 3, p. 100686, Mar., 2025. DOI: 10.1016/j.modpat.2024.100686.

18

Nature-Based Carbon Capture and Ecological Solutions

18.1 Introduction to Nature-Based Carbon Removal

18.1.1 Fundamentals and Benefits of Nature-Based Carbon Removal

Nature-based carbon removal refers to a group of climate mitigation approaches that use natural ecosystems and biological processes to remove carbon dioxide (CO_2) from the atmosphere and store it in vegetation, soils, sediments, and marine environments. These methods rely on the Earth's natural carbon cycle, where plants absorb CO_2 during photosynthesis and transfer carbon into biomass and organic matter. Over time, portions of this carbon become stored in long-lived ecosystems such as forest soils, coastal sediments, and deep ocean systems [1]. Nature-based removal strategies are increasingly recognized as important complements to engineered carbon capture and storage (CCS) technologies.

Nature-based solutions differ from industrial CCS methods, which capture carbon from exhaust gases or directly from the air and store it in geological formations. Instead, nature-based solutions focus on restoring, protecting, or enhancing ecosystems that already act as carbon sinks [2]. A carbon sink is any natural or artificial system that absorbs more carbon dioxide from the atmosphere than it releases. These activities include afforestation and reforestation, the restoration of wetlands and mangroves, sustainable land and agricultural management practices to increase soil carbon, and the conservation of seagrass meadows and salt marshes [3, 4]. Seagrass meadows are permanently submerged marine plant ecosystems, while salt marshes are intertidal coastal wetlands that are intermittently flooded, both functioning as highly efficient blue-carbon carbon sequestration systems through sediment carbon burial [5]. Additionally, emerging marine-based approaches, such as ocean alkalinity enhancement (OAE), aim to bolster natural chemical processes that absorb and store atmospheric CO_2 [6].

Nature-based carbon removal offers benefits that extend beyond mitigating climate change. Healthy forests play a crucial role in managing water flow, reducing soil erosion, and protecting wildlife [7]. Coastal wetlands and mangroves serve as essential habitats for marine life. Additionally, increasing soil carbon enhances soil quality, boosts food production, and helps farmland become more resilient to drought [8]. These added advantages make nature-based solutions particularly valuable, as they combat climate change while simultaneously supporting ecosystems and communities.

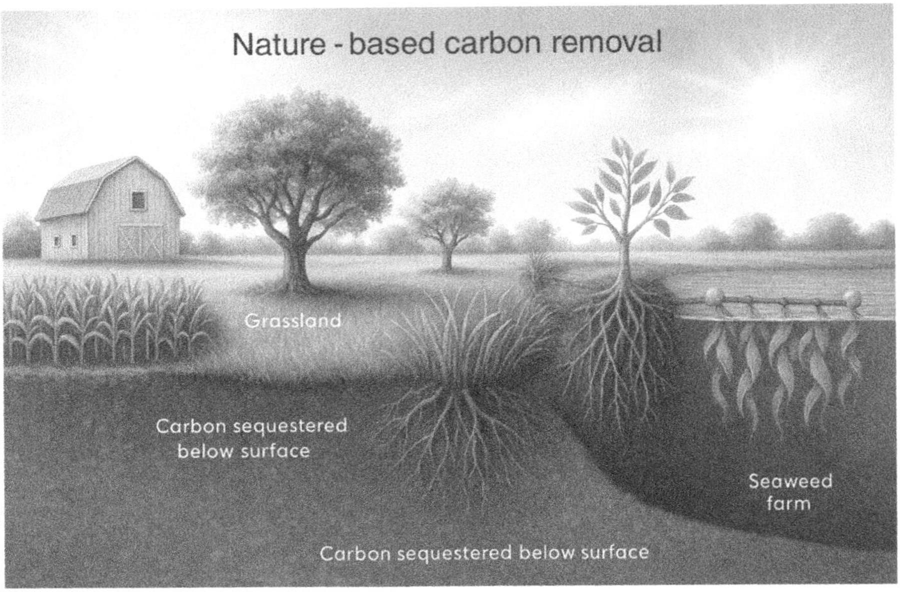

Figure 18.1 Nature-based carbon capture and removal.

Figure 18.1 illustrates the concept of nature-based carbon capture and removal, where natural ecosystems absorb and store atmospheric carbon dioxide through biological processes. Plants and trees capture CO_2 and convert it into chemical energy and organic matter through photosynthesis. Carbon is stored in their biomass (leaves, trunks, and roots) as well as in soils below the surface. Grasslands and agricultural systems increase soil carbon, improving plant growth and productivity, while coastal ecosystems such as mangroves and seaweed farms capture and retain carbon in underwater vegetation and sediments. Sunlight is shown as the primary energy source driving photosynthesis and enabling these natural carbon uptake processes.

18.1.2 Limitations, Measurement Challenges, Social and Policy

Although nature-based carbon removal has many benefits, it also has significant limitations. Carbon stored in plants and soils is not permanent. Forest fires, pests, droughts, land clearing for human use, such as constructing pipelines, or poor land management, such as cutting trees in forests, can quickly release stored carbon back into the atmosphere. Soil carbon can decrease if sustainable farming practices are stopped, and coastal ecosystems can be damaged by pollution or construction [2]. Because of these risks, carbon stored in living ecosystems is less secure than carbon captured and stored deep underground using engineered CCS. For this reason, nature-based solutions cannot replace CCS for long-term climate protection. Instead, they should be used together as part of a balanced carbon management strategy [9].

Measurement and verification present additional challenges. Unlike industrial CCS, where captured and injected CO_2 volumes can be measured precisely, ecosystem carbon

Figure 18.2 Integrated field and remote-sensing monitoring of ecosystem carbon sequestration using tree biomass measurements and drone observations to quantify carbon storage in the tree.

captures are more difficult to quantify. However, scientists use satellite imagery, field sampling, modeling, and remote sensing technologies to estimate carbon uptake and long-term storage. The amount of capture depends on the dynamic factors such as species growth rates, climate conditions, and soil composition [10]. Figure 18.2 shows an integrated ecosystem carbon monitoring approach combining field measurements and remote sensing technologies. Ground surveys of tree biomass are supported by drone-based aerial observations and satellite imagery to assess vegetation growth, health, and disturbances.

The social and ethical aspects of carbon removal are also important. Many projects take place on land used by Indigenous peoples, farmers, and coastal communities. It is essential to protect land rights, share benefits fairly, and involve local people in decision-making. Protecting land rights ensures that local people keep legal control and ownership of their land and are not forced to relocate. Relocation can occur when large land areas are claimed for carbon projects. Sharing benefits fairly means that communities receive direct advantages from projects, such as income, jobs, training, or infrastructure, rather than profits going only to companies running natural carbon capture projects. Involving local people in decision-making means that communities actively participate in planning, approving, and managing projects that affect their land and livelihoods [11].

From a policy perspective, nature-based carbon removal is now part of many national climate plans and international agreements. International mechanisms support funding for ecosystem restoration while also setting rules to protect the environment and local communities [11]. Another critical issue is scale. The amount of carbon that can be removed by natural processes is limited. Using land for carbon capture may compete with other important land uses. Large areas of land are needed to grow food for people, provide homes for communities, and protect wildlife habitats [12]. If land is converted to tree plantations or bioenergy crops for carbon sequestration, it may reduce space for agriculture or replace natural ecosystems that support wildlife.

In summary, nature-based carbon removal is an important part of climate solutions. Restoring forests and managing land responsibly, meaning using farming, forestry, and livestock management practices, are essential for nature-based carbon removal. These practices protect soil health, prevent deforestation, maintain natural ecosystems, and protect marine environments. Nature can capture carbon from the air while also supporting wildlife, food production, clean water, and local communities. However, because this carbon storage is not always permanent and can be difficult to measure accurately, nature-based methods cannot replace engineered CCS. Instead, both approaches should be used together to achieve strong, fair, and long-lasting climate protection.

18.2 Forest-Based Carbon Capture Systems

18.2.1 Principles and Benefits of Forest-Based Carbon Capture

Forests play one of the most important roles in natural carbon capture, as shown in Figure 18.3. Through photosynthesis, trees absorb carbon dioxide (CO_2) from the atmosphere and convert it into biomass, storing carbon in trunks, branches, leaves, roots, and

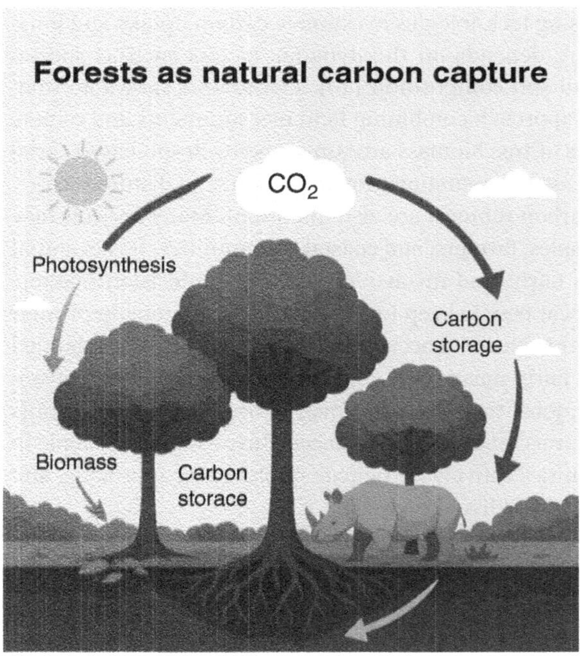

Figure 18.3 Forests as natural carbon capture systems. Trees absorb CO_2 from the atmosphere through photosynthesis, storing it as biomass and in the soil, making forests vital for climate regulation and ecosystem balance.

soils. Globally, forests already act as major carbon sinks, removing billions of tons of CO_2 each year. Forest-based carbon capture systems aim to strengthen this natural function by expanding forests and protecting existing forests to maximize CCS while supporting ecosystem health [13].

The most common forest-based strategies are afforestation and reforestation. Afforestation refers to establishing forests on lands that historically did not exist, while reforestation restores tree cover on lands where forests were previously removed by logging or agriculture. Both approaches promote carbon sequestration through tree growth [14]. Well-designed programs plant trees that match the local climate and ecosystem. These trees support wildlife and plant diversity by providing food, shelter, and breeding spaces for birds, insects, and animals. Their leaves, fruits, and seeds feed many species, while branches and trunks offer nesting and hiding places. In addition, trees improve water flow by helping rain slowly soak into the soil instead of quickly running off the surface, stabilize soil by holding it in place with their roots, and reduce erosion by protecting the ground from wind and heavy rainfall that would otherwise wash or blow soil away [15].

Forest projects must be planned carefully. Planting only one fast-growing tree species to store carbon can increase carbon uptake quickly, but it can also damage ecosystem health. These single-species forests are more vulnerable to pests, diseases, droughts, and wildfires, which can destroy trees and release stored carbon back into the atmosphere. In contrast, forests with many different tree species are more resilient and stable [16].

Another important forest-based approach is forest conservation. Protecting existing forests often prevents more carbon loss than planting new trees, because mature forests already store large amounts of carbon in their trees and soil. When forests are cleared or damaged, this stored carbon is quickly released into the atmosphere [17]. Therefore, stopping deforestation is one of the most effective ways to reduce climate change. Actions such as controlling illegal logging, protecting land rights, using sustainably sourced products, and enforcing environmental laws are essential for forest conservation [18]. Protecting land rights ensures that local and Indigenous communities legally own and manage their lands, which helps prevent outside groups from clearing forests without consent [19]. Using sustainably sourced products reduces demand for timber taken from destroyed forests and supports responsible forestry practices. Enforcing environmental laws ensures that regulations designed to protect forests are followed and that violations are punished.

18.2.2 Challenges, Monitoring, and Governance

Monitoring and verification are important challenges in forest-based carbon capture. Measuring stored carbon requires field surveys of trees, soil testing, and the use of satellite images and aerial scanning. However, uncertainty remains due to variations in tree species growth rates, climate influences, fire risks, and land-use changes [20]. Carbon performance in nature-based carbon capture refers to how well a natural system removes and stores carbon over time. It is a measure of the amount, stability, and durability of carbon stored in ecosystems. Carbon permanence is a major challenge for forest-based carbon removal. Events such as wildfires, droughts, storms, pests, or land clearing can quickly release carbon that has been stored for many years. Climate change increases these risks by making forests more vulnerable to damage [21].

Community involvement is very important for successful forest carbon projects. Many forests are managed by Indigenous peoples and small farmers, so protecting land rights and respecting local traditions and involving them in decision-making helps ensure forests are cared for by respecting their owners and managers. Economic considerations supporting forest-based carbon projects include government support and encouragement programs, private investment partnerships in forestry, and carbon markets. Payments for ecosystem services reward landowners for maintaining forest cover and ecosystem health [22].

When designed responsibly, forest-based carbon capture systems provide many additional benefits. Restoring forests helps wildlife recover, protects water sources, and makes landscapes more resilient to extreme weather. Tree-planting programs can reduce flooding and improve soil quality [22, 23]. Ecotourism and forest-management jobs also provide new income opportunities for rural communities. These wide-ranging benefits are summarized in Figure 18.4, highlighting the critical ecosystem services provided by healthy forests.

Although forests are important for climate action, they cannot solve the problem alone. There is limited land for planting new forests, and much land is needed to grow food. In addition, carbon stored in trees is not permanent because it can be released by fires or forest loss [21]. For these reasons, forest-based solutions must be used together with industrial emission reductions and technological CCS systems.

In summary, forest-based carbon capture is one of the most effective and practical natural climate solutions. By planting new forests, restoring damaged forests, and protecting

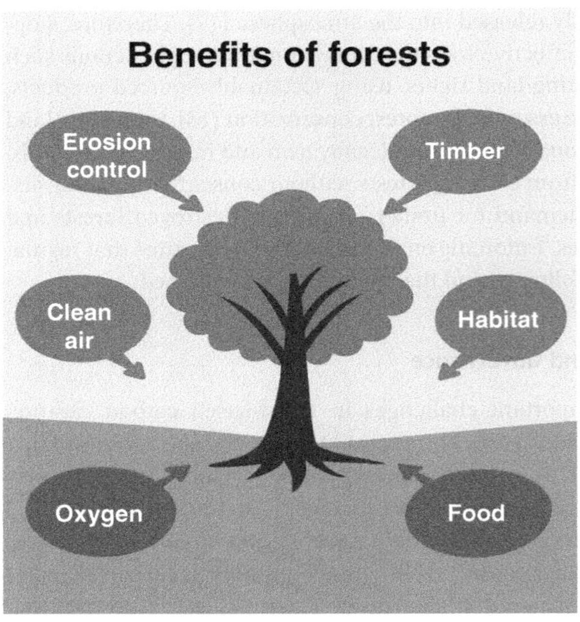

Figure 18.4 Key benefits of forests. Forests contribute to clean air, oxygen production, erosion control, biodiversity habitat, timber resources, and food security, demonstrating their value beyond carbon capture.

existing forests, forests can remove large amounts of CO_2 [13]. However, these projects must be well planned, carefully monitored, and governed transparently to protect both the environment and local people. When combined with engineered carbon capture to reduce emissions, forests become a strong and reliable part of climate change solutions [20, 22].

18.3 Wetlands, Mangroves, and Blue Carbon Ecosystems

Wetlands, mangroves, seagrass meadows, and tidal salt marshes, often referred to as blue carbon ecosystems, are among the most powerful natural carbon sinks on Earth. These coastal and marine ecosystems store carbon not only in plant biomass but also in deep water sediments that can retain organic carbon for centuries or even millennia. Although blue carbon ecosystems occupy a relatively small percentage of global land area, they capture carbon at rates significantly higher than most terrestrial forests, making them critical parts of nature-based climate mitigation strategies [24].

Mangrove forests grow along tropical and subtropical coastlines where seawater meets land. Their thick roots slow down waves and water currents. Because the water moves more slowly, mud, sand, and dead leaves or plants (called organic matter) settle and stay around the roots instead of being washed away. This organic matter builds up in the soil and slowly turns into carbon-rich sediment. Tidal marshes and freshwater wetlands work in a similar way. Plants grow and die, and their remains collect in wet, low-oxygen soil where decomposition happens very slowly. This allows a large amount of carbon to stay locked in the soil for long periods. Seagrass meadows grow underwater. They take in carbon dioxide (CO_2) directly from the water during photosynthesis. Like mangroves and wetlands, they also trap and store carbon in the sediments beneath them, helping reduce the amount of carbon in coastal waters [25]. Figure 18.5 illustrates blue carbon ecosystems, including mangrove forests, tidal wetlands, and seagrass meadows, that capture and store significant amounts of carbon in their biomass and marine sediments.

Figure 18.5 Blue carbon ecosystems, including mangrove forests, tidal wetlands, and seagrass meadows.

In addition to helping the climate, blue carbon ecosystems also provide many other important benefits for nature and people. Wetlands act like natural water filters. They trap pollutants and extra nutrients, which helps keep water cleaner [26]. Mangroves and salt marshes protect coastlines from storms and waves. They reduce erosion, slow down storm surge, and help keep shorelines stable. These ecosystems are essential nursery habitats for fish populations and contribute to food security for millions of coastal communities. Their restoration strengthens biodiversity while supporting fisheries and tourism economies [24].

Blue carbon ecosystems, such as mangroves, wetlands, and seagrass meadows, are some of the most threatened environments in the world. Many have been damaged or destroyed because of coastal cities' growth, fish farms, farming activities, and construction projects. When wetlands or mangroves are drained or cleared, often due to agriculture, construction, or development, the water that typically saturates the soil is lost, and the plants die. Water loss allows air and oxygen to penetrate the soil, leading to rapid decomposition of dead plants and the organic matter they store. As a result, carbon dioxide (CO_2) and methane are released into the atmosphere. Consequently, these areas cease to act as carbon sinks and instead become sources of greenhouse gas emissions that contribute to climate change [24, 27].

Restoring wetlands and mangroves is one of the most effective and affordable ways to reduce climate change. Restoration means helping damaged ecosystems return to their natural state so they can store carbon again. This work includes reopening blocked water channels to restore natural water flow, replanting mangroves where trees were lost, stabilizing loose soil and sediments using vegetation along the edges to prevent erosion, and removing plant species that stop native plants from growing [24, 27].

Monitoring and measurement are crucial for ensuring credible blue carbon accounting. Carbon stocks are assessed through soil sampling, biomass measurements, satellite mapping, and hydrological modeling. Recent advances in drone surveys and remote sensing enhance tracking of vegetation recovery and sediment dynamics [28].

Social inclusion and fair economic support are essential for successful wetland restoration. Coastal communities depend on wetlands for livelihoods and protection, so projects must involve local people, provide jobs, and respect traditional land rights. Community-run restoration programs have proven successful by protecting fisheries, supporting eco-tourism, and creating employment [29]. Financial support comes from governments such as climate funds and carbon credit programs to reward ecosystem protection. In addition, strong regulation and are needed to prevent "green grabbing," where local people lose land or benefits to outside investors [30].

Blue carbon approaches also face limitations. Ocean warming, sea-level rise, and pollution stress ecosystems and may reduce long-term sequestration effectiveness. Restoration capacity is constrained by coastal land availability and sometimes in conflict with urban development [31]. Importantly, blue carbon does not replace engineered CCS but complements it. Coastal ecosystems provide rapid carbon removal and immediate ecological recovery benefits, while geological CCS provides long-term, permanent carbon storage at large industrial scales. Integrating both solutions creates diversified climate mitigation portfolios. In conclusion, wetlands, mangroves, and blue carbon ecosystems are powerful natural allies in the fight against climate change.

18.4 Ocean Alkalinity and Marine-Based Approaches

Ocean-based carbon removal strategies focus on enhancing the natural ability of marine systems to absorb and store carbon dioxide (CO_2). The ocean already plays a major role in global climate conditions, absorbing nearly one-quarter of human-produced CO_2 emissions each year. However, rising atmospheric CO_2 has led to increased ocean acidity, threatening coral reefs, organisms, and marine food chain. Marine-based approaches aim both to strengthen ocean carbon uptake and to mitigate harmful acidification effects [32].

One of the most researched methods is OAE. This approach involves adding alkaline minerals, such as limestone or basalt into coastal waters or open-ocean settings. These minerals dissolve and increase seawater alkalinity, allowing additional atmospheric CO_2 to be chemically converted into stable dissolved bicarbonate and carbonate forms. This process stores carbon in the ocean for centuries or longer. Importantly, increasing alkalinity also counteracts ocean acidification, providing protective benefits for marine organisms [33].

Figure 18.6 shows an ocean carbon capture field experiment where a research vessel releases alkaline material into the water to enhance CO_2 absorption. Sensors, buoys, drones, and underwater vehicles monitor how the added alkalinity spreads and increases the ocean's ability to capture and store carbon. The orange swirls represent the mixing zone where CO_2 reacts with alkalinity to form more stable, long-term carbon storage.

Another marine strategy is enhancing phytoplankton growth. Phytoplankton are microscopic, photosynthetic organisms that form the foundation of the aquatic food web and produce a significant portion of the Earth's oxygen. Phytoplankton absorb CO_2 through photosynthesis and form the base of marine food chains. When these microscopic organisms die, a portion of their carbon sinks into deep ocean waters or sediments. Some proposals explore iron fertilization, in which trace amounts of iron nutrients stimulate plankton

Figure 18.6 Ocean carbon capture process involving alkalinity enhancement.

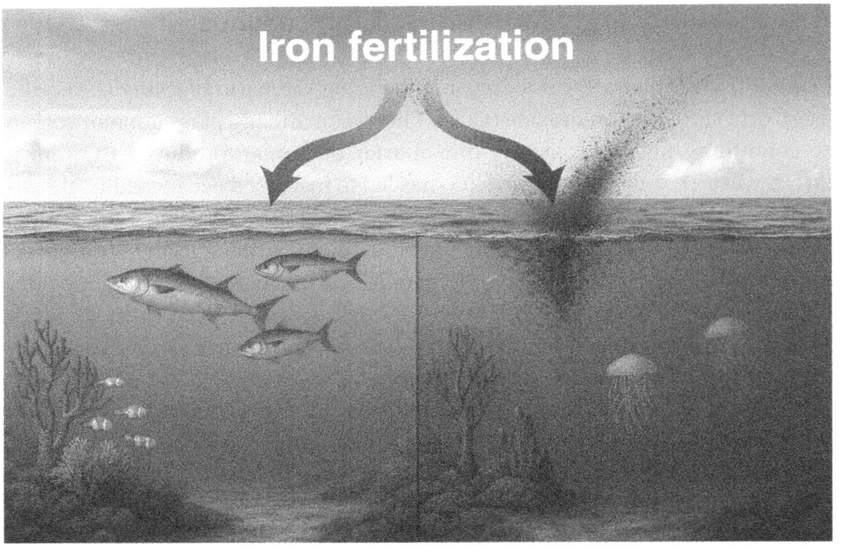

Figure 18.7 Illustration of ocean iron fertilization; adding iron to surface waters stimulates phytoplankton growth, potentially enhancing CO_2 uptake, but may also alter marine ecosystems and species balance.

growth [34]. However, ecological risks, including changes to plankton populations and oxygen depletion, have raised concerns, leading to strict regulatory limitations on ocean fertilization experiments.

Figure 18.7 illustrates the concept of ocean iron fertilization, a method where iron particles are added to surface waters to stimulate phytoplankton growth. The left side shows a healthy marine ecosystem with fish and corals, representing the intended positive effects of iron fertilization. The right side shows the iron being released into the water, leading to increased plankton activity but also potential ecological changes, such as shifts in species pattern changes, like jellyfish dominance. The split design highlights both the potential benefits and uncertainties of iron fertilization as a carbon-removal strategy.

Marine vegetation restoration is also essential. Seagrass meadows, salt marshes, mangroves, and kelp forests absorb carbon through biomass accumulation and sediment burial. These systems were discussed in the blue carbon section (previous section) [35]. Monitoring marine carbon removal presents measurement challenges. Ocean chemistry is complex, and accurately verifying carbon uptake requires combining satellite data, buoy-based sensors, water sampling, and oceanographic modeling. Researchers must track changes in alkalinity, dissolved carbon concentrations, sediment burial rates, and biological productivity [36].

Potential risks must be carefully managed. Improper mineral dosing injection into the water could harm marine organisms or disrupt localized ecosystems. Extensive plankton growth could alter food webs and oxygen levels. As with all nature-based and climate technologies, environmental impact and risk assessments are essential before scaling operations. Ocean-based approaches also raise questions of governance and global responsibility. Most marine ecosystems exist in international waters governed by different legal rules.

Large-scale ocean carbon interventions must be regulated and controlled through international agreements to prevent not only ecological harm but also geopolitical conflict [37].

Compared with engineered CCS, marine carbon capture offers both strengths and limitations. Ocean-based methods have the potential to absorb very large amounts of CO_2 naturally and provide important ecosystem benefits, such as shoreline protection. However, they are sensitive to environmental disturbances, including pollution, coastal development, extreme storms, and other human activities that can damage marine ecosystems and reduce carbon storage capacity. In addition, measuring and verifying captured CO_2 in marine systems is more difficult than in geological CCS, where storage volumes are well-contained and directly monitored. CCS provides reliable, measurable long-term storage but does not offer ecosystem recovery benefits. Together, marine approaches and geological CCS complement each other as part of a diversified climate mitigation strategy [31, 38].

In summary, OAE and related marine-based approaches represent promising tools in climate mitigation science. These methods potentially provide long-lasting carbon storage while reducing ocean acidification impacts.

18.5 Soil-Carbon Sequestration in Agriculture

18.5.1 Mechanisms, Practices, Benefits, and Measurement

Soil-carbon sequestration refers to the process of increasing the amount of organic carbon stored in agricultural soils. Plants remove carbon dioxide (CO_2) from the atmosphere during photosynthesis and transfer part of this carbon to the soil through roots and crop residues. When farming practices promote healthy soils and reduce disruption of natural biological processes, carbon becomes stabilized in soil structures and can remain stored for decades or longer. Soil-carbon sequestration is one of the most accessible nature-based climate solutions because it can be implemented across millions of agricultural hectares while improving food production systems [39].

Farming practices that promote healthy soils focus on protecting and feeding the soil instead of damaging it. Farmers plant cover crops to protect bare soil from erosion and add nutrients when the plants break down. They rotate crops from season to season so the soil does not lose the same nutrients year after year. Compost and natural plant wastes are added to enrich the soil and support beneficial organisms living in it. Farmers reduce heavy plowing to avoid breaking soil structure. Chemical fertilizers and pesticides are used less often to protect soil life that naturally improves fertility [40].

Increasing soil organic carbon means adding more natural plant material to the soil, which makes the soil healthier. Healthy soil can hold water better, so crops survive more easily during droughts and heavy rain. Rich soil also provides natural nutrients to plants, which reduces the need for chemical fertilizers and lowers pollution while improving crop growth. When soil is healthy, tiny organisms such as bacteria and worms increase, helping keep ecosystems balanced. Farmers benefit as well because healthier soil leads to better harvests and more reliable income over time [41].

Despite these advantages, soil-carbon sequestration is subject to important limitations. The amount of carbon that soils can store is finite; once soils reach saturation levels,

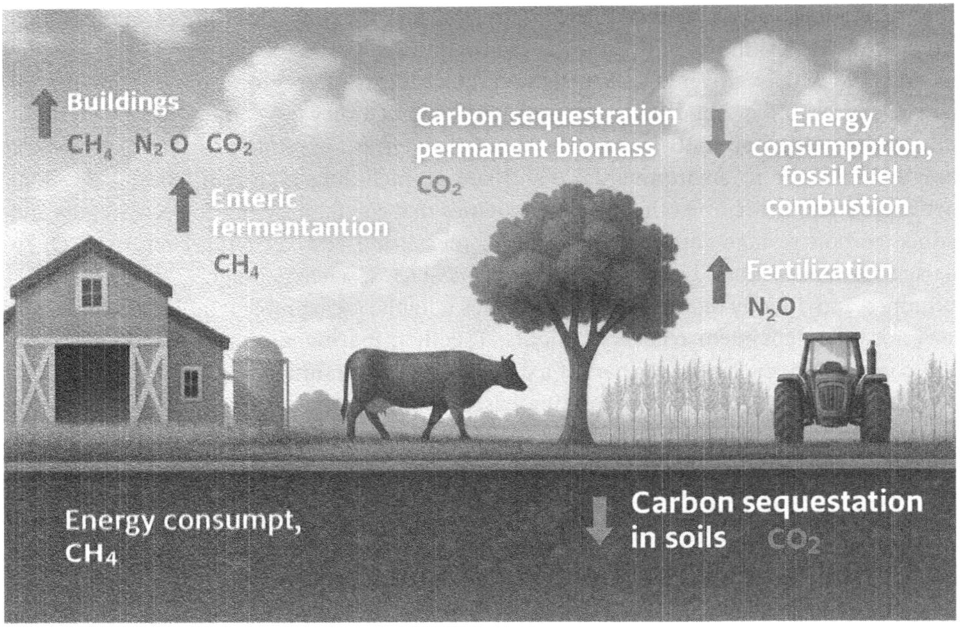

Figure 18.8 Buildings and machinery on farms release greenhouse gases: CO_2 from burning fuel in tractors and heaters, CH_4 from manure and animal housing, and N_2O from fertilized soils and stored waste.

additional storage slows. Soil-carbon stocks are also vulnerable to rapid loss if management practices change or extreme weather events occur [42]. For example reverting to deep tillage in a few seasons can release stored carbon back into the atmosphere. Likewise, droughts and erosion can reduce soil-carbon retention.

Figure 18.8 shows how different parts of a farm contribute to greenhouse gas emissions and carbon sequestration. Buildings and machinery release CO_2, CH_4, and N_2O, while livestock produce methane through enteric fermentation. Fertilizer use emits nitrous oxide. CO_2 (carbon dioxide) is released when tractors, trucks, generators, and heaters burn diesel, gasoline, or natural gas for fuel. CH_4 (methane) can leak from stored manure and animal housing. N_2O (nitrous oxide) is released from fertilized soils and from manure that is stored near barns and sheds. At the same time, trees and crops absorb CO_2 and store it as biomass, and soils capture additional carbon underground. The image highlights both sources of agricultural emissions and natural processes that remove carbon from the atmosphere.

Accurate measurement and verification remain significant challenges. Soil carbon varies across fields, soil depths, climates, and land uses. Direct soil sampling combined with laboratory analysis provides reliable measurements but can be expensive and labor-intensive. Remote sensing (using satellite and aerial images), computer models (which simulate soil carbon changes over time), and digital farm management platforms (which track field practices such as crop rotation, fertilizer use, and tillage) help estimate carbon gains across large areas, but some uncertainty still remains. Reliable soil-carbon accounting depends on consistent soil sampling [43].

18.5.2 Policy, Socioeconomic Factors, and Integration with CCS

Soil-carbon systems refer to farming practices that increase the amount of carbon stored in soils, such as planting cover crops, reducing plowing, improving crop rotations, and adding organic matter. These methods help remove carbon dioxide from the atmosphere while improving soil health. However, soil-carbon programs are closely linked to food security and land-use priorities, because farmland must continue to produce enough food for a growing population. If soil-carbon policies are poorly designed, they could shift farmland away from food production toward carbon-focused activities [40–42].

Social and economic factors strongly affect whether farmers adopt new climate-friendly farming practices, such as reduced tillage, cover cropping, or changes in fertilizer use [39, 40]. Many farmers face financial challenges, including the cost of buying new equipment, the risk of lower crop yields while learning and adjusting to new farming methods, and limited access to training and advice from agricultural extension services, universities, or technical consultants. Payments for ecosystem services and income from carbon-credit programs reward farmers for storing carbon in their soils, while cost-sharing programs with governments or agricultural organizations help pay for new equipment [44]. Free technical support also helps farmers learn how to apply these practices correctly. These programs encourage farmers to participate in sustainable agriculture and carbon-farming initiatives. At the same time, rules are needed to protect small farmers from unstable agricultural and carbon-credit markets. Land use refers to how land is used, such as for farming, forests, housing, or industry [44, 45]. Poorly designed soil-carbon strategies can negatively affect land use if they encourage farmers to stop growing food and instead convert farmland into carbon-only projects.

Soil-carbon storage should be seen as a partner to engineered CCS, not a replacement for it. Soils can store carbon at low cost and bring many short-term benefits, such as healthier soils and better crops. However, carbon stored in soils can be lost again through farming changes or climate impacts. In contrast, geological CCS stores carbon deep underground for very long periods and can handle much larger amounts of carbon safely over time. Using both approaches together is important: soil carbon delivers quick environmental improvements, while engineered CCS provides long-term, reliable carbon removal for climate protection [46].

In summary, soil-carbon storage in agriculture is one of the most practical and socially positive ways to address climate change today. By using suitable farming practices, farmland can store carbon while also improving food production, supporting farmer incomes, and protecting ecosystems. Although challenges remain, such as keeping carbon stored long term, accurately measuring soil carbon, and encouraging farmers to adopt new practices. When combined with nature-restoration efforts, clean energy growth, and engineered CCS, soil-carbon programs play an important role in building a more sustainable global climate solution [46].

18.6 Synergy Between Nature-Based and Engineered CCS Solutions

Effective climate mitigation relies on a variety of complementary approaches rather than a single solution. Nature-based solutions and engineered CCS systems each fulfill unique but somehow similar roles in tackling the accumulation of atmospheric carbon dioxide (CO_2).

Instead of competing against each other, these strategies are most effective when integrated into comprehensive carbon management systems that harness the advantages of both biological ecosystems and technological innovations [47].

Nature-based solutions, such as forest restoration, wetland rehabilitation, soil-carbon sequestration, and marine ecosystem enhancement, can deliver fast, large-scale, and low-cost carbon removal while providing important additional benefits. These co-benefits include improved biodiversity (providing food and habitat for animals), water management, stronger food systems, job creation in rural communities, and climate protection services such as flood and storm mitigation, reduced soil erosion, and improved land stability. Because these solutions can be applied across many regions, they can be implemented quickly at a global scale [1].

However, biological carbon storage also has important limitations. Forest expansion is restricted by land availability for farming, housing, or industrial use, ocean-based carbon uptake reaches an acidity condition, and all living ecosystems are vulnerable to climate disturbances. In addition, carbon stored in vegetation or soils is not permanent and can be rereleased into the atmosphere through fire, droughts, pests, or land-use changes [21]. For these reasons, nature-based solutions are valuable for near-term climate action but cannot provide permanent carbon storage on their own.

CCS solutions, in contrast, provide permanent and more reliable carbon storage. Capture systems installed on industrial sources or operating as direct air capture units can capture CO_2 molecules and inject them into stable geological formations, potentially sequestering carbon safely for thousands of years. CCS provides large-scale, measurable carbon removal, especially for industrial sectors where completely eliminating emissions is technically difficult. However, CCS involves high upfront costs and significant energy use, and it offers fewer direct environmental or ecosystem benefits compared to nature-based solutions [38].

When integrated thoughtfully, the strengths of each approach compensate for the other's limitations. For example industrial carbon capture clusters may pair with adjacent land restoration projects. In farming areas, soil-carbon programs can work together with biomass or bioenergy plants that use carbon capture and storage (BECCS). Farmers grow crops or biomass that improve soil carbon while also supplying material to produce heat, electricity, or low-emission energy. The carbon released during energy production is captured and stored underground, creating a closed-loop system that supports both climate protection and sustainable land management [48].

Figure 18.9 shows an integrated Bioenergy with Carbon Capture and Storage (BECCS) system, linking a biomass production and processing facility with a dedicated carbon capture plant. The left side illustrates biomass feedstock storage, digesters, and agricultural infrastructure, representing carbon absorbed naturally through plant growth. The right side depicts the carbon capture unit, where CO_2 released during biomass conversion is captured, purified, and prepared for geological storage. The separation of the two areas reflects real engineering practice, with distinct zones for biomass handling and high-pressure capture operations. Together, the layout demonstrates how BECCS produces energy while achieving net-negative emissions. The layout demonstrates how BECCS links nature-based carbon uptake from biomass with engineered CCS systems to create a net-negative emissions pathway.

Figure 18.9 BECCS links nature-based carbon uptake from biomass with engineered CCS systems to create a net-negative emissions pathway.

From a policy perspective, using a mix of carbon-removal approaches helps reduce risk. If natural carbon storage is damaged by droughts, fires, or other climate events, permanent storage through CCS can continue to protect the climate. On the other hand, if CCS projects are delayed or remain in smaller scale, nature-based solutions can provide immediate carbon-reduction benefits. Using both approaches together ensures steady progress toward climate goals, even when environmental, political, and economic conditions are uncertain [47].

New technologies help connect natural and engineered climate solutions. Digital monitoring systems track carbon stored in both ecosystems and underground CCS sites [38]. Financial systems are increasingly supporting both nature-based solutions and CCS together. Climate funds and programs, and international development banks now invest in projects that combine ecosystem restoration with CCS [49]. In addition, carbon-credit systems also recognize both natural carbon storage and permanent underground storage [50]. Social protection rules are included in both CCS and nature-based projects to make sure that local communities benefit from climate change mitigation projects.

Integration must adhere to strict ethical and governance principles. CCS projects should not be used to justify the expansion of fossil fuel-based facilities, such as power plants. Furthermore, nature-based carbon capture activities should not delay essential decarbonization efforts, which include both CCS and the development of renewable energy. The advancement of CCS must not divert focus or funding from urgent ecosystem protection initiatives. Investment strategies should prioritize direct emissions reductions above all else.

Public engagement further strengthens synergy. Communities benefit most when projects jointly deliver jobs, ecosystem protection, air quality improvements, and educational opportunities. Clear communication helps people understand how climate projects really work and what benefits they bring [51]. In summary, synergy between nature-based solutions and engineered CCS represents a balanced and strong climate mitigation strategy. Biological systems offer near-term carbon removal with essential ecological and social benefits, while technological CCS provides the permanence required for long-term atmospheric stabilization. Together, these approaches maximize mitigation effectiveness, diversify risk, enhance co-benefits, and support development goals.

18.7 Key Highlights (Chapter Summary)

◇ Chapter 18—Key Highlights:

1) Nature-based carbon capture approaches use forests, wetlands, oceans, and soils to capture CO_2 naturally while delivering major ecological and social benefits.
2) Forest restoration through afforestation, reforestation, and conservation represents one of the most accessible large-scale carbon sinks.
3) Blue carbon ecosystems, including mangroves, wetlands, salt marshes, and seagrass, capture and store carbon extremely efficiently in biomass and deep sediments.
4) Ocean-based approaches such as alkalinity enhancement and phytoplankton stimulation are used to amplify natural marine carbon uptake while addressing ocean acidification.
5) Soil-carbon programs through suitable agricultural approaches increase organic soil matter while improving food security and soil fertilizing efficiency.
6) Nature-based carbon capture solutions provide food security for human and animals, water management, soil quality improvement, and community employment, beyond carbon removal alone.
7) Measurement and verification of ecosystem carbon storage remain technically challenging.
8) Carbon stored in nature is vulnerable to reversal from fires, drought, land-use change, pollution, and extreme weather events.
9) Ethical aspects of nature-based carbon capture are essential. These include ensuring meaningful community participation in decision-making, protecting the rights of people who live in and depend on forests and natural ecosystems, and guaranteeing fair benefit-sharing, such as access to income, jobs, and other economic opportunities created by restoration projects.
10) The strongest climate change mitigation outcomes occur when nature-based approaches are integrated with engineered CCS, combining near-term removal with permanent geological storage.

Keywords: Nature-Based Solutions, Carbon Removal, Forest Restoration, Afforestation, Reforestation, Deforestation Avoidance, Blue Carbon, Wetland Restoration, Mangrove

Ecosystems, Seagrass Meadows, Salt Marshes, Marine Carbon Capture, Ocean Alkalinity Enhancement, Phytoplankton Fertilization, Soil-Carbon Sequestration, Regenerative Agriculture, Cover Cropping, Carbon Permanence, Ecosystem Restoration, Biodiversity Co-Benefits, Coastal Resilience, Carbon Monitoring And Verification, Environmental Safeguards, Community Stewardship, Land Use Governance, Integrated CCS Strategies

References

1 J. Samaniego *et al.*, *Nature-Based Solutions and Carbon Dioxide Removal*. Santiago, Chile: ECLAC, 2023.

2 H. D. Matthews, K. Zickfeld, M. Dickau, *et al.*, "Temporary nature-based carbon removal can lower peak warming in a well-below 2 °C scenario," *Communications Earth & Environment*, vol. 3, p. 65, 2022. DOI: 10.1038/s43247-022-00391-z.

3 N. Gruber, D. C. E. Bakker, T. DeVries, *et al.*, "Trends and variability in the ocean carbon sink," *Nature Reviews Earth & Environment*, vol. 4, pp. 119–134, 2023. DOI: 10.1038/s43017-022-00381-x.

4 S. Ruehr, T. F. Keenan, C. Williams, *et al.*, "Evidence and attribution of the enhanced land carbon sink," *Nature Reviews Earth & Environment*, vol. 4, pp. 518–534, 2023. DOI: 10.1038/s43017-023-00456-3.

5 S. C. Johannessen, "How can blue carbon burial in seagrass meadows increase long-term, net sequestration of carbon? A critical review," *Environmental Research Letters*, vol. 17, no. 9, p. 093004, Sep., 2022. DOI: 10.1088/1748-9326/ac8ab4.

6 K. R. Martin, A. Nickoloff, L. Moffat, A. J. Weaver, and M. Eby, "Assessing the effectiveness of ocean alkalinity enhancement on carbon sequestration and ocean acidification," *FACETS*, Sep., 2025. DOI: 10.1139/facets-2024-0171.

7 L. Marx, C. Zimmerman, T. Ontl, and M. Janowiak, *Healthy Forests for Our Future: A Management Guide to Increase Carbon Storage in Northeast Forests*. Arlington, VA, USA: The Nature Conservancy; Madison, WI, USA: Northern Institute of Applied Climate Science, 2021, pp. 1–40.

8 R. Das *et al.*, "Soil carbon sequestration for soil quality improvement and climate change mitigation," in *Advances in Carbon Capture and Utilization*, D. Pant, A. K. Nadda, K. K. Pant, and A. K. Agarwal, Eds, Energy, Environment, and Sustainability. Singapore: Springer, 2021, pp. 77–98, DOI: 10.1007/978-981-16-0638-0_4.

9 F. A. S. Islam, "The convergence of AI and nature: Advancing carbon dioxide capture, removal, and storage technologies through integrated ecosystem-based strategies," *International Journal of Applied and Natural Sciences*, 2024. DOI: 10.61424/ijans.

10 H. Keith, M. Vardon, C. Obst, V. Young, R. A. Houghton, and B. Mackey, "Evaluating nature-based solutions for climate mitigation and conservation requires comprehensive carbon accounting," *Science of the Total Environment*, vol. 769, 144341, May, 2021. DOI: 10.1016/j.scitotenv.2020.144341.

11 B. K. Sovacool, C. M. Baum, and S. Low, "Reviewing the sociotechnical dynamics of carbon removal," *Joule*, vol. 7, no. 1, pp. 57–82, Jan. 18, 2023. DOI: 10.1016/j.joule.2022.12.004.

12 A. Gurgel, J. Morris, M. Haigh, A. D. Robertson, R. van der Ploeg, and S. Paltsev, "Land-use competition in 1.5 °C climate stabilization: Is there enough land for all potential needs?," *Frontiers in Environmental Science*, vol. 12, Nov., 2024. DOI: 10.3389/fenvs.2024.1393327.
13 M. D. Hurteau, "The role of forests in the carbon cycle and in climate change," in *Climate Change (Third Edition): Observed Impacts on Planet Earth*, 3rd ed.. Amsterdam, Netherlands: Elsevier, 2021, pp. 561–579, DOI: 10.1016/B978-0-12-821575-3.00027-X.
14 I. Ménard, E. Thiffault, W. A. Kurz, and J. F. Boucher, "Carbon sequestration and emission mitigation potential of afforestation and reforestation of unproductive territories," *New Forests*, vol. 54, pp. 1013–1035, 2023. DOI: 10.1007/s11056-022-09955-5.
15 A. Raj, M. K. Jhariya, and N. Khan, "The importance of forest for soil, food, and climate security in Asia," in *Biodiversity, Conservation and Sustainability in Asia*, M. Öztürk, S. M. Khan, V. Altay, R. Efe, D. Egamberdieva, and F. O. Khassanov, Eds. Cham, Switzerland: Springer, 2022, pp. 37–58, DOI: 10.1007/978-3-030-73943-0_3.
16 A. B. Salisbury, J. W. Miesbauer, and A. K. Koeser, "Long-term tree survival and diversity of highway tree planting projects," *Urban Forestry & Urban Greening*, vol. 73, p. 127574, Jul., 2022. DOI: 10.1016/j.ufug.2022.127574.
17 I. H. Luby, S. J. Miller, and S. Polasky, "When and where to protect forests," *Nature*, vol. 609, pp. 89–93, 2022. DOI: 10.1038/s41586-022-05096-z.
18 S. Winter, "Stop rainforest deforestation," in *3 Degrees More*, K. Wiegandt, Ed. Cham, Switzerland: Springer, 2024, pp. 93–108, ch. 6, DOI: 10.1007/978-3-031-58144-1_6.
19 J. S. Sze, D. Z. Childs, L. R. Carrasco, and D. P. Edwards, "Indigenous lands in protected areas have high forest integrity across the tropics," *Current Biology*, vol. 32, no. 22, pp. 4949–4956.e3, Nov., 2022. DOI: 10.1016/j.cub.2022.10.014.
20 C. Huang, W. Gong, and Y. Pang, "Remote sensing and forest carbon monitoring: A review of recent progress, challenges and opportunities," *Journal of Geodesy & Geoinformation Science*, vol. 5, no. 2, pp. 65–77, 2022.
21 A. W. Dye, R. M. Houtman, P. Gao, *et al.*, "Carbon, climate, and natural disturbance: A review of mechanisms, challenges, and tools for understanding forest carbon stability in an uncertain future," *Carbon Balance and Management*, vol. 19, 35, 2024. DOI: 10.1186/s13021-024-00282-0.
22 E. B. Ewane, "Understanding community participation in tree planting and management in deforested areas in Cameroon's Western Highlands," *Environmental Management*, vol. 73, pp. 274–291, 2024. DOI: 10.1007/s00267-023-01902-0.
23 K. B. Dsouza, E. Ofosu, J. Salkeld, R. Boudreault, J. Moreno-Cruz, and Y. Leonenko, "Assessing the climate benefits of afforestation: Processes, methods, and frameworks," arXiv preprint, arXiv:2407.14617, 2024.
24 G. Doolan and S. Hynes, "Ecosystem service valuation of blue carbon habitats: A review for saltmarshes and seagrasses," *Journal of Ocean and Coastal Economics*, vol. 10, no. 1, 2, 2023.
25 D. M. Alongi, "Functional role of mangrove forests along the subtropical and tropical coasts of China," *Current Chinese Science*, vol. 1, no. 1, pp. 73–86, Jan., 2021. DOI: 10.2174/2665997201999200729164528.
26 A. Nayak and B. Bhushan, "Wetland ecosystems and their relevance to the environment: Importance of wetlands," in *Handbook of Research on Monitoring and Evaluating the Ecological Health of Wetlands*, P. K. Shit, P. S. Roy, and R. Avtar, Eds, 2022, pp. 1–16, DOI: 10.4018/978-1-7998-9498-3.ch001.

27 K. Palit, S. Rath, S. Chatterjee, and S. . Das, "Microbial diversity and ecological interactions of microorganisms in the mangrove ecosystem: Threats, vulnerability, and adaptations," *Environmental Science and Pollution Research*, vol. 29, pp. 32467–32512, 2022. DOI: 10.1007/s11356-022-19048-7.

28 J. Liu, P. Failler, and D. Ramrattan, "Blue carbon accounting to monitor coastal blue carbon ecosystems," *Journal of Environmental Management*, vol. 352, 120008, Feb., 2024. DOI: 10.1016/j.jenvman.2023.120008.

29 C. C. Pricillia, M. P. Patria, and H. Herdiansyah, "Social consideration for blue carbon management," *IOP Conference Series: Earth and Environmental Science*, vol. 755, 012025, 2021. DOI: 10.1088/1755-1315/755/1/012025.

30 C. Esteve-Jordà and A. Scheidel, "Do laws provoke or prevent green grabbing? A systematic review," *The Journal of Peasant Studies*, vol. 52, no. 7, pp. 1493–1517, 2025. DOI: 10.1080/03066150.2025.2483262.

31 N. Mengis, A. Paul, and M. Fernández-Méndez, "Counting (on) blue carbon—challenges and ways forward for carbon accounting of ecosystem-based carbon removal in marine environments," *PLOS Climate*, Aug., 2023. DOI: 10.1371/journal.pclm.0000148.

32 M. H. Delval, N. Thonemann, P. J. G. Henriksson, S. E. Tanzer, and P. Behrens, "Life cycle assessment of ocean-based carbon dioxide removal approaches: A systematic literature review," *Renewable and Sustainable Energy Reviews*, vol. 224, 116091, Dec., 2025. DOI: 10.1016/j.rser.2025.116091.

33 J. Hartmann *et al.*, "Stability of alkalinity in ocean alkalinity enhancement (OAE) approaches—consequences for durability of CO_2 storage," *Biogeosciences*, vol. 20, no. 4, pp. 781–802, 2023. DOI: 10.5194/bg-20-781-2023.

34 J. Nishioka, H. Obata, T. Hirawake, *et al.*, "A review: Iron and nutrient supply in the subarctic Pacific and its impact on phytoplankton production," *Journal of Oceanography*, vol. 77, pp. 561–587, 2021. DOI: 10.1007/s10872-021-00606-5.

35 E. Asbridge and P. Dwyer, "Marine vegetation management strategies: A framework for estuary-wide prioritization of protection and rehabilitation," *Restoration Ecology*, vol. 31, no. 7, p. e13819, 2023. DOI: 10.1111/rec.13819.

36 A. E. A. Blomberg, I.-K. Waarom, C. Totland, and E. Eek, "Marine monitoring for offshore geological carbon storage—a review of strategies, technologies and trends," *Geosciences*, vol. 11, no. 9, p. 383, 2021. DOI: 10.3390/geosciences11090383.

37 A. Westholm and G. Argüello, "Dynamic ocean management in areas beyond national jurisdiction," *Ocean Development & International Law*, vol. 54, no. 4, pp. 448–468, 2023. DOI: 10.1080/00908320.2023.2296392.

38 K. Sotoodeh, *Piping, Valves, and Process Engineering in Carbon Capture Systems: Applications, Sustainability, and Environmental Protection*, 1st ed.. Amsterdam, Netherlands: Elsevier, 2026. ISBN: 978–0443510786.

39 C. I. D. Rodrigues, L. M. Brito, and L. J. R. Nunes, "Soil carbon sequestration in the context of climate change mitigation: A review," *Soil Systems*, vol. 7, no. 3, p. 64, 2023. DOI: 10.3390/soilsystems7030064.

40 H. R. A. Alsammarai and S. A. Y. Al-Ajeeli, "The wrong agriculture practices of field crop farmers leading to land agricultural deterioration in Samarra District–Salah Al-Din Governorate," *IOP Conference Series: Earth and Environmental Science*, vol. 1262, 082031, 2023. DOI: 10.1088/1755-1315/1262/8/082031.

41 S. A. Wood and J. C. Blankinship, "Making soil health science practical: Guiding research for agronomic and environmental benefits," *Soil Biology and Biochemistry*, vol. 172, 108776, Sept., 2022. DOI: 10.1016/j.soilbio.2022.108776.

42 G. Y. K. Moinet, R. Hijbeek, D. P. van Vuuren, and K. E. Giller, "Carbon for soils, not soils for carbon," *Global Change Biology*, vol. 29, no. 9, pp. 2384–2398, 2023. DOI: 10.1111/gcb.16570.

43 J. A. Reijneveld, M. J. van Oostrum, K. M. Brolsma, and O. Oenema, "Soil carbon check: A tool for monitoring and guiding soil carbon sequestration in farmer fields," *Frontiers of Agricultural Science and Engineering*, vol. 10, no. 2, 2023.

44 C. T. Barbato and A. L. Strong, "Farmer perspectives on carbon markets incentivizing agricultural soil carbon sequestration," *npj Climate Action*, vol. 2, 26, 2023. DOI: 10.1038/s44168-023-00055-4.

45 M. Dustova and S. Ochilova, "Foreign experience of financial support of agricultural enterprises," *The Innovation Economy*, vol. 1, no. 2, pp. 79–86, Jul., 2023.

46 D. Pant, K. K. Shah, S. Sharma, *et al.*, "Soil and ocean carbon sequestration, carbon capture, utilization, and storage as negative emission strategies for global climate change," *Journal of Soil Science and Plant Nutrition*, vol. 23, pp. 1421–1437, 2023. DOI: 10.1007/s42729-023-01215-5.

47 H. Jain, "Synergizing carbon reduction technologies and nature-based solutions for a climate-neutral future," *Mitigation and Adaptation Strategies for Global Change*, vol. 30, 42, 2025. DOI: 10.1007/s11027-025-10237-6.

48 Y. Zhang *et al.*, "Energy and CO_2 emission analysis of a bio-energy with CCS system: Biomass gasification—solid oxide fuel cell—mini gas turbine—CO_2 capture," *Fuel Processing Technology*, vol. 238, 107476, Dec., 2022. DOI: 10.1016/j.fuproc.2022.107476.

49 K. Mintz-Woo and J. Lane, "Why and where to fund carbon capture and storage," *Science and Engineering Ethics*, vol. 27, 70, 2021. DOI: 10.1007/s11948-021-00344-3.

50 N. Lokuge and S. Anders, "Carbon-credit systems in agriculture: A review of literature," SSRN Scholarly Paper, Apr. 2022. Available: https://ssrn.com/abstract=4096174

51 F. Mulyasari, A. K. Harahap, A. O. Rio, R. Sule, and W. G. A. Kadir, "Potentials of the public engagement strategy for public acceptance and social license to operate: Case study of carbon capture, utilisation, and storage Gundih pilot project in Indonesia," *International Journal of Greenhouse Gas Control*, vol. 108, 103312, June, 2021. DOI: 10.1016/j.ijggc.2021.103312.

19

Education, Skills Development, and Public Capacity-Building for Carbon Capture

19.1 Importance of Knowledge and Skills in CCS Expansion

The successful expansion of carbon capture and storage (CCS) depends not only on technical innovation and financial investment but also on the development of a skilled and knowledgeable workforce. As CCS is deployed more widely across power generation, biomass facilities, hydrogen production, and other energy sectors, the demand for trained professionals grows rapidly [1]. This section explains why human capacity-building is a core pillar of CCS implementation and highlights the knowledge and skills required to support a safe, efficient, and socially responsible carbon management system.

CCS is a multidisciplinary field that combines elements of engineering, chemistry, geology, environmental science, digital technology, policy, safety management, and community engagement. Each stage of the CCS process, including capture, compression, transport, injection, monitoring, and long-term stewardship, requires specialized expertise. Engineers must understand thermodynamics and solvent chemistry. They also need knowledge of pipeline design, materials selection, and reservoir dynamics [2]. Technicians and operators must be skilled in operating equipment and troubleshooting issues. Geologists and geophysicists are responsible for identifying safe storage formations. These diverse responsibilities emphasize the necessity for advanced training systems that prepare individuals for real-world CCS environments.

A strong knowledge base is essential for ensuring safety and reliability in carbon capture and geological storage. CCS processes involve handling high pressures, corrosive chemicals, and subsurface carbon dioxide injection operations that must be carefully managed [3]. Skilled operators can reduce the risk of leaks by understanding potential operational issues before they become critical, thereby preventing equipment failures and environmental harm. Well-trained monitoring teams ensure that CO_2 remains safely contained in the storage area beneath the ground surface for decades to centuries [4]. Without proper education and technical training, CCS projects risk operational failures that could undermine public trust and hinder long-term adoption.

A key reason why knowledge and skills are essential is the rapid technological evolution in CCS. Innovations such as new capture materials to prevent corrosion or improve carbon capture, digitalization, autonomous inspection robots, and advanced seismic imaging technologies are transforming the operations of facilities [5]. To keep up with these emerging technologies, workers need ongoing learning opportunities.

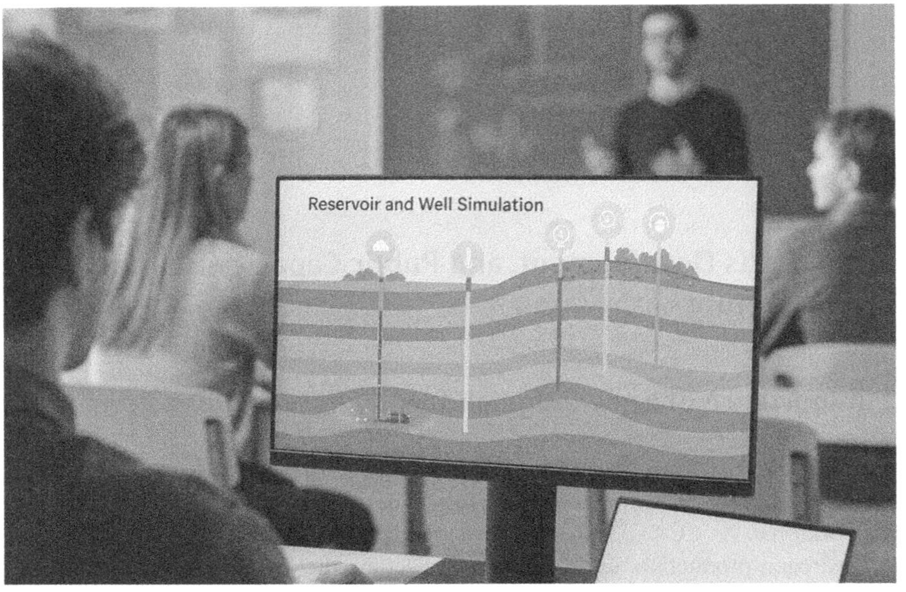

Figure 19.1 Classroom-based CCS training course with focus on reservoir and well engineering simulation. This training helps students understand CO_2 injection, subsurface storage of CO_2, and long-term geological storage that are key skills for future CCS professionals.

Figure 19.1 shows a classroom where students are learning about CCS through reservoir and well simulation. A computer monitor in the foreground displays a simplified geological cross-section with different subsurface layers and an injection well. This visual model helps students understand how CO_2 moves underground, how reservoirs hold injected carbon, and how monitoring systems track long-term storage. In the background, the instructor is teaching while other students participate in the training session. This highlights the importance of education and hands-on training in developing skilled CCS professionals.

The energy transition presents both opportunities and challenges for workforce mobility. As fossil fuel industries decline, many skilled workers, such as plant operators and pipeline specialists, face the risk of job loss. However, CCS provides a pathway for these workers to transition into low-carbon sectors where their skills are still valuable. For instance, expertise in oil and gas drilling is highly applicable to CO_2 injection operations for long-term storage, while experience in pipeline maintenance can effectively support CO_2 transport pipeline networks [6].

Knowledge and skills extend beyond technical disciplines. Successful CCS implementation requires understanding policy, economics, permitting processes, environmental justice considerations, and community engagement strategies [7]. In addition, public awareness is particularly important because CCS infrastructure, including pipelines and injection wells, often intersects with local communities. Workers must communicate with local people and be capable of explaining CCS concepts clearly, addressing concerns, and building transparent relationships with stakeholders [8].

Knowledge also plays a central role in long-term stewardship. Geological storage sites require monitoring and maintenance for decades after injection ends [9]. This means future generations of professionals must be equipped with the skills needed to perform safety assessments and interpret monitoring data. International cooperation further increases the importance of knowledge expansion. As CCS expands globally, countries with limited technical experience may depend on knowledge transfer from regions with mature CCS industries.

Building human skills and training supports long-term sustainability. Skilled workers drive innovation, reduce the risk of failures and environmental impacts, and strengthen local communities, especially in regions where CCS projects are developed. This highlights that CCS is not only a technical climate solution but also part of a broader strategy for developing knowledge, expertise, and local capacity. In summary, the expansion of CCS depends fundamentally on the availability of well-trained professionals who understand the technological, environmental, and social dimensions of carbon management.

19.2 University Programs and Technical Training Pathways

The deployment of CCS at scale requires a highly trained workforce equipped with strong scientific, technical, and practical skills. Universities, technical colleges, and vocational training centers play a central role in preparing this workforce by offering specialized programs that cover the broad range of disciplines involved in CCS [10]. This section explores the educational programs that support workforce readiness, the key competencies emphasized in academic programs, and the importance of clear and flexible training models for long-term CCS success.

Universities provide the foundation for advanced CCS knowledge through degree programs in engineering, environmental science, geology, chemistry, materials science, economics, and public policy. Many institutions now offer specialized courses or concentrations in carbon capture or energy systems [11]. These programs introduce students to topics such as solvent chemistry, membrane technologies, digital plant operations, and geological storage processes. Courses also cover safety standards and life-cycle assessment (LCA), skills essential for real-world CCS projects.

Technical colleges and vocational schools complement university education by offering training for technicians, operators, and workers. These programs focus on practical skills such as equipment operation, failure detection and repair, pipeline maintenance, control room monitoring, welding, electrical works, and industrial safety. Nearly every stage of CCS operations from CO_2 capture to compression to pipeline integrity checks and injection well operation requires skilled technicians who understand both the technical and safety dimensions of carbon handling facilities [10, 11].

As CCS technology evolves rapidly, educational institutions need to update their curricula to include emerging technologies. Digitalization, artificial intelligence, robotics, and predictive analytics are becoming increasingly integrated into CCS systems [12]. In response, universities are creating interdisciplinary courses that combine engineering fundamentals with data science, automation, and digital twin modeling. Likewise, technical colleges are introducing modules on drone operation, remote sensing tools, and automated

Figure 19.2 Digital-twin-based monitoring of a carbon capture plant using real-time operational data.

inspection technologies. These revised curricula are designed to prepare students for modern CCS workplaces that heavily rely on digital tools.

Figure 19.2 illustrates the use of digitalization in carbon capture systems. A digital twin of the plant is displayed on a laptop and used to monitor plant process performance and analyze operating conditions. This reflects realistic modern industrial practice, where digital tools are used for online monitoring, process analysis, and decision-making regarding operation and maintenance activities.

Accessibility of training is essential. "Accessibility of training" means that workers can easily get the education they need to work in CCS. This includes having courses that are affordable and flexible in schedule. "Flexible in schedule" means that the courses are offered in ways that fit workers' busy lives [13]. For example, workers may still have full-time jobs, shift work, or family responsibilities. Flexible training options allow them to participate without disturbing their daily routines and busy schedules. In addition, many workers and engineers already have basic skills from conventional energy industries (like oil and gas), and they only need complementary programs or short courses to adapt those skills to CCS. Examples include learning about CO_2 transportation and well engineering used to inject CO_2 for long-term storage.

International training programs are also important. Universities work together through joint degrees, student exchange programs, and research partnerships focused on CCS. Exchanged students between universities in two countries can join internships or field visits at CCS sites in other countries. These experiences help share knowledge between regions and countries, particularly between different levels of CCS development [14]. International cooperation builds a larger global workforce and helps emerging economies prepare for CCS more quickly.

An important part of CCS training is learning across different fields, not just engineering. Professionals need to understand how government rules and permitting processes affect

CCS projects. They also need to understand environmental impacts, such as how CO_2 injection interacts with groundwater and ecosystems. Social understanding is equally important: workers must know how to engage communities, address concerns, and build trust with public. Economic knowledge matters as well because CCS projects involve large investments.

Practical training facilities make CCS education much stronger [15]. Many universities and training centers now use small pilot capture units where students can see how CO_2 is actually separated from flue gas. Geoscience laboratories allow learners to study real rock samples and understand how storage formations hold CO_2 underground. Simulated control rooms teach students how to operate equipment. Together, these hands-on tools help learners connect what they study in class to how CCS works in real life.

In summary, university programs and technical training are essential for developing the workforce needed for CCS. Through interdisciplinary education, practical training, reskilling programs, and international collaboration, these approaches ensure that workers possess the knowledge and technical skills required to operate and manage CCS systems effectively.

19.3 Interdisciplinary Competencies for Future Workforce

CCS is not a single field. It is a system that depends on many areas of knowledge working together. An important part of CCS is industrial facilities, where engineers design equipment to separate CO_2 from flue gas. This requires expertise in chemical and mechanical engineering and materials science because the equipment must handle high temperatures, corrosive fluids, and mechanical loads. In addition, CO_2 transportation after capture requires multidisciplinary collaboration among several fields, including pipeline engineers (typically mechanical engineers), as well as process and materials engineers [2].

Geological storage introduces an entirely different discipline. Geologists and geophysicists evaluate underground formations, study rock properties, and predict how CO_2 will flow and store deep beneath the surface. Environmental scientists assess possible CO_2 subsurface storage impacts on groundwater, ecosystems, and surrounding land [9]. Figure 19.3 shows the basic concept of carbon capture and geological storage. CO_2 is captured at an industrial facility, transported to an injection well, and stored deep underground in a porous geological formation. A low-permeability caprock above the storage layer prevents the CO_2 from migrating upward, ensuring long-term containment.

Monitoring experts use seismic imaging, satellite data, sensors, and digital tools to track CO_2 after injection. Seismic surveys help specialists "see" underground and detect how the CO_2 plume moves through rock layers over time. Satellite observations provide surface-level information such as ground uplift or subtle CO_2 leaks that may not be visible on-site. Downhole sensors measure pressure, temperature, and chemical changes inside the reservoir, giving continuous, real-time feedback on storage performance. Digital-twin models and advanced software combine all these data sources to simulate how CO_2 behaves deep underground [16].

Systems thinking is a core competency. CCS projects are not standalone. They operate within the broader energy context. Engineers must understand how capture systems

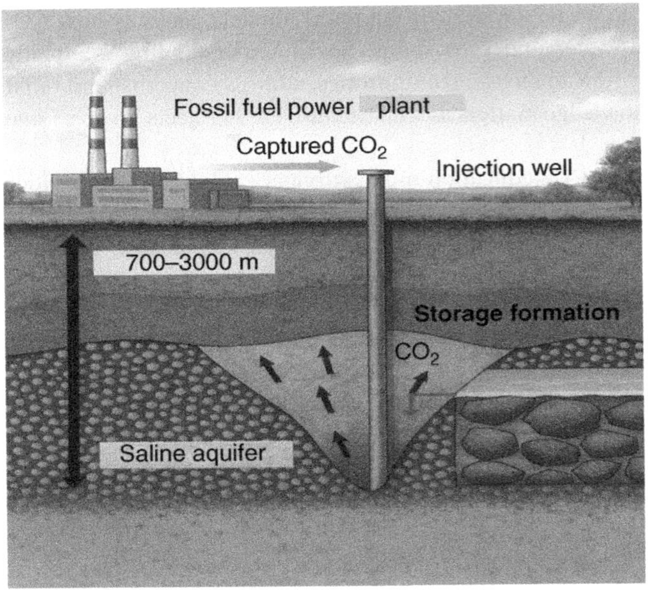

Figure 19.3 Schematic illustration of CO_2 capture at a power plant and injection into a deep geological storage formation.

increase energy demand through heat and electricity. Reducing this demand requires approaches such as wastewater use and heat recovery from surrounding industries. Digital technologies play an important role in lowering operating costs while improving safety and reliability. These improvements also influence overall emissions. Geologists must understand regulatory requirements, public concerns, and long-term responsibility for storage sites. Systems thinking enables clear understanding of interactions across disciplines.

Another important competency is cross-disciplinary communication. CCS projects involve many different professional backgrounds. Different departments must work together at technical interfaces. Well engineers often need support from flow assurance specialists to simulate CO_2 injection into wells. Pipeline engineers rely on materials specialists to select suitable materials and to manage corrosion and integrity issues. Effective collaboration across disciplines ensures that interface problems are identified early and addressed correctly [2]. Figure 19.4 shows a multidisciplinary training session where professionals from different engineering backgrounds discuss key aspects of CCS. The focus is on CO_2 transportation through pipelines and CO_2 injection into underground reservoirs. This picture highlights the importance of cross-disciplinary understanding and collaboration in CCS projects.

The CCS workforce must develop a strong understanding of regulations. Permitting requirements as part of regulations are needed to start a CCS project, and they strongly influence project design. The permitting process gives approval to investors and project developers. It defines what data must be collected and how risks must be managed. If these requirements are not understood early, projects can face serious delays [17]. For example, a storage site may be technically sound but still be denied a permit if the monitoring plan

Figure 19.4 Multidisciplinary training session addressing CO_2 transport through pipelines and injection into geological reservoirs in CCS projects.

is incomplete. Early regulatory awareness allows these issues to be addressed during the design phase rather than after construction planning has begun.

Environmental literacy is an essential skill for the CCS workforce. CCS projects affect the natural environment and nearby communities. Project members must understand how projects can impact land, water, and ecosystems. This knowledge helps reduce environmental damage and supports responsible project development [18]. It also helps ensure that CCS projects respect community interests and long-term sustainability goals.

Digital skills are important for the entire CCS project team, including engineers, operators, managers, and technical specialists. Operators need digital skills to run and monitor systems. Engineers use digital tools for design, analysis, and optimization. Project managers rely on digital information for decision-making and risk control. When the whole team has digital understanding, system efficiency improves, and digital systems are used correctly across the project [19]. Figure 19.5 shows a professional in a CCS plant using digital tools to monitor and manage industrial process systems. The image highlights how digital skills support real-time operation, data analysis, and decision-making across engineering and operation roles in CCS facilities.

Another important competency is risk assessment and safety culture. CCS projects involve safety-critical operations. These include high-pressure systems and long-term storage responsibilities [20]. Project teams must understand how risks are identified and managed. They must also be prepared to respond to emergencies. A strong safety culture supports early hazard recognition and consistent compliance with safety requirements.

Economic and financial literacy is also essential for CCS projects. CCS deployment depends on both capital investment and operating costs. Project financing is often linked to carbon pricing and carbon markets. Professionals must understand basic cost modeling

Figure 19.5 Digital tools support CCS professionals, including engineers and operators, by enabling real-time monitoring and system optimization (Courtesy: Shutterstock).

and project economics. This knowledge supports informed technical decisions and realistic project planning [21]. In addition to technical and financial skills, future CCS professionals must develop social and community engagement competencies. Ethical aspects are equally important. CCS raises questions of long-term responsibility for CO_2 storage and respect for community rights. In summary, interdisciplinary competencies are critical for the future CCS workforce. Beyond technical expertise, workers must understand systems thinking, environmental literacy, economy and policy awareness, digital network, risk assessment, ethics, and community engagement.

19.4 Public Awareness and Climate Education Programs

Public awareness and climate education are crucial for the successful implementation of CCS technology. Even if a CCS project is technically sound and environmentally safe, it cannot progress without public understanding and acceptance. Communities need to feel informed, respected, and confident about how CCS operates, the benefits it offers, and how any potential risks are managed. This section outlines the importance of public education and how effective engagement strategies can enhance social acceptance and support the long-term development of CCS [10].

CCS technologies are complex and unfamiliar to many people. They involve pipelines, underground storage, injection wells, and monitoring systems. Without clear public education, opinions may be shaped by misunderstanding or fear. Climate education helps communities understand how CCS works and why it is used. It shows how CCS fits within wider climate efforts such as emission reduction and industrial decarbonization. Clear explanations reduce confusion and support informed public discussion [22].

A key goal of public awareness is transparency. People are more likely to trust CCS projects when information is open and easy to access. This includes safety measures, monitoring systems, and emergency planning. For example, public trust increases when people can clearly see how CO_2 is stored safely deep underground. Showing what safeguards are in place, and how problems would be handled (emergency response) if they occur, makes the project easier to understand [23].

Educational programs must also explain the local benefits of CCS. CCS projects create jobs throughout the project lifecycle. Companies involved in carbon capture often offer employment related to construction activities, equipment maintenance, and environmental monitoring. These activities create new economic opportunities within the region. CCS also helps existing industries, such as power plants, reduce emissions while continuing to operate. This protects current jobs, especially in regions where a sudden shift to renewable energy could lead to job losses. At the same time, CCS supports cleaner industrial activity. Reduced emissions can improve local air quality and environmental conditions [24]. Clear explanation of these benefits helps communities understand how CCS can contribute to long-term local development.

Schools and universities play an essential role in climate and CCS education. Teaching basic climate science helps students understand why emission prevention matters. Learning about the carbon cycle explains how CO_2 moves between the atmosphere, land, and oceans, as shown in Figure 19.6. Introducing low-carbon technologies, including CCS, shows students how emissions can be reduced in practice. On the land side, plants absorb CO_2 from the atmosphere through photosynthesis and store it in plant biomass. When plants die, part of this carbon is transferred into the soil, forming soil carbon. Some carbon is also released back into the atmosphere. On the ocean side, CO_2 from the atmosphere dissolves into surface waters. Marine plants and phytoplankton use this carbon for

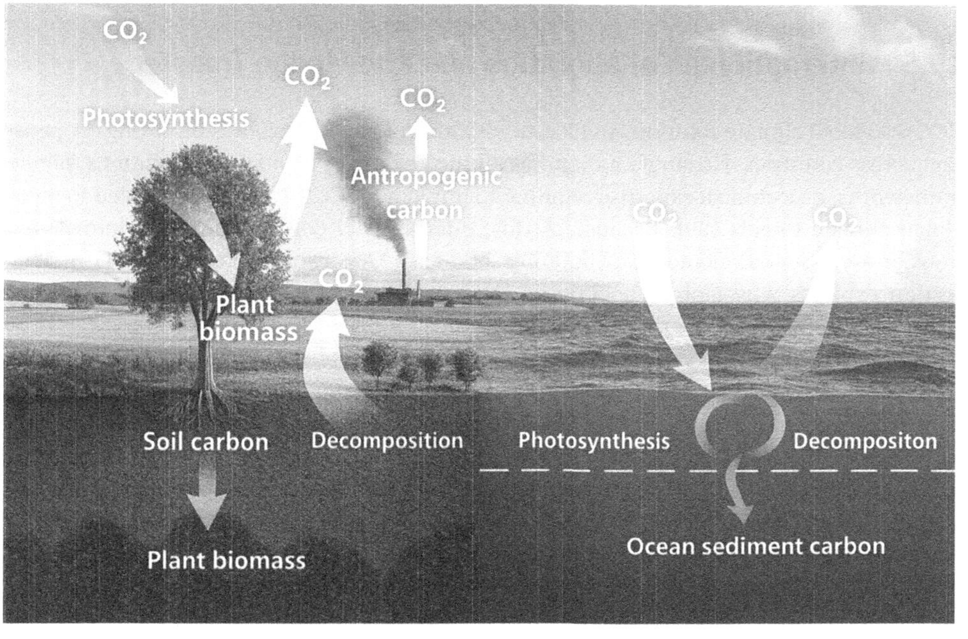

Figure 19.6 Simplified illustration of the global carbon cycle showing carbon exchange between the atmosphere, soils, oceans, and marine sediments.

photosynthesis. When marine organisms die, some of the carbon sinks and becomes stored in ocean sediments where it can remain for long periods. At the same time, decomposition and ocean–atmosphere exchange return part of the CO_2 back to the atmosphere. When these topics are included in primary and secondary education, students develop early awareness and understanding [25].

Another important part of public awareness is explaining how CCS fits into broader climate policy. Many people believe CCS is used to extend fossil fuel use [26]. Education programs can clarify that CCS is mainly applied to industries that are difficult to decarbonize. These include sectors such as cement, steel, and fertilizers. In these industries, renewable energy or full electrification is not yet practical [27]. CCS helps reduce emissions from these sources while longer-term solutions such as renewable energies continue to develop.

Public awareness initiatives must avoid oversimplification. Oversimplification means presenting CCS as a quick or perfect solution without explaining its limits. Education should explain both the benefits and the challenges of CCS in an honest way. This includes discussing risks, uncertainties, costs, and the need for other climate solutions alongside CCS [28]. Open and balanced discussion helps people make informed judgments rather than feeling persuaded in one direction.

In summary, public awareness and climate education are essential for successful CCS deployment [10]. Clear and accessible information helps people understand how CCS works and why it is used. Open communication supports transparency and allows concerns to be discussed honestly. When communities are informed and involved, trust in CCS projects increases. This social understanding and acceptance form a strong foundation for the long-term success of CCS projects.

19.5 International Collaboration and Knowledge Transfer

CCS is a global climate solution, and its successful deployment depends on strong cooperation across countries. No single nation has all the resources, geological formations, policy frameworks, or technical expertise required to implement CCS at the scale needed to meet global climate targets [29]. Because of this, international collaboration and knowledge transfer are essential elements of building long-term CCS capacity worldwide [14]. This section explores why global cooperation matters, how knowledge is shared, and what structures support effective international learning.

The first reason collaboration is crucial is that CCS technologies and geological storage conditions vary widely across regions. Some countries have advanced CCS technologies and industrial experience, but lack suitable geological formations for large-scale CO_2 storage. Conversely, other nations may have ample geological capacity for CO_2 storage but limited technical expertise in CCS. Knowledge-sharing partnerships allow countries to combine strengths. For example, nations with well-developed CCS projects, such as Norway, Canada, Australia, Japan, and the United States, often serve as training sources for countries beginning their CCS journeys. These collaborations accelerate learning and reduce costly trial-and-error processes for countries with less capability for CCS technology development [30].

International collaboration helps align technical and safety standards. Different regulatory bodies may apply different design rules, safety factors, and approval processes.

International cooperation helps bring these frameworks closer together. For example, ASME and ISO both publish standards for pressure equipment and pipelines used in CCS projects. ASME and ISO collaboration can be driven by international cooperation between countries and the need for integrated regulatory frameworks. In international CCS projects, regulatory approval can be delayed when different standards are required. For example a project may be designed according to ASME standards, while the approving authority requires compliance with ISO standards. This mismatch can lead to additional reviews, redesign, or requalification of equipment. Greater alignment between ASME and ISO standards helps reduce such barriers, making approval processes smoother and reducing delays in international CCS projects [31].

Knowledge transfer between countries takes place through several channels. Research partnerships allow universities and laboratories from different countries to collaborate on CCS-related topics. International conferences and technical workshops bring together leading specialists and managers to share results and discuss common challenges. Student exchange programs between universities in different countries help young professionals gain experience in diverse CCS environments. These activities strengthen international cooperation and support the development of global expertise in CCS.

Another important form of collaboration is capacity-building support. High-income countries and banks often provide funding, training, and technical assistance to countries that want to develop CCS but lack experience or infrastructure. Training programs may include geological courses, operational safety, or digital monitoring. This support helps countries build local expertise and prepare for safe and effective CCS deployment [32].

Policy collaboration is also important for CCS development. Governments work together through international climate agreements. For example, under the Paris Agreement, countries share climate strategies and report how technologies such as CCS contribute to emission reduction targets [33]. In carbon markets, such as the European Union Emissions Trading System (EU ETS), countries align rules on carbon pricing, monitoring, and verification, which directly affect how CCS projects are supported. To clarify, carbon markets are based on measuring emissions and verifying emission reductions or excess emissions [34]. For these markets to function properly, emissions must be measured and verified in a clear and uniform way. This is especially important in international or regional systems such as emissions trading schemes.

Another growing form of collaboration is cross-border industrial clustering. In some regions, countries develop shared CO_2 transport and storage networks. These networks allow CO_2 captured in one country to be transported and stored in another. A key example is the Northern Lights project in Norway. It allows CO_2 from industrial facilities in other European countries to be shipped and stored offshore in Norwegian geological formations [35]. Shared infrastructure reduces costs and limits environmental impact by avoiding unnecessary construction and duplication of facilities [36]. It also supports cooperation between industries across national borders.

Figure 19.7 illustrates a cross-border CCS system in which multiple countries cooperate by sharing CO_2 transport and storage infrastructure. Industrial facilities in different countries (Country A, Country B, and Country C) capture CO_2 at the source. The captured CO_2 is then transported across national boundaries using both pipelines and ships, depending on distance and available infrastructure. All transported CO_2 is delivered to a shared

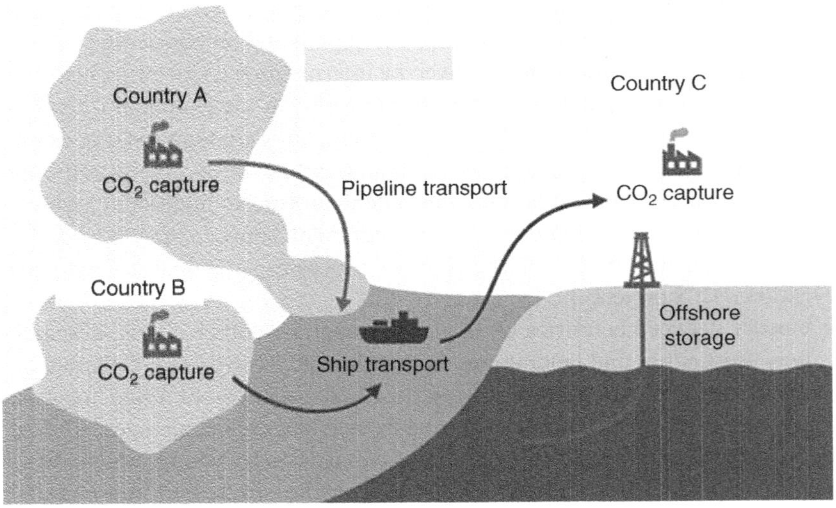

Figure 19.7 Conceptual illustration of cross-border CCS cooperation, showing CO_2 captured in multiple countries and transported by pipeline and ship to a shared offshore geological storage site.

offshore geological storage site, where it is injected deep underground for long-term storage. The diagram highlights how international cooperation allows countries to use common storage resources, reduce infrastructure duplication, and lower overall costs.

International collaboration also strengthens public trust. When global scientific organizations validate CCS technologies, communities gain greater confidence in the safety and reliability of carbon storage [37]. However, effective international collaboration must be fair and inclusive. Developing countries should have equal access to training, technology, and funding. Finally, international cooperation encourages innovation by exposing engineers, students, and researchers to science and technology from different perspectives and national contexts [38]. In summary, international collaboration and knowledge transfer are foundational pillars of global CCS development.

19.6 Strategies for Long-Term Capacity-Building

Long-term capacity-building is essential for ensuring that CCS can be deployed and managed effectively over many decades. Long-term success of the CCS project requires a steady supply of trained professionals, informed communities, a strong education system, and robust governance supportive of policy toward CCS. This section presents strategic approaches for strengthening capacity-building to support widespread CCS development.

19.6.1 Strengthening Educational Infrastructure

A long-term CCS workforce depends on strong education and training. Universities need to offer programs that teach carbon capture and environmental technology. Technical colleges should provide practical courses on industrial operations and digital monitoring.

Cooperation between schools and industry helps ensure that students learn skills that are needed in real projects. Laboratories and pilot plants should give students hands-on experience and prepare them for work in CCS projects.

19.6.2 Promoting Long-Term Learning and Workforce Reskilling

As CCS technologies change, workers need to keep learning and updating their skills. Lifelong learning programs such as short courses and online training help people stay up to date. Lifelong training means continuous learning throughout a person's career, not only during formal education. Reskilling is especially important for workers moving from coal, oil, and gas industries into low-carbon jobs. Reskilling means learning new skills so a person can move into a different role or industry. Support from governments and industry can make this transition easier by offering training programs and financial assistance.

19.6.3 Building Strong Institutional Organs

In this context, institutions are organizations that have official responsibilities for regulating, managing, and supporting CCS activities. Institutions include regulatory agencies, government departments, and research organizations. Capacity-building also depends on strong institutions. These institutions need clear responsibilities so each organization knows its role in CCS development. They also need enough financial and technical resources to perform their tasks effectively. Trained staff is essential to enforce regulations and respond and act correctly to technical or environmental issues.

19.6.4 Enhancing Community Understanding and Participation

Communities are important to the success of CCS projects. Community awareness includes helping people understand climate change, carbon management, and how CCS works safely. Public meetings allow residents to ask questions and share concerns. When people feel informed and respected, trust increases, and projects gain acceptance.

19.6.5 Strengthening International Knowledge Exchange

Global collaboration helps build CCS knowledge and experience, expanding by allowing countries to learn from each other. International training programs help share skills and experience. Joint research and open data make learning faster and reduce repeated mistakes. Global organizations support this cooperation by bringing countries together and coordinating knowledge sharing. Countries with more CCS experience can support others by providing training and practical experience.

19.6.6 Supporting Innovation and Research Capacity

Innovation is important for improving CCS over the long term. This requires a strong research capacity. Governments and industry should support universities and research centers that study CCS technologies. Testing new ideas in pilot projects and research facilities can help improve performance and safety. When universities and companies work together, new solutions develop faster, and theoretical ideas can be aligned better with practice.

19.7 Key Highlights (Chapter Summary)

◇ Chapter 19–Key Highlights:

1) Successful long-term deployment of carbon capture and storage (CCS) depends on building strong human capacity alongside technological and infrastructure development.
2) Universities, technical colleges, and institutions play an essential role in educating engineers, scientists, and technicians for CCS systems.
3) Workforce reskilling programs enable workers from fossil fuel industries to transition into CCS roles, supporting energy transition.
4) Interdisciplinary competencies combining engineering, environmental science, policy, digital skills, and communication with public are essential for future CCS professionals.
5) Public awareness and climate education programs strengthen social acceptance and trust by improving their understanding of CCS safety, benefits, and limitations.
6) Community training and engagement around CCS sites enhance transparency, local participation, and CCS project success.
7) Digital and data-related skills are increasingly important due to automation, AI, and smart monitoring technologies in modern CCS systems.
8) International collaboration and knowledge transfer accelerate CCS development through shared research in conferences and training programs.
9) Strong regulatory expertise is required to handle CCS design, operation, and providing permission.
10) Long-term capacity-building strategies must emphasize long-term learning even after graduation, equitable access to education, and sustained investment in research and development.

Keywords: Capacity-Building, Climate Education, Workforce Development, Technical Training, Interdisciplinary Skills, Public Awareness, Community Engagement, Knowledge Transfer, International Collaboration, CCS Literacy, Lifelong Learning, Digital Skills, Reskilling Programs, Institutional Capacity, Environmental Education, Social Acceptance, Just Transition, Policy Literacy, Sustainable Workforce Development

References

1 M.-A. Pourmoosavi and T. Amraee, "Low carbon generation expansion planning with carbon capture technology and coal phase-out under renewable integration," *International Journal of Electrical Power & Energy Systems*, vol. 128,art. no.106715, June 2021, 10.1016/j.ijepes.2020.106715.

2 A. Zauner *et al.*, "Multidisciplinary assessment of a novel carbon capture and utilization concept including underground sun conversion," *Energies*, vol. 15, no. 3, p. 1021, 2022. DOI: 10.3390/en15031021.

3 F. Tamburini, F. Zanobetti, S. Bonvicini, and V. Cozzani, "The crucial role of safety in carbon capture and storage: Addressing knowledge gaps for a sustainable future," *Chemical Engineering Transactions*, vol. 116, pp. 205–210, 2025.

4 P. D. Patil, A. S. Al-Qasim, A. I. Al-Zayani, and S. L. Kokal, "End-to-end surface and subsurface monitoring and surveillance for the onshore and offshore CCS or CCUS projects," *Proc. Offshore Technol. Conf.*, Houston, TX, USA, May 2023, paper OTC-32172-MS. DOI: 10.4043/32172-MS

5 A. Kumar, "Evolution of carbon capture, utilization, and storage technology: Past, present, and future developments," in *Clean Energy*, 2024, pp. 154–171.

6 M. Armiento, M. Lelli, C. Andrews, B. Idini, and R. Ruff, "Analysing the global workforce dynamics of the energy transition: Main findings from the world energy employment report 2023," *Discover Sustainability*, vol. 6, 35, 2025. DOI: 10.1007/s43621-024-00773-7.

7 Y.-C. Ko, K. Zigan, and Y.-L. Liu, "Carbon capture and storage in South Africa: A technological innovation system with a political economy focus," *Technological Forecasting and Social Change*, vol. 166, 120633, May, 2021. DOI: 10.1016/j.techfore.2021.120633.

8 H. J. Buck, "Social science for the next decade of carbon capture and storage," *The Electricity Journal*, vol. 34, no. 7, 107003, Aug.–Sept., 2021. DOI: 10.1016/j.tej.2021.107003.

9 S. Sunjay, "Carbon capture geological storage," in Proc. 12th Congress of the Balkan Geophysical Society, May 2024, vol. 2024, pp. 1–5, DOI: 10.3997/2214-4609.202449BGS39

10 Y. Vasilev, A. Cherepovitsyn, A. Tsvetkova, and N. Komendantova, "Promoting public awareness of carbon capture and storage technologies in the Russian federation: A system of educational activities," *Energies*, vol. 14, no. 5, p. 1408, 2021. DOI: 10.3390/en14051408.

11 A. M. Hernandez, D. A. Pacheco Rojas, and D. B. Villaverde, "Carbon lock-in and contradictions—applied guide to academic teaching of Mexico's energy transition," *Applied Sciences*, vol. 11, no. 18, 8289, 2021. DOI: 10.3390/app11188289.

12 P. Bielka, "The future of carbon capture and storage technology: Innovative approach with digital twin," *Journal of Geotechnology and Energy*, vol. 40, no. 2, 2023. DOI: 10.7494/jge.2023.40.2.5413.

13 M. López-Morado, L. S. Caamaño, and V. Díaz Casás, "Bridging the skills gap of workers in the offshore renewable energies industry by creating a set of guidelines to promote innovative training," in Proc. 2024 IEEE Global Engineering Education Conference (EDUCON), 2024, DOI: 10.1109/10578932

14 M. Nooraiepour, P. Gladysz, and E. Melaas, "Norwegian—polish carbon capture and storage network: Bilateral collaboration for European climate action," *Energy Research and Social Science*, vol. 126, p. 104106, 2025. DOI: 10.1016/j.erss.2025.104106.

15 A. M. Laleye, "Practical and technological skills: An inevitable social engineering tool for sustainable development," *European Journal of Education and Pedagogy*, vol. 3, no. 2, 2022. DOI: 10.24018/ejedu.2022.3.2.318.

16 M. Fawad and N. H. Mondol, "Monitoring geological storage of CO_2: A new approach," *Scientific Reports*, vol. 11, no. 1, pp. 1–14, 2021. DOI: 10.1038/s41598-021-85346-8.

17 N. Ada, D. Ilic, and M. Sagnak, "A framework for new workforce skills in the era of industry 4.0," *International Journal of Mathematical, Engineering and Management Sciences*, vol. 6, no. 3, pp. 771–786, 2021. DOI: 10.33889/IJMEMS.2021.6.3.046.

18 H. Sudhakar, "Sustainability education: Cultivating environmental awareness and action through campus carbon sequestration studies," *International Journal of Sustainability in Higher Education*, vol. 26, no. 6, pp. 1450–1472, 2025. DOI: 10.1108/IJSHE-09-2024-0641.

19 E. Obode, T. Osunrinde, and O. Ifalade, "Skills for the energy transition and beyond: Human resource in the Industry 4.0 energy sector," in Proc. Abu Dhabi Int. Petroleum Exhibition & Conf. (ADIPEC), Abu Dhabi, UAE, Oct. 2023, Paper No. SPE-216637-MS, DOI: 10.2118/216637-MS

20 M. Shourideh, S. Yasseri, and H. Bahai, "Safety culture influence on safety performance of a post-combustion carbon capture facility," *Heliyon*, vol. 10, no. 14,Art.no.e34640, Jul. 2024, 10.1016/j.heliyon.2024.e34640.

21 S. P. Filippov, "The economics of carbon dioxide capture and storage technologies (review)," *Thermal Engineering*, vol. 69, pp. 738–750, 2022. DOI: 10.1134/S0040601522100020.

22 F. S. Khatibi, A. Dedekorkut-Howes, M. Howes, and E. Torabi, "Can public awareness, knowledge and engagement improve climate change adaptation policies?," *Discover Sustainability*, vol. 2, 18, 2021. DOI: 10.1007/s43621-021-00024-z.

23 P. R. Lima, A. A. M. Pereira, G. L. D. Chaves, and A. P. Meneguelo, "Environmental awareness and public perception on carbon capture and storage (CCS) in Brazil," *International Journal of Greenhouse Gas Control*, vol. 111, 103467, Oct., 2021. DOI: 10.1016/j.ijggc.2021.103467.

24 M. Xu, X. Zhang, S. Shen, S. Wei, and J.-L. Fan, "Assessment of potential, cost, and environmental benefits of CCS-EWR technology for coal-fired power plants in Yellow River Basin of China," *Journal of Environmental Management*, vol. 289,Art. no.112717, 2021, 10.1016/j.jenvman.2021.112717.

25 D. Newsome, K. B. Newsome, and S. A. Miller, "Teaching, learning, and climate change: Anticipated impacts and mitigation strategies for educators," *Behavior and Social Issues*, vol. 32, pp. 494–516, 2023. DOI: 10.1007/s42822-023-00129-2.

26 M. G. Fikru and N. Nguyen, "Factors shaping public support for more carbon capture and storage projects in the United States," *Environmental Management*, vol. 74, pp. 425–438, 2024. DOI: 10.1007/s00267-024-02000-5.

27 T. Zhang, M. Zhang, and L. Jin, "Advancing carbon capture in hard-to-abate industries: Technology, cost, and policy insights," *Clean Technologies and Environmental Policy*, vol. 26, pp. 2077–2094, 2024. DOI: 10.1007/s10098-024-02810-5.

28 S. M. Farouq Ali and M. Y. Soliman, "Limitations and fallacies of carbon capture and storage (CCS) and impact on oil and gas production," in Proc. SPE Annual Technical Conference and Exhibition (ATCE), San Antonio, TX, USA, Oct. 2023, Paper SPE-214950-MS, DOI: 10.2118/214950-MS

29 E. H. Oelkers and S. R. Gíslason, "Carbon capture and storage: From global cycles to global solutions," *Geochemical Perspectives*, vol. 12, no. 2, pp. 179–180, Oct., 2023. DOI: 10.7185/geochem_persp.12.2.179.

30 U. Tanveer, S. Ishaq, and T. G. Hoang, "Enhancing carbon trading mechanisms through innovative collaboration: Case studies from developing nations," *Journal of Cleaner Production*, vol. 482, 144122, Dec., 2024. DOI: 10.1016/j.jclepro.2024.144122.

31 K. Sotoodeh, *Piping, Valves, and Process Engineering in Carbon Capture Systems: Applications, Sustainability, and Environmental Protection*, 1st ed.. Oxford, UK: Elsevier, 2026. ISBN: 978–0–443-51078-6.

32 K. Neuhoff *et al.*, "Structuring international financial support to support domestic climate change mitigation in developing countries," *Climate Strategies*, 2022.

33 L. C. King and J. C. J. M. van den Bergh, "Potential carbon leakage under the Paris agreement," *Climatic Change*, vol. 165, 52, 2021. DOI: 10.1007/s10584-021-03082-4.

34 K. Kulovesi, "EU emissions trading scheme: Preventing carbon leakage before and after the Paris agreement," in *Research Handbook on EU Energy Law and Policy*, R. Leal-Arcas, Ed. Cheltenham, UK: Edward Elgar Publishing, 2024, pp. 586–599, DOI: 10.4337/9781035328024.00037.

35 P. Slorach, R. Lesemann, and P. Leung, "How an autonomous offshore power system can transform the ocean economy: A hypothetical case study utilizing an autonomous offshore power system in Northern lights carbon capture and storage project," in Proc. Offshore Technology Conf. (OTC), Houston, TX, USA, May 2023, Paper OTC-32664-MS, DOI: 10.4043/32664-MS

36 K. Sotoodeh, "Material failure and corrosion behavior of CO_2–H_2 hybrid pipelines in high-pressure transport: Mechanisms, risks, and mitigation," *Journal of Failure Analysis and Prevention*, vol. 26, no. 1, pp. 160–174, 2025. DOI: 10.1007/s11668-025-02344-y.

37 E. LoPreiato, "The public trust as an international solution to climate inaction," *German Law Journal*, vol. 25, no. 3, pp. 481–496, 2024.

38 N. Pandey, H. de Coninck, and A. D. Sagar, "Beyond technology transfer: Innovation cooperation to advance sustainable development in developing countries," *Wiley Interdisciplinary Reviews: Energy and Environment*, vol. 11, no. 2, e422, 2022.

20

Ethics, Governance, and Long-Term Stewardship of Carbon Storage

20.1 Ethical Considerations in Long-Term Carbon Storage

Long-term carbon storage is not only a technical and environmental challenge but also a deeply ethical one. Unlike many industrial activities, carbon storage creates responsibilities that extend far beyond the lifetime of a single project, company, or generation. Carbon dioxide injected into geological formations is expected to remain safely stored for decades to centuries. Therefore decisions made today can affect people and ecosystems far into the future [1]. The exceptionally long timescales associated with geological CO_2 storage extend responsibility beyond current stakeholders, raising ethical questions about obligations to future generations who will inherit both the benefits and potential risks of today's decisions [2, 3].

One of the most important ethical issues in carbon capture and storage (CCS) is intergenerational responsibility [2, 3]. Current generations rely on industrial activities that produce carbon emissions to support modern society and economic development. However, future generations may inherit the responsibility of living with stored carbon underground. Ethical carbon storage requires ensuring that future societies are not exposed to unmanaged risks, such as CO_2 leakage [4]. This means designing storage systems that are as safe as possible, maintaining long-term monitoring, and ensuring that storage sites remain secure long after operations have ended.

Another key ethical principle is accountability. Carbon storage projects must clearly define responsibility for monitoring, maintenance, and corrective actions in the event of operational problems [5]. Without clearly assigned accountability, long-term risks such as CO_2 leakage may be unfairly transferred to individuals or communities, particularly future generations, who were not involved in the original activities and have no authority or power to prevent or manage the consequences [6].

The concept of liability is closely linked to accountability. Ethical management requires that individuals or companies that benefit from carbon storage, such as industrial operators, also take on financial responsibility if there are storage failures or negative impacts. This responsibility, known as liability, includes covering the costs for repairing CO_2 leaks, restoring the environment, and addressing any health impacts on humans. Fair liability frameworks encourage the prevention of the transfer of risk to vulnerable populations or future generations [7].

Justice and equity are also central ethical concerns. Justice refers to fairness in how benefits, risks, and responsibilities are shared among people and communities. In CCS, justice means that no group, particularly those who live near industrial facilities and carbon storage, is unfairly exposed to environmental risks or excluded from decision-making. Equity focuses on recognizing differences between groups and ensuring that those with fewer resources or greater vulnerability receive appropriate support. In CCS, equity means giving special attention to communities that may be more affected or have less capacity to influence project decisions [8].

Figure 20.1 shows a visual comparison between equity and justice using a simple everyday example of people watching a baseball game. In the equity panel, individuals of different heights stand on boxes so that everyone can see over a barrier, illustrating how support is adjusted according to individual needs while the barrier itself remains. In contrast, the justice panel removes the barrier entirely, allowing all individuals to view the game equally without additional support, highlighting how justice addresses the root causes of inequality rather than managing its effects.

Informed consent and public participation are essential ethical principles. Communities living near storage sites or transport infrastructure have the right to be informed about the potential risks and benefits of carbon capture projects being developed nearby. They also have the right to openly express their concerns and receive informative and honest answers from project developers and operators [9]. It is unethical for carbon capture projects to move forward in a top-down manner; instead, they should be developed through transparent consultation processes that respect the rights, knowledge, and opinions of local residents.

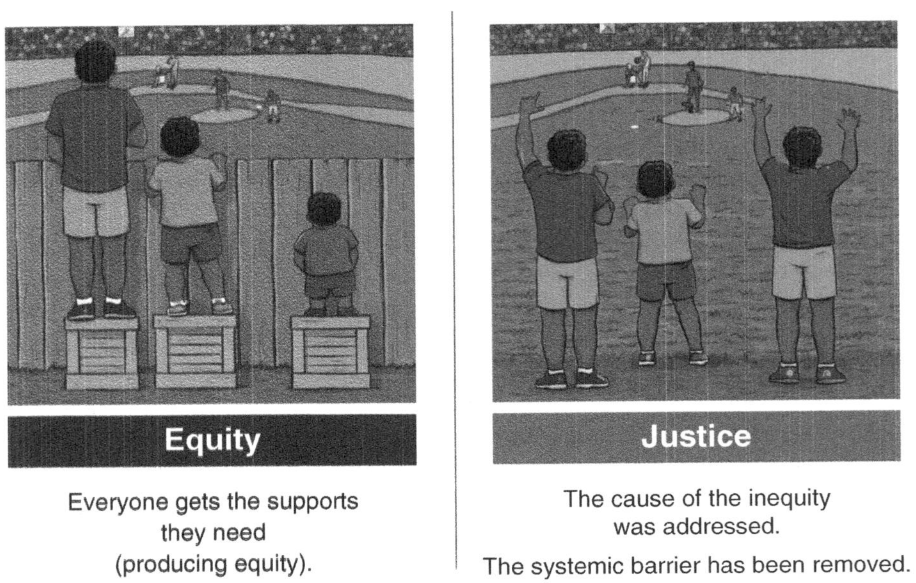

Figure 20.1 Conceptual illustration comparing equity and justice: equity provides unequal support to achieve equal outcomes, while justice removes systemic barriers to create fair access for all.

An important ethical consideration is recognizing the risks and uncertainties associated with projects. No engineered system is completely risk-free, and geological storage carries uncertainties related to subsurface storage over long periods. Ethical responsibility entails honestly acknowledging these uncertainties with local people instead of exaggerating certainty or hiding potential risks [10].

Environmental protection is a fundamental ethical responsibility [11]. Carbon storage must not harm groundwater resources or the surrounding environment. To ethically mitigate environmental risks associated with carbon dioxide storage, it is essential to conduct careful site selection, perform environmental impact assessments, and implement long-term monitoring of ecosystems [12].

Ethics contains global responsibility [13]. Countries and industries that have historically emitted high levels of carbon, particularly those in more developed nations, have a moral obligation to establish strong governance frameworks, promote technology sharing, and enhance capacity in areas where carbon storage is underdeveloped. Ethical climate action acknowledges this global responsibility while also respecting local authority and rights [14].

Finally, ethical carbon storage must avoid becoming a justification for delaying emissions reductions. While storage is a valuable climate tool, it should not be used to excuse continued high emissions or prevent investment in renewable energy and efficiency improvements. In summary, ethical considerations in long-term carbon storage center on global responsibility, fairness and justice, transparency with communities, and care for future generations.

20.2 Governance Models and Institutional Responsibilities

Effective long-term carbon storage depends not only on reliable technology and ethical principles but also on clearly defined institutional roles. Institutions play a central role in CCS decision-making. These include regulators, operating companies, and public authorities. Together, they determine how responsibilities are assigned and how rules are applied and enforced over time. Institutions set rules by defining how CCS projects must be designed, operated, and monitored [15]. For example, a national regulatory authority may require a storage company to submit a CO_2 storage monitoring plan before a permit is approved. It should be noted that when an operator is described as operating a CO_2 storage site, this refers specifically to the active injection phase and the associated management of wells and reservoir conditions, and does not include long-term stewardship after site closure [16].

Figure 20.2 illustrates the institutional governance framework for long-term carbon storage, highlighting the roles of public authorities (governments), regulators, operators, and independent monitoring bodies across the project lifecycle. The upper part of the figure represents surface-level governance, where government institutions define policy and regulators enforce technical and environmental standards. Operators manage active injection and monitoring, and independent bodies provide verification and auditing. The lower part shows the subsurface storage system, linking governance responsibilities to physical infrastructure such as injection wells, monitoring wells, caprock, and the CO_2 storage reservoir. Together, the figure emphasizes that safe and transparent carbon storage depends on coordinated institutional roles that extend from operational control to long-term stewardship.

Figure 20.2 Institutional governance for carbon storage, showing the interaction between government authorities, regulators, and operators, in relationship to subsurface CO_2 injection, monitoring, and long-term responsibility.

The two puzzle pieces are empty because the exact institutions, rules, and responsibilities that will govern carbon storage far into the future do not yet exist or cannot be fully defined today. Carbon storage operates on timescales of decades to centuries, while current legal systems, companies, and regulatory frameworks are designed for much shorter horizons. As a result, governance beyond the post-closure phase remains partially undefined and must be developed by future societies. The empty puzzle pieces may eventually be filled by international or cross-border governance structures or future monitoring and verification technologies. The injection well is used to place CO_2 into the subsurface reservoir, while monitoring wells do not inject CO_2 and are used solely to observe pressure, fluid behavior, and CO_2 movement in the reservoir to verify CO_2 storage safety and detect leakage.

One of the primary goals of governance in carbon storage is to clearly assign roles and responsibilities across the entire project lifecycle. Operating companies are typically responsible for day-to-day operations, monitoring, and maintenance. Regulatory authorities handle project approval and ensure compliance with safety and environmental standards. Independent monitoring and verification bodies may operate within regulatory agencies or as separate organizations. Their role is to audit project performance and validate carbon storage data, along with other relevant information. Independent verification helps ensure transparency, accuracy, and trust in CCS operations [17].

Several governance models are used around the world for carbon storage projects. One common approach is operator-led governance under state regulation. In this model, private or state-owned companies design, build, and operate carbon storage sites. Government authorities are responsible for approving projects, setting safety and environmental rules, and checking that these rules are followed. During the operational phase, the operating company (either governmental or private) carries out monitoring and maintenance of the storage site, while regulators supervise performance and enforce safety requirements. This governance model can encourage investment and technological innovation, but it depends on strong and capable regulators to protect public safety and the environment [18].

Another widely used model is public–private partnership (PPP) governance. In this approach, governments and private companies work together to develop and manage carbon storage projects, for example, both cooperate on design phase of the project. Private companies often provide technical expertise, project development, and daily operations, while governments support the project through regulation, funding mechanisms, or infrastructure development. Responsibilities and risks, such as long-term monitoring, liability, and financial costs, are shared according to clearly defined agreements. PPP governance can reduce financial risk for private companies and increase public confidence, as governments remain actively involved in long-term responsibility for carbon storage safety [19].

Some countries use a state-led or public ownership model for long-term carbon storage. In this model, private companies are responsible for designing, constructing, and operating storage sites during the active injection and after that for a specific period. Early post-closure phases because they possess the technical expertise, operational experience, and financial capacity needed to develop and manage complex storage infrastructure efficiently. Once a storage site has been closed and demonstrated stable and secure performance for a defined period, responsibility is transferred to the government [20]. This transfer recognizes that long-term risks extend beyond the lifespan of private companies and that public institutions are better suited to provide permanent monitoring and stewardship.

Regardless of the governance model, regulatory institutions play a central role in ensuring the safety, transparency, and long-term safe storage of carbon storage projects. They are responsible for approving projects, setting technical and environmental standards, reviewing and auditing monitoring data, and enforcing compliance with rules and regulations throughout the project lifecycle [21].

Clear governance frameworks must also address liability transfer mechanisms, which define how responsibility for a carbon storage site is transferred over time. These mechanisms explain who is legally responsible if problems occur, such as CO_2 leakage or environmental damage, during different phases of the project [7]. For example, the operating company may be responsible during the active injection phase and the early post-closure storage period, while responsibility may later be transferred to the government once the storage site has been demonstrated to be stable and secure over a defined period.

Community engagement as shown in Figure 20.3 is another important institutional responsibility within carbon storage governance systems. Institutions must establish clear

Figure 20.3 Community engagement in carbon storage project. Formal public engagement processes, including information meetings, community consultation, and mechanisms for raising and addressing public concerns during the planning and operation of carbon storage projects.

and formal processes to inform the public, consult local communities, and allow public participation in decision-making. This includes public meetings, access to information, and mechanisms that allow public concerns or complaints to be raised and addressed [9]. In addition, cross-border and offshore storage projects introduce additional governance complexity. In such cases, international coordination is required to align regulatory standards, monitoring protocols, and emergency response procedures [22].

Governance systems must also be adaptive. Over decades, scientific understanding, monitoring technologies, and climate conditions will evolve. Governance frameworks should allow regulatory updates [23]. Finally, governance effectiveness depends on institutional capacity. Regulators and public agencies must have adequate technical expertise and financial resources to perform their roles. In summary, governance models for carbon storage must clearly define institutional roles, including public authorities, private companies, and regulatory bodies.

20.3 Long-Term Monitoring and Liability Frameworks

Long-term monitoring and liability are essential to ensure the safety, reliability, and ethical commitment of carbon dioxide (CO_2) storage. Unlike many industrial activities, carbon storage must remain secure for decades to centuries after injection operations have ended. This extended timeframe creates unique challenges related to how storage sites are monitored and who is responsible for potential failures. Effective monitoring and liability work together to protect the environment and safeguard communities [24].

Long-term monitoring refers to the continuous observation and assessment of CO_2 storage sites to confirm that injected carbon dioxide remains safely contained within geological formations. Monitoring of a CO_2 storage site usually begins before injection, when baseline measurements are collected to understand the natural conditions of the reservoir and the surrounding environment. This baseline monitoring helps assess whether the reservoir is suitable for long-term CO_2 storage and provides reference information that allows engineers and regulators to identify potential CO_2 leakage and its impacts on the environment [25].

Monitoring then continues during the active injection phase to track CO_2 movement, pressure changes, and well integrity. After injection has stopped and the wells are closed, monitoring is maintained during the post-closure phase to verify long-term stability and ensure that the stored CO_2 remains safely contained over extended periods of time. In carbon storage projects, the operator is the main organization responsible for managing the storage site, including CO_2 injection, safety, and monitoring. Monitoring may be carried out by the operator itself or by a specialized monitoring company hired by the operator. In both cases, the operator remains responsible for the monitoring results and for reporting them to regulatory authorities [24, 25].

Monitoring frameworks should be designed for long-term use and remain adaptable to changes over time, such as advances in monitoring technologies. Although monitoring methods may evolve, the primary goal remains the same: to ensure that CO_2 is safely stored and to identify potential problems as early as possible [26]. Regulatory authorities typically define minimum monitoring requirements, but ethical best practices often encourage operators to go beyond these basic standards [27]. Continuous collection of monitoring data can also be used to improve understanding of long-term CO_2 storage through ongoing research and analysis. This information helps improve engineering design, operational practices, and risk assessment for future carbon storage projects.

Monitoring is closely linked to measurement, reporting, and verification (MRV) systems, which are used to track and confirm the performance of carbon storage projects. Monitoring involves collecting data from the storage site, such as pressure, CO_2 movement, and well integrity. Measurement refers to quantifying this data in a consistent and reliable way. Reporting means documenting and submitting the results to regulatory authorities and other stakeholders. Verification involves independent checks to confirm that the reported data are accurate and that CO_2 is being stored safely as intended [28].

Figure 20.4 presents a conceptual cross-sectional view of a geological carbon storage system, illustrating the main components involved in CO_2 injection, containment, monitoring, and potential leakage risks. Captured CO_2 is injected through a dedicated well into a deep subsurface reservoir with sufficient porosity to store the gas. An overlying cap rock forms a low-permeability sealing layer that prevents upward migration of CO_2 and ensures long-term containment. During and after injection, surface and subsurface monitoring systems are used to track CO_2 movement, pressure changes, and well integrity. The figure also highlights potential but unlikely risk possibilities, such as CO_2 leakage along faults or reactivated fractures, which are key targets for monitoring. Overall, the illustration emphasizes how geological barriers and monitoring work together to ensure the safe and secure storage of CO_2 over a long period.

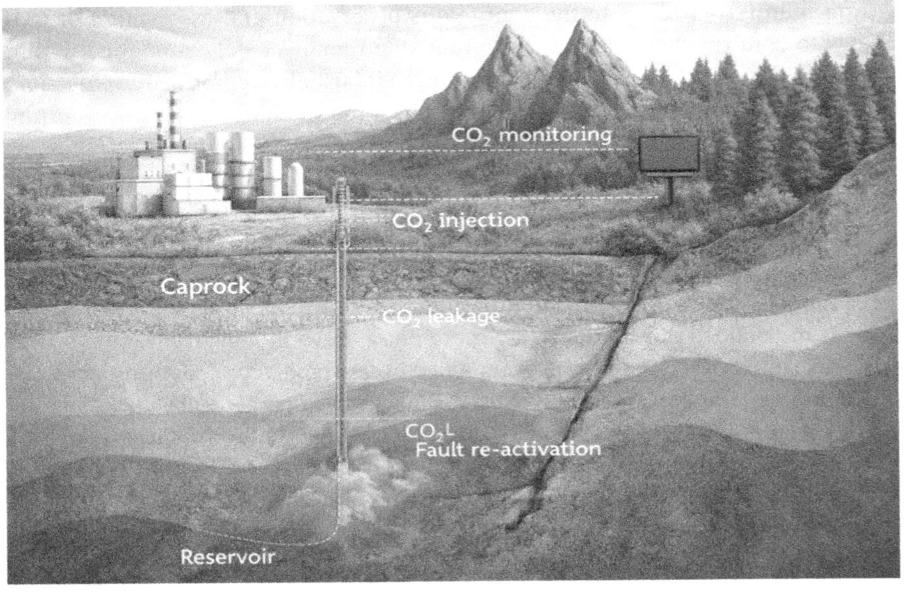

Figure 20.4 Conceptual cross-sectional schematic of a geological carbon storage system showing CO_2 injection into a deep reservoir, containment by an overlying cap rock, and monitoring of potential leakage paths such as faults.

Liability defines who is legally and financially responsible if a carbon storage site does not perform as expected. This includes situations such as CO_2 leakage, environmental damage, or the need for repairs and corrective actions. Liability rules clarify whether responsibility lies with the operating company, the government, or another institution during different stages of the project. Clear liability definition helps ensure that problems are addressed properly and that costs are not unfairly transferred to the public or future generations [7].

Liability exists at all stages of a carbon storage project, including during site development, active CO_2 injection, and early post-closure. However, liability becomes more complex during the post-closure phase because responsibilities extend over very long time periods, often beyond the lifetime of the operating company. During this period, the original operating company may no longer exist, technologies and regulations may change, and it can be difficult to clearly identify the cause of any future problems [29].

Financial support is essential in carbon storage projects because monitoring, maintenance, and corrective actions in case of CO leakage cost money and may be needed many years after CO_2 injection has ended. Without secured financial provisions, problems such as CO_2 leakage or environmental damage may not be properly addressed, leaving losses and financial costs to local communities or future generations [30]. To avoid this situation, operators are usually required to provide financial support during the active phase of the project.

Another key element is remediation responsibility. If monitoring detects leakage or unexpected migration, liability frameworks must specify corrective actions and timelines. Prompt remediation prevents minor issues from escalating. Long-term monitoring and liability frameworks must also consider community protection and compensation [31]. International projects add further complexity. Cross-border CO_2 transport or offshore storage requires harmonized liability rules and shared monitoring responsibilities [32]. In summary, long-term monitoring and liability are essential parts of ethical carbon storage systems.

20.4 Transparency and Community Accountability

Transparency and community accountability are key elements of ethical carbon storage governance. Because CO_2 storage projects operate over very long time periods and are often located near communities, public trust is just as important as technical performance. Transparency means that information about carbon storage activities, risks, and monitoring results is shared openly and in a way that people can understand. Community accountability ensures that project developers, operators, and governing institutions remain responsible and responsive to the concerns of local communities who may be affected by these projects [33].

Transparency in carbon storage begins with providing clear and understandable information about the project. This includes access to monitoring reports, technical documents such as risk assessments and emergency response plans, as well as relevant financial information [34]. Communities, regulators, and other stakeholders should be able to understand where CO_2 is stored, how it is monitored, what risks exist, and how those risks are managed.

Monitoring data play a central role in transparency. Measurements related to reservoir pressure, CO_2 movement, groundwater quality, and surface conditions should be reported regularly and clearly [35]. While technical details may be complex, summaries should be presented in plain language so that non-experts can understand system performance.

Community accountability goes beyond information sharing. It refers to the responsibility of operators and authorities to respond to community concerns. Meaningful public participation is especially important during key stages of a carbon storage project, including site selection, permitting, operation, and post-closure planning. During these phases, communities should have opportunities to ask questions and express concerns about CO_2 storage before major decisions are made by project developers. Early and ongoing participation helps identify local issues, improves project design, and increases public trust and acceptance of carbon storage projects [33, 34].

For example, local communities may express concerns about the location of a proposed CO_2 storage site, especially if it is close to residential areas or water resources. In response to these concerns, project developers may adjust the site location or introduce additional safety layers in their design and operation [36]. In other cases, community feedback may focus on the risk of CO_2 leakage and emergency preparedness [37]. This can lead project developers to use stronger monitoring systems, improved well designs, or enhanced emergency response measures. Such changes show how public participation can directly improve project safety and design.

Environmental justice is a crucial aspect of ethical governance in CCS [38]. Historically, large-scale industrial and energy infrastructure projects have often been located in or near

Figure 20.5 Symbolic representation of environmental injustice, illustrating how unequal social, economic, and governance conditions can lead to very different environmental outcomes for different communities.

marginalized and economically disadvantaged communities. These communities typically have limited political influence and face challenges in making decisions regarding such projects [39]. Figure 20.5 symbolically illustrates how similar communities can experience very different environmental outcomes depending on social, economic, and governance conditions. Effective governance and ethical frameworks for CCS must actively work to prevent the continuation of these historical inequities.

Education and capacity-building help make carbon storage projects more transparent and trustworthy. Public education activities, such as community workshops, open meetings, and site visits, can make carbon storage less mysterious and easier to understand for the public. When communities have a basic understanding of climate change, underground geology, and monitoring technologies, they are better able to understand what is happening and ask informed questions [40].

In international or offshore carbon storage projects, transparency becomes even more important because activities may involve multiple countries, regulatory systems, and stakeholders [41]. In international or offshore carbon storage projects, strong project management systems are also essential. Clear definition of roles and responsibilities helps ensure that everyone understands who is responsible for decision-making, monitoring, reporting, and emergency response. Well-structured management systems reduce confusion and improve cooperation between parties. In summary, transparency and community accountability (responsibility) are essential for ethical and sustainable carbon storage [42]. Open access to information, meaningful public participation, and clear project management structures are essential for the success of carbon capture projects.

20.5 International Legal and Regulatory Perspectives

Carbon dioxide (CO_2) storage increasingly operates within an international context, particularly when CO_2 is transported across national borders or stored in offshore and shared marine environments. As a result, long-term carbon storage is not governed solely by national laws but is also shaped by international legal and regulatory frameworks [43]. These frameworks aim to ensure environmental protection, clarify responsibility, and promote cooperation between countries while supporting global climate mitigation goals.

One of the primary challenges in governing international carbon storage is the issue of jurisdiction. Carbon capture may occur in one country, the transport of CO_2 may pass through another, and the actual storage may take place in offshore geological formations located either under international waters or within the jurisdiction of a different state. Therefore, clear legal rules are essential to define which country has the authority over permitting, monitoring, and liability. Without international coordination, legal gaps or conflicts could jeopardize the success of the project [44].

Several international agreements influence carbon storage activities. The United Nations Framework Convention on Climate Change (UNFCCC) and the Paris Agreement provide the climate policy by encouraging emissions reduction and carbon management strategies, including CCS [45]. While these agreements do not regulate CCS operations directly, they shape national commitments that influence how carbon storage is governed.

Marine carbon storage is particularly influenced by international maritime law. The United Nations Convention on the Law of the Sea (UNCLOS) establishes rules for protecting the marine environment and territorial waters. Offshore CO_2 storage projects must comply with UNCLOS obligations to prevent marine pollution and protect ecosystems [46]. In addition, the London Convention and London Protocol play a critical role in regulating the disposal of substances at sea [47]. Amendments to the London Protocol allow the subseabed storage of CO_2 streams from carbon capture processes, provided that strict environmental safeguards are met.

Regional regulatory frameworks further support international coordination. In the European Union, the EU CCS Directive establishes comprehensive rules for site selection, operation, monitoring, closure, and post-closure responsibility [48]. International legal perspectives also address liability and compensation. In cross-border projects, questions arise about which country is responsible if CO_2 leakage causes environmental harm or economic damage. International law generally requires states to prevent transboundary environmental harm and cooperate in resolving disputes [29]. Figure 20.6 provides a simplified illustration of the international cooperation between countries for offshore carbon storage projects.

Another important aspect is data sharing and transparency. International regulatory bodies encourage consistent reporting of emissions reductions and storage performance to support global climate change prevention. These records help ensure that stored CO_2 is captured and stored accurately toward climate commitments [49].

Despite recent progress, several challenges still exist. For example, different countries may apply different safety standards or permitting procedures for the same type of project, making coordination difficult. Some countries also lack the technical staff or financial

Figure 20.6 Conceptual overview of offshore and cross-border CO_2 storage, illustrating the role of international legal and regulatory frameworks in project governance.

resources needed to properly enforce regulations. In addition, international agreements can take many years to be approved, as they often require political negotiations and parliamentary approval in multiple countries [43]. Together, these issues can significantly delay or complicate cross-border projects. In summary, international legal and regulatory frameworks are essential for governing carbon storage in a globalized energy system. International agreements, maritime law, and regional directives together shape how CO_2 storage is regulated beyond national borders.

20.6 Designing Durable Stewardship Frameworks

Designing durable stewardship frameworks is essential to ensure that carbon dioxide (CO_2) storage remains safe, secure, and ethically governed over very long time periods. Unlike most industrial activities, carbon storage must be managed for decades to centuries, long after injection operations have ended. Durable stewardship refers to the institutions, systems, and clearly defined responsibilities that support continued monitoring of stored CO_2 and the maintenance or remediation of infrastructure such as wells [50]. This section outlines the key elements needed to establish effective stewardship frameworks capable of withstanding technological change, political transitions, and evolving societal conditions.

A key principle of effective stewardship is long-term institutional continuity [50]. Carbon storage sites must remain under responsible oversight even if the original operating company no longer exists. For this reason, governments play a central role as long-term stewards, either by taking direct responsibility after site closure or by supervising independent public stewardship organizations.

Clear and well-defined mechanisms for transferring responsibility are another critical element of effective stewardship. During the operational phase, private operators are usually responsible for CO_2 injection, monitoring, and safety management [51]. After injection has ended and a defined post-closure period has passed, responsibility may transfer to the government for long-term monitoring and safety considerations. This transfer process must be transparent and carefully structured to ensure that any remaining technical risks are identified, managed, and not unfairly shifted to the public [52].

Financial sustainability is essential for stewardship quality and durability [53]. Long-term monitoring and emergency response in case of issues with CO_2 storage require funding that remains available far into the future. Knowledge storage and preservation are another crucial element. Information about storage sites, such as geological data, monitoring records, CO_2 injection histories, and safety risk assessments must remain accessible for future regulators and scientists [54].

Long-term stewardship frameworks must preserve community accountability to ensure that public trust is maintained over time. Community engagement remains important even after injection ends. Local communities should continue to have access to monitoring results [6, 9].

International considerations further affect stewardship design. Offshore and cross-border storage projects require coordination between countries to manage shared responsibilities and potential transboundary risks. International agreements and regional cooperation frameworks support consistency in monitoring standards, data sharing reports, and emergency response. For example, in offshore carbon storage projects in the North Sea, participating countries apply harmonized monitoring requirements, such as regular pressure measurements, seismic surveys to track CO_2 movement, and standardized reporting formats [18].

Resilience is a defining feature of effective stewardship frameworks. Stewardship systems must remain stable and functional even when political leadership changes, economic conditions worsen, or government institutions are reorganized. For example, a carbon storage site must continue to be monitored even if a new government with different political priorities comes to power. A new government may choose to reduce funding for climate programs or change environmental policies. Similarly, government reorganization can occur when governmental organizations are merged, renamed, or restructured, which may interrupt oversight if responsibilities are unclear [55]. Figure 20.7 shows long-term stewardship of a carbon storage site, with inspectors monitoring infrastructure in a natural landscape. Simple visual elements represent importance of oversight, funding, knowledge preservation, and community engagement.

In summary, designing durable stewardship frameworks requires long-term institutional commitment, clear responsibility transfer, secure financing, knowledge preservation, and sustained community engagement.

Figure 20.7 Durable stewardship frameworks ensure long-term safety, accountability, and trust in CO_2 storage beyond site closure.

20.7 Key Highlights (Chapter Summary)

◇ **Chapter 20—Key Highlights:**

1) Long-term carbon storage creates ethical responsibilities that extend across decades and centuries, requiring long-term accountability and care.
2) Ethical governance of carbon storage is built on clear principles of responsibility, fairness, transparency, and environmental protection.
3) Strong governance models define institutional roles across the full project lifecycle, from site selection and operation to post-closure stewardship.
4) Public–private partnerships can support CCS deployment but require careful design to prevent risk transfer to communities.
5) Accountability and liability frameworks ensure that operators remain legally and financially responsible for storage safety and environmental impacts.
6) Continuous long-term monitoring and robust measurement, reporting, and verification (MRV) systems are essential to confirm storage integrity.
7) Transparent communication and community accountability strengthen public trust and social acceptance of carbon storage projects.
8) Community rights and participation are essential for ethical deployment, especially for populations living near storage or transport infrastructure.
9) Public participation and environmental justice considerations help prevent unequal distribution of risks and benefits of carbon storage projects.

10) International legal frameworks and cross-border cooperation are applicable to offshore and transnational carbon storage projects.

Keywords: Ethics, Carbon Storage, Long-Term Stewardship, Governance Frameworks, Accountability, Liability, Intergenerational Responsibility, Environmental Justice, Transparency, Community Accountability, Long-Term Monitoring, Measurement, Reporting, And Verification, Institutional Responsibility, Public–Private Partnerships, Regulatory Oversight, Post-Closure Management, Financial Assurance, Risk Management, Knowledge Preservation, Adaptive Governance, International Environmental Law, Cross-Border Carbon Storage, Offshore Storage Governance, Climate Responsibility, Sustainable Climate Mitigation

References

1 V. Jones-Bassett, R. Bassett, E. C. Rowe, E. Tipping, D. Yumshavey, and J. Davies, "Changes in carbon storage since the pre-industrial era: A national scale analysis," *Anthropocene*, vol. 34, 100289, Jun., 2021. DOI: 10.1016/j.ancene.2021.100289.

2 F. Medvecky, J. Lacey, and P. Ashworth, "Examining the role of carbon capture and storage through an ethical lens," *Science and Engineering Ethics*, vol. 20, no. 4, pp. 1111–1128, 2014. DOI: 10.1007/s11948-013-9474-z.

3 P. Boucher and C. Gough, "Mapping the ethical landscape of carbon capture and storage," *Poiesis & Praxis*, vol. 9, no. 3–4, pp. 249–270, 2012. DOI: 10.1007/s10202-012-0117-2.

4 S. White *et al.*, "A risk-based approach to evaluating the area of review and leakage risks at CO_2 storage sites," *International Journal of Greenhouse Gas Control*, vol. 93, 102884, Feb., 2020. DOI: 10.1016/j.ijggc.2019.102884.

5 F. Vidler, "Accountability for CO_2 climate crisis by carbon majors' emissions in oil and gas production?," *SSRN Electronic Journal.*, Mar., 2023. DOI: 10.2139/ssrn.4403449.

6 C. Oltra, P. Upham, and A. Lis, "Public responses to CO_2 storage sites: Lessons from five European cases," *Public Understanding of Science*, vol. 23, no. 2–3, pp. 227–248, 2012. DOI: 10.1260/0958-305X.23.2-3.227.

7 E. J. Wilson, A. B. Klass, and S. Bergan, "Assessing a liability regime for carbon capture and storage," *Energy Procedia*, vol. 1, no. 1, pp. 4575–4582, Feb., 2009. DOI: 10.1016/j.egypro.2009.02.277.

8 C. Vale, R. Averill, J. Hall, H. Forgasz, and G. Leder, "Equity, social justice, and ethics," in *Research in Mathematics Education in Australasia 2016–2019*, J. Way, C. Attard, J. Anderson, J. Bobis, H. McMaster, and K. Cartwright, Eds. Singapore: Springer, 2020, pp. 161–182, ch. 8, DOI: 10.1007/978-981-15-4269-5_8.

9 F. Mulyasari, A. K. Harahap, A. O. Rio, R. Sule, and W. G. A. Kadir, "Potentials of the public engagement strategy for public acceptance and social license to operate: Case study of carbon capture, utilisation, and storage Gundih pilot project in Indonesia," *International Journal of Greenhouse Gas Control*, vol. 108, 103312, Jun., 2021. DOI: 10.1016/j.ijggc.2021.103312.

10 B. W. Terwel, F. Harinck, N. Ellemers, and D. L. Daamen, "Going beyond the properties of CO_2 capture and storage (CCS) technology: How trust in stakeholders affects public acceptance of CCS," *International Journal of Greenhouse Gas Control*, vol. 5, no. 2, pp. 181–188, Mar., 2011. DOI: 10.1016/j.ijggc.2010.10.001.

11 M. Soleimanpour Omran, "The effect of educating environmental ethics on behavior and attitude to environment protection," *European Online Journal of Natural and Social Sciences*, vol. 3, no. 3(S), p. 141, 2014.

12 K. Damen, A. Faaij, and W. Turkenburg, "Health, safety and environmental risks of underground CO_2 storage—overview of mechanisms and current knowledge," *Climatic Change*, vol. 74, no. 1–3, pp. 289–318, 2006. DOI: 10.1007/s10584-005-0425-9.

13 A. M. González, "Ethics in global business and in a plural society," *Journal of Business Ethics*, vol. 44, no. 1, pp. 23–36, 2003. DOI: 10.1023/A:1023230222707.

14 A. R. Pearson, C. G. Tsai, and S. Clayton, "Ethics, morality, and the psychology of climate justice," *Current Opinion in Psychology*, vol. 42, pp. 36–42, Dec., 2021. DOI: 10.1016/j.copsyc.2021.03.001.

15 I. L. de Araujo, Y. Marcelle, Z. Makuch, and H. K. de Medeiros Costa, "Designing Brazilian institutional framework for carbon capture, use and storage (CCUS) activities: The role of competent authority," *Revista Jurídica da Presidência*, vol. 26, no. 138, pp. 44–74, 2024.

16 M. Osazuwa-Peters and M. Hurlbert, "Analyzing regulatory framework for carbon capture and storage (CCS) technology development: A case study approach," *The Central European Review of Economics and Management (CEREM)*, vol. 4, no. 1, pp. 107–148, 2020.

17 S. I. Plasynski, J. T. Litynski, H. G. McIlvried, D. M. Vikara, and R. D. Srivastava, "The critical role of monitoring, verification, and accounting for geologic carbon dioxide storage projects," *Environmental Geosciences*, vol. 18, no. 1, pp. 19–34, Mar., 2011. DOI: 10.1306/eg.06231010008.

18 S. Vajjhala, J. Gode, and A. Torvanger, "An international regulatory framework for risk governance of carbon capture and storage," Policy Note, CICERO—Center for International Climate and Environmental Research, Oslo, Norway, 2007. [Online]. Available: http://urn.nb.no/URN:NBN:no-3647

19 J. Zhu, Q. Shi, C. Zhang, J. Yuan, Q. Li, and X. Wang, "Investigating cooperative strategies in low-carbon public–private partnership projects through evolutionary game," *Engineering Construction and Architectural Management*, vol. 31, no. 2, pp. 789–811, 2024. DOI: 10.1108/ECAM-04-2022-0324.

20 J. R. Zadick, "The public pore space: Enabling carbon capture and sequestration by reconceptualizing subsurface property rights," *William & Mary Environmental Law and Policy Review*, vol. 36, pp. 257–302, 2011.

21 C. Vannan and J. C. Gemmell, "The role of regulators in reducing regulatory risk: Using scenario planning to assess the regulatory framework for carbon capture and storage," *Risk Management*, vol. 14, no. 1, pp. 27–41, 2012. DOI: 10.1057/rm.2011.11.

22 H. Pór Hilmarsson, "Managing risks in cross border energy projects in emerging markets," Central and Eastern European Online Library (CEEOL), EDİTURA ASE, 2013. [Online]. Available: https://www.ceeol.com/search/article-detail?id=732146

23 S. Ulli-Beer, *Dynamic Governance of Energy Technology Change*. Berlin, Germany: Springer-Verlag, 2016.

24 A. B. Klass and E. J. Wilson, "Climate change and carbon sequestration: Assessing a liability regime for long-term storage of carbon dioxide," *The Emory Law Journal*, vol. 58, pp. 103–180, 2008.

25 D. Li, S. Peng, Y. Guo, Y. Lu, and X. Cui, "CO_2 storage monitoring based on time-lapse seismic data via deep learning," *International Journal of Greenhouse Gas Control*, vol. 108, 103336, Jun., 2021. DOI: 10.1016/j.ijggc.2021.103336.

26 M. Fawad and N. H. Mondol, "Monitoring geological storage of CO_2: A new approach," *Scientific Reports*, vol. 11, 5942, 2021. DOI: 10.1038/s41598-021-85346-8.

27 S. E. Schütz and A. Omar, Report on regulations and technological capabilities for monitoring CO_2 storage sites, Tech. Rep., 2023.

28 M. N. Rahman, "Regulatory and standard insights on transboundary CO_2 in the context of MRV," *Future Sustainability*, vol. 3, no. 4, Nov., pp. 1–11, 2025.

29 H. Bjerke, State responsibility and liability in the context of carbon capture and storage, Master's thesis, Faculty of Law, UiT The Arctic University of Norway, Tromsø, Norway, 2023. [Online]. Available: https://hdl.handle.net/10037/35912

30 W. Li, "Energy finance and carbon finance: Key roles of the financial community in addressing climate change," *International Journal of Social Sciences and Public*, vol. 2, no. 2, pp. 1–10, 2024. DOI: 10.62051/ijsspa.v2n2.43.

31 F. Zheng, D. Morgan, M. Ma, and B. Chen, "Leakage remediation strategy at carbon storage reservoir," in Proc. SPE Reservoir Simulation Conf., Galveston, TX, USA, Mar. 2025, Paper SPE-223860-MS. DOI: 10.2118/223860-MS

32 S. Gola and K. Noussia, "From CO_2 sources to sinks: Regulatory challenges for transboundary trade, shipment and storage," *Resources, Conservation and Recycling*, vol. 179, 106039, Apr., 2022. DOI: 10.1016/j.resconrec.2021.106039.

33 J. A. E. Nielsen, K. Stavrianakis, and Z. Morrison, "Community acceptance and social impacts of carbon capture, utilization and storage projects: A systematic meta-narrative literature review," *PLoS One*, vol. 17, no. 8, e0272409, Aug., 2022. DOI: 10.1371/journal.pone.0272409.

34 Y. Vasilev, A. Cherepovitsyn, A. Tsvetkova, and N. Komendantova, "Promoting public awareness of carbon capture and storage technologies in the Russian Federation: A system of educational activities," *Energies*, vol. 14, no. 5, 1408, Mar., 2021. DOI: 10.3390/en14051408.

35 G. Fibbi, M. Del Soldato, and R. Fanti, "Review of the monitoring applications involved in the underground storage of natural gas and CO_2," *Energies*, vol. 16, no. 1, 12, Jan., 2023. DOI: 10.3390/en16010012.

36 M. G. Fikru and N. Nguyen, "Factors shaping public support for more carbon capture and storage projects in the United States," *Environmental Management*, vol. 74, pp. 425–438, 2024. DOI: 10.1007/s00267-024-02000-5.

37 A. Hansson, J. Anshelm, M. Fridahl, and S. Håkold, "The underworld of tomorrow? How subsurface carbon dioxide storage leaked out of the public debate," *Energy Research and Social Science*, vol. 90, 102606, Aug., 2022. DOI: 10.1016/j.erss.2022.102606.

38 B. K. Sovacool, "Expanding carbon removal to the Global South: Thematic concerns on systems, justice, and climate governance," *Energy and Climate Change*, vol. 4, 100103, Dec., 2023. DOI: 10.1016/j.egyc.2023.100103.

39 H. N. Brehm and D. N. Pellow, "Environmental justice: Pollution, poverty and marginalized communities," in *Routledge Handbook of Global Environmental Politics*, G. Kütting and Z. R. Taylor, Eds. London, UK: Routledge, 2022, pp. 348–361.

40 M. Khan, D. Mfitumukiza, and S. Huq, "Capacity building for implementation of nationally determined contributions under the Paris Agreement," in *Making Climate Action More Effective*, 1st ed.. London, UK: Routledge, 2021, pp. 1–12, DOI: 10.4324/9781003162407-8.

41 D. Honig, R. Lall, and B. C. Parks, "When does transparency improve institutional performance? Evidence from 20,000 projects in 183 countries," *American Journal of Political Science*, vol. 67, no. 4, pp. 1096–1116, Oct., 2023. DOI: 10.1111/ajps.12698.

42 H. M. Uddin, S. A. Mustaqeem, and F. A. Siddique, "Challenges in carbon capture and storage projects: A strategic and project management perspective," in Proc. SPE/PAPG Pakistan Section Annu. Tech. Symp. Exhib., Islamabad, Pakistan, Feb. 2025, Paper SPE-226837-MS, DOI: 10.2118/226837-MS

43 R. A. Duraiswami, "Global carbon capture and storage efforts: Challenges and opportunities," *Journal of the Geological Society of India*, vol. 101, no. 2, pp. 143–148, Feb., 2025. DOI: 10.17491/jgsi/2025/174076.

44 N. Krisch, "Jurisdiction unbound: (Extra)territorial regulation as global governance," *European Journal of International Law*, vol. 33, no. 2, pp. 481–514, May, 2022. DOI: 10.1093/ejil/chac028.

45 W. Burns, "Governance of ocean-based carbon dioxide removal research under the United Nations Convention on the Law of the Sea," *Maine Law Review*, vol. 75, p. 37, 2023.

46 B. E. Klerk, "The ITLOS advisory opinion on climate change: Revisiting the relationship between the United Nations Convention on the Law of the Sea and the Paris agreement," *Review of European, Comparative & International Environmental Law*, vol. 34, no. 1, pp. 181–193, Apr., 2025. DOI: 10.1111/reel.12588.

47 C. Vivian and L. Del Savio, "The London convention and protocol: Adapting to address the ocean-climate crisis," *The International Journal of Marine and Coastal Law*, vol. 39, no. 3, pp. 519–553, Aug., 2024. DOI: 10.1163/15718085-bja10141.

48 R. Maddahi and K. Talus, "Carbon dioxide utilization and storage through mineralization under European Union law: A legal analysis of the EU CCS directive and the EU emissions trading system," *The Journal of World Energy Law & Business*, vol. 18, no. 4, pp. 1, jwaf023–24, 2025. DOI: 10.1093/jwelb/jwaf023.

49 S. D. Comello, J. Reichelstein, and S. Reichelstein, "Corporate carbon reporting: Improving transparency and accountability," *One Earth*, vol. 6, no. 7, pp. 803–810, Jul., 2023. DOI: 10.1016/j.oneear.2023.07.006.

50 J. G. Meng, C. M. PMP, and C. M. EngD, "Committed to long-term stewardship," in Proceedings of the International Annual Conference of the American Society for Engineering Management (ASEM), 2023, pp. 1–9.

51 D. Tanase and J. Tanaka, "Progress of CO_2 injection and monitoring of the Tomakomai CCS demonstration project," in Proc. 15th Greenhouse Gas Control Technologies Conf. (GHGT-15), Mar. 2021, pp. 1–12. [Online]. Available: SSRN. Accessed: Apr. 1, 2021.

52 K. Romank and T. Dixon, "CO_2 storage guidelines and the science of monitoring: Achieving project success under the California Low Carbon Fuel Standard CCS Protocol and other global regulations," *International Journal of Greenhouse Gas Control*, vol. 113, p. 103523, Jan., 2022. DOI: 10.1016/j.ijggc.2021.103523.

53 Fakhrullah, D. Xiao, N. Jan, S. Khan, and M. Suplata, "Environmental stewardship and economic prosperity: A comprehensive assessment of CO_2 emissions and sustainable development goals in European countries," *Journal of the Knowledge Economy*, vol. 16, pp. 5572–5593, 2025. DOI: 10.1007/s13132-024-02169-4.

54 S. Shah, T. Modi, J. Kanakia, E. Shah, K. Shah, and A. Nanade, "A comprehensive approach to unify carbon registries data access," in Proc. 2024 OITS Int. Conf. Inf. Technol. (OCIT), Dec. 2024, pp. 525–529, DOI: 10.1109/OCIT.2024.10973666

55 M. Shen *et al.*, "Carbon capture and storage (CCS): Development path based on carbon neutrality and economic policy," *Carbon Neutrality*, vol. 1, p. 37, 2022. DOI: 10.1007/s43979-022-00039-z.

Index

f

Printed and bound by CPI Group (UK) Ltd, Croydon, CR0 4YY
17/09/2026
14964270-0004